Vienna Circle Institute Yearbook

Institute Vienna Circle, University of Vienna
Vienna Circle Society, Society for the Advancement of Scientific World Conceptions

Volume 26

Editorial Address
Wiener Kreis Gesellschaft
Universitätscampus, Hof 1
Spitalgasse 2-4, A-1090 Wien, Austria
Tel.: +431/4277 46501 (international) or 01/4277 46501 (national)
Email: vcs@univie.ac.at
Homepage: http://univie.ac.at/vcs/

More information about this series at http://www.springer.com/series/6669

Christian Damböck • Adam Tamas Tuboly
Editors

The Socio-Ethical Dimension of Knowledge

The Mission of Logical Empiricism

Springer

Editors
Christian Damböck
Institute Vienna Circle
University of Vienna
Wien, Austria

Adam Tamas Tuboly
Institute of Philosophy
Research Centre for the Humanities
Budapest, Hungary

ITD, Medical School
University of Pécs
Pécs, Hungary

ISSN 0929-6328 ISSN 2215-1818 (electronic)
Vienna Circle Institute Yearbook
ISBN 978-3-030-80362-9 ISBN 978-3-030-80363-6 (eBook)
https://doi.org/10.1007/978-3-030-80363-6

This Springer imprint is published by the registered company Springer Nature Switzerland AG
The registered company address is: Gewerbestrasse 11, 6330 Cham, Switzerland

Contents

Part I
The Socio-ethical Dimension of Knowledge – The Mission of Logical Empiricism

Chapter 1
Introduction: The Socio-ethical Concern of Logical Empiricism

Adam Tamas Tuboly

Abstract In this introduction, various notions of socio-political and ethical engagements are introduced. Some examples from logical empiricism are discussed along with the summaries of the volume's chapters.

Keywords Socially engaged philosophy of science · Rudolf Carnap · Vienna circle · Otto Neurath · Scientific world-conception

1.1 Introduction

On December 11 and 12, 2017, the Institute of Philosophy at the Hungarian Academy of Sciences organized a conference on "The Socio-Ethical Dimension of Knowledge: The Mission of Logical Empiricism". The participants (most of whose presentations are reproduced here) discussed various topics, both closely and loosely related to the theme, as suggested by the conference's broad title. In this short introduction, we will present the context and aim of the conference and try to clarify certain basic notions.

A. T. Tuboly (✉)
Institute of Philosophy, Research Centre for the Humanities, Budapest, Hungary

ITD, Medical School, University of Pécs, Pécs, Hungary

C. Damböck, A. T. Tuboly (eds.), *The Socio-Ethical Dimension of Knowledge*,
Vienna Circle Institute Yearbook 26, https://doi.org/10.1007/978-3-030-80363-6_1

1.2 *Ways* of Socio-ethical Issues in Logical Empiricism

Whenever one looks at the socio-ethical and political engagements of scientists and philosophers, the web of ideas that can be considered is quite broad and perhaps unexpectedly complex. In discussing the political significance of Sir Arthur Eddington's famous popularizing works, Matthew Stanley noted that

> [i]n order to better understand the relationship between scientists and their political activities, it is critical to expand our expectations of where to find politics and what it should look like: the values underlying political engagements can have many different sources and allegiances, and we must carefully search out the interrelationships that quietly suffuse politics throughout other activities. (Stanley 2008, 193)

One aim of this volume is to show that the same is also true of logical empiricism, and thus of a significant part of early analytic philosophy. But what exactly do we have in mind?

The terms "social", "ethical" and "political" are anything but well-defined; they are vague notions, and perhaps for good. These terms – and the ideas they express – are context-relative and could take different forms of realization among various (spatially and temporally) local webs of relations. We will thus not attempt to define them here; for the purposes of this introduction it suffices to note that by "ethical" and "moral" we mean the characterization of certain non-epistemic or non-cognitive values of a person's worldview. Consequently, whoever pursues ethics or moral philosophy is concerned with the place, role and nature of these non-cognitive values, which are related (loosely or strongly, depending on one's particular approach) to the well-being of the individual and of society.[1] By "social" or "social concern" we mean an individual's relation to the cultural and human-made environment and how this is made explicit. This scope is intentionally general so as to cover both activism and theoretical approaches towards the subject, as well as theoretical reflections on primary scientific fields; the aim is also to include philosophy of the social sciences, since the possible approaches towards this field will also determine people's views on whether philosophy and science should be allowed to shape the world at all, and how. Finally, by "political" we mean the general concern of citizens for their lives as well as power relations, instead of party politics in a more restricted sense.

As the history of twentieth-century (analytic) philosophy has shown, there are numerous *ways* to pursue such a form of philosophy, which is not only sensitive to internal-theoretical issues, but also tries to keep an eye on the broader questions of human life. Thus, if we intend to do so, we can presumably find *a possible way* to pursue the goals that fit our mentality and temper. As Rudolf Carnap, Hans Hahn and Otto Neurath famously claimed in their jointly-authored manifesto,

> Of course, not every single adherent of the scientific world-conception will be a fighter. Some, glad of solitude, will lead a withdrawn existence on the icy slopes of logic; some

[1] For a broad list of social and political values, and a discussion of how they intertwined in the case of logical empiricist philosophy of science, see Richardson (2003).

> may even disdain mingling with the masses and regret the 'trivialization' which is inevitable when popularizing. (In Stadler and Uebel 2012, 90)

Independently of our theoretical or practical orientation, our "achievements take their place in the historical development" (ibid.). This characterization seems to trivialize socio-ethical and political concerns, since whatever we do, we do it firstly in a socio-cultural environment, and secondly, we thereby document a socio-ethical commitment. As Neurath (1921/1973, 158) famously said, "politics are action", and "'non-action' is also an action" (1913/1983, 2), so that supposedly non-political moves are, in fact, also political moves. Either we think this is trivial or not, what matters is that many of these social engagements and our place in the socio-historical development have to be made explicit – and that is not trivial at all.

One way or another, people are going to find channels in which their commitments will manifest themselves in a specific form of philosophy. For some people, this takes a rather conventional form, namely that of *moral philosophy*. Despite the expectations and common perceptions of many observers, a significant part of the works that appeared under the aegis of logical empiricism in general, and that of the Vienna Circle in particular, is what might be called moral philosophy (as Anne Siegetsleitner shows in her chapter). To give a few examples, this includes Karl Menger's *Morality, Decision and Social Organization* (1934/1974); Viktor Kraft's *Foundations for a Scientific Analysis of Value* (1937/1981); and Moritz Schlick's book on the wisdom of life (*Lebensweisheit*, 1908) as well as his well-known *Problems of Ethics* (1930/1939) and, of course, Rudolf Carnap's "Theoretical Questions and Practical Decisions" (1934).[2]

Besides publishing their own works, however, members of the circle also accepted in their series works by other authors that concerned questions of ethics. Of note are the inclusion of John Dewey's *Theory of Valuation* (1939) in Neurath's *International Encyclopedia of Unified Science*, and that of Otto Kant's *Zur Biologie der Ethik* (1932) in Schlick and Philipp Frank's *Schriften zur wissenschaftliche Weltauffassung*. Though the group's in-house journal *Erkenntnis* never featured any papers on ethics or moral philosophy, the logical empiricists supported such inquiries wherever they could. As can be gleaned from this not at all complete list, questions of ethics and moral philosophy were very much in the foreground of the logical empiricists' attention, even at monograph length.

But that is just one of the possible ways of engaging with socio-ethical issues. Others gave voice to their social and moral concerns not by contributing to the field of ethics, but by taking a step back and considering *(the philosophy of) the social sciences*, and the roles of science, philosophy and society vis-à-vis one another. The often-cited locus classicus of this field is Carl Gustav Hempel's somewhat misleading article on "logical positivism and the social sciences" (1969). Hempel discussed Carnap's *Der logische Aufbau der Welt* at length in his work, but he focused mainly

---

[2]Although he was a member of the Berlin Group (led by Hans Reichenbach), Wolfgang Köhler (1938) also published a long treatise on *The Place of Value in a World of Facts*. For more on the ethical issues in and the moral philosophy of logical empiricism, see Siegetsleitner (2010, 2014).

on concept formation in psychology and on some of Neurath's broad physicalist ideas.[3] Therefore, Hempel simply missed or ignored all of those works that explicitly and directly engaged with the social sciences.

Felix Kaufmann, on the other hand, nicely fits the description of this type of engagement with his *Methodenlehre der Sozialwissenschaften* (1936/2014), as does Otto Neurath's overwhelming work in the social sciences, which covers various fields from economics to pedagogy, and from museology to sociology (Cat and Tuboly 2019). But even Philipp Frank comes to mind as an example of someone who was highly interested in the sociology of knowledge and therefore broadened the theoretical scope of social matters, by incorporating and investigating non-cognitive values in theory building and concept formation (Frank 2021; Uebel 2000; Nemeth 2003; Tuboly 2017). Finally, Edgar Zilsel should also be noted. Besides his socio-historical work on the cult of genius (described by Günther Sandner in this volume), Zilsel produced important material on the socio-history of science and the possible ways of socializing scientific knowledge (Krohn and Raven 2000; Nemeth 2011).[4]

There are other matters; as Thomas Uebel (2007, 250) noted, "philosophy of social science is unlikely to figure in many people's judgment as a field in which logical empiricism effected great progress. If anything, impressions run to the contrary." Nonetheless, Uebel discusses various issues in the context of physicalism, encyclopedism and the general conception of unified science that are – as one might say, indirectly – relevant for the actual practice and methods of the social sciences.

The logical empiricists' work on the social sciences, however, took quite an interesting form and context, a fact that was recognized already in the 1930s. In 1936, at the *Second International Congress for the Unity of Science* in Copenhagen, John Sommerville – a philosopher from Columbia University, founder of institutes on Marxist and Soviet philosophy and the relation between philosophy and society – argued that "logical empiricism needs social science" in order to justly claim that unified science indeed embraces *all* the sciences, and that "social science needs logical empiricism." What exactly did Sommerville have in mind? He argued that there are fields – such as physics and mathematics – that have already collected a lot of data and knowledge, that have a "relatively advanced state of development, [and that are] characterized by a fairly well crystallized methodology." Work in these fields aims for "ultimate digestion, which arises out of the very super-abundance of the harvest" (Sommerville 1936, 406). But the situation is quite different in the case of economics and sociology:

> For in this case the philosopher comes upon the field, not after the harvest has been gathered, but even before the ground has been properly prepared. He is confronted with the kind of science in which even the most fundamental concepts are vague and loose, in which specialized instruments and a complex apparatus can hardly be said to exist, in which there

---

[3] For an updated view on "logical positivism and the social sciences", see Uebel (2007, 2010, 2019).

[4] One might note here Ernest Nagel as well, who worked out quite specific concerns about the social sciences (mainly during his co-taught seminar with the sociologist Paul Lazarsfeld). On these matters see Neuber (2022).

> is little cooperation of one group of investigators with another, or of one generation of workers with the preceding and following. He is dealing with the kind of science wherein the outstanding characteristic of the methodology is not its stability and fertility, but its instability and relative sterility. In this case, accordingly, his general problem is not digestion, but production. It is a problem which arises, not out of a plethora, but out of a scarcity. [...] Logical empiricism cannot afford to place itself in the over-modest position of waiting patiently about for social science to appear, because in the first place, it might not appear at all [...]. (Sommerville 1936, 406–407)

Sommerville's narrative has some well-known lessons to offer: logical empiricists, as they are usually presented, did not consider the social sciences, which they viewed as less-developed fields of inquiry and thus ignored in order to gather the fruits of the natural sciences. As the social sciences required much greater effort and different types of work and attitude, the logical empiricists simply disregarded them. Neurath indeed spent a lot of his time arguing that, contrary to the natural sciences, the social sciences are often characterized by a certain amount of instability and by a lack of consensually agreed principles and methods. Nevertheless, Sommerville's argumentation had a normative element. Logical empiricists should contribute to the field in order to secure it in the first place. On the other hand, he seems to have forgotten that Schlick, Frank and Hans Reichenbach did their best physics-related work at a moment when physics itself was becoming quite instable. These logical empiricists thus had a certain amount of freedom in their attempts to incorporate the latest findings from the physical sciences into their conceptual frameworks as philosophers. If there was a difference between the logical empiricist approaches towards the natural and the social sciences, it was, consequently, mainly a question of degree.

Besides pursuing moral philosophy and philosophy of the social sciences as theoretical enterprises, some logical empiricists also became substantially involved in *political activism*. Even though there were obvious connections between Neurath and the Austrian Social Democratic Party (one of its leaders, Otto Bauer, gave a talk at the Ernst Mach Society), for the logical empiricists political activism did *not* mean party politics. Rather, they were quite busy delivering public lectures, writings pamphlets explaining the latest results in science and philosophy, establishing educational institutes for laymen, and organizing seminars at adult education centers – that is, they did politics in a very broad sense (Romizi 2012). Though for some members, these activities were motivated by financial and thus existential reasons (e.g. for Friedrich Waismann, who did not have any academic position at the university), others were galvanized by substantial democratic motivations (like Carnap, Zilsel, Hahn and Neurath, perhaps even Schlick to a certain degree, as this volume will testify). They aimed to emancipate and empower the masses in order to assess the newly emerged possibilities of modern life and to consciously avoid the deteriorative webs of oppressing ideologies and propaganda.

Though we do not presuppose a direct link between the theoretical content of philosophy and political activism – that is, we will not take at face value any claims that political activism in itself is a form of philosophy – it can be argued that the majority of Neurath's scholarly activism was induced by social concerns expressed as activism. Therefore, in his case, the relation between practice (activism) and theory (systematic philosophical claims) was quite direct and substantial (see Cartwright, Cat, Fleck, and Uebel 1996; Howard 2019). In fact, Neurath often argued that logical empiricism *as a form of philosophy* is directly connected to certain broadly taken political values, so that the "encyclopedism of logical empiricism [...] with the unified science encyclopedia are the children of the tolerant approach of democratic cooperation [...] [they are] anti-totalitarian through and through" (Neurath 1946/1983, 242).

Political activism is engagement for society within society. This may take philosophical forms, as we have seen in the previous paragraph, but philosophy done in a society for society does not have to be political activism from the first instance. Though Rudolf Carnap (1963, 23) famously claimed that everyone in the Vienna Circle was "strongly interested in social and political progress," and that they were even socialists, they "liked to keep [their] philosophical work separated from [their] political aims." As such, the logical empiricists, as concerned and endangered citizens, actively took important steps in social matters (most of these are described by George Reisch (2005), at least regarding Carnap's American period), but as philosophers, they did their work independently of and with a neutral stance regarding social concerns. While Neurath excluded certain scientific inquiries and questions *because* of their *possible* totalitarian and dangerous connotations, Carnap (1963, 23) argued that other Circle members "defended the right to examine objectively and scientifically all processes or alleged processes without regard for the question of whether other people use or misuse the results."

The authors of the group's manifesto (most probably Neurath) described this attitude as "falling back on the icy slopes of logic", namely to such fields and forms of philosophy where they could sit back and do their job without taking any risks or social engagements. History proved otherwise, however, and many people recognized the implicit socio-political power of logical empiricism. Take, for example, the *Seventh International Congress for Philosophy*, which was held in Prague in 1934. After the Congress, the Hungarian philosopher István Boda described his experience of the event in general, and regarding the sections on logic in particular, in the Hungarian journal *Athenaeum*. He claimed that those who expected any guidance or deeper wisdom from philosophy would have been disappointed by the logic sections: "most of the logisticians are completely alien towards the reality of life: their whole field of problems lies outside the world of 'reality'" (Boda 1935, 76–77). For a congress that purported to address questions of democracy and its enemies in the 1930s, it seems surprising that a significant part of the event was limited to abstract and otherworldly concerns. More importantly, at least for our present considerations, Boda also drew some social conclusions:

> The huge number of logistic presentations, as well as the *extraordinary* and surprisingly big attention towards them, could already be interpreted in itself as a characteristic sign of the times (especially as a tendency that is contrary to the official [socio-political] goals of the congress): the philosophical concern of the most modern today does not *dive* into the reality of life in order to reform it in accordance with the products of wisdom and philosophy, but wants *to get ever further away* from the sad world of today – and instead dives into the

> abstract world of "cold" signs and equations of mathematical logic that is alien to any life and "reality". (Boda 1935, 77; my translation, original emphases)

Boda seems to imply that the turn towards the "cold signs of mathematics" (or, as they were called in the manifesto, "the icy slopes of logic"), was itself an expression of socio-political considerations. If logical empiricism, as a logistic movement, entailed such a withdrawal from public life, then it was already making a move into the space of social concerns. As Neurath pointed out many times in his correspondence with Carnap (and presumably also at the Circle meetings), avoiding social engagement is also a form of social engagement.

However, following recent historical investigations into the background of the young Carnap, as well as into the philosophical toolkits of the mature Carnap, we know that there were times when he acted socio-politically not just as a citizen, but as a philosopher. We know that as a student, he actively participated in the so-called German Youth Movement that had a quite stratified social dimension (see the essays in Damböck, Sandner, and Werner, 2021), and that he later connected his *Aufbau* to the modernist movements in architecture, education, science, and the arts. There were thus times when Carnap's philosophy explicitly addressed practical and social matters (Damböck 2017). Even if sometimes in hidden form, "Carnap intended his philosophy to make a difference to everyday life: it was to be *political* in the broadest sense", as Thomas Uebel (2012, 133, original emphasis) has claimed.

One might argue that the young Carnap was merely idealistic, and as soon as he found his real voice in the "logic of the sciences" program and fell back onto the icy slopes of logic, he excluded his social and political concerns from his philosophical ones. Nonetheless, A. W. Carus (2007) has argued in great detail that the mature Carnap's ideal of explication could also be read (at least with a grain of salt) as an approach to settling practical questions. Carus has shown how, through the dialectics of explication, Carnap was able to render the supposedly sharp theory-practice distinction into a sophisticated machinery of decision processes (for a further elaboration of this idea, see Dutilh Novaes 2020 and Pearson 2017).

But it was not Carus who first spotted this political (or social) dimension of Carnap's late philosophy. Though he mentioned that this idea goes back to his teacher, Howard Stein, it had already been recognized by mid-century. In his short article on the subject, titled "The 'Political' Philosophy of Logical Empiricism", Warner A. Wick, then a colleague of Carnap at Chicago, argued that even though logical empiricists do not have any explicit and sharp social or political philosophy, their philosophy "is [...] what may be properly called a political or social philosophy" (Wick 1951, 49). The main example and source of reference for Wick was Carnap's "Empiricism, Semantics, and Ontology", which had recently been published and in which Carnap famously scrutinized his philosophy of language, his ideas on ontology and their relation to one another. The core argument of the paper is that ontological questions take different levels and forms, and in order to make them theoretically relevant and meaningful, they need to be relativized to what Carnap called "linguistic frameworks". Within these fixed (but still flexible) frameworks, certain rules and principles of construction determine the ontological questions (either analytically or empirically). Outside these frameworks, absolutized ontological questions – that is, questions along the lines of "are there *really* such and such entities?" – are cognitively meaningless as they cannot be answered by any cognitive means.

Nonetheless, Carnap was not a hardheaded eliminativist; he was not afraid to postulate entities and to raise ontological issues. He thus argued that most of these seemingly cognitive, but in fact absolutized questions are practical decisions in disguise. When we ask the question whether there really are numbers, the only meaningful question that we can, in fact, ask fruitfully is whether we should introduce a framework through which we are able to discuss the question of numbers; that is, whether such a number framework is expedient for *our given aims* or not. After we *decide* to introduce a new framework and determine its rules, we can also raise theoretical-cognitive issues. Consequently, there is an obvious route from decisions to theories and knowledge.

As Wick (1951, 53) argued, "it is evident that we have here a view according to which – to use a rather antique mode of speech – practical reason (if indeed it be a reason) and its maxims are constitutive of the realm of empirical truth as well as moral action." By bringing in "moral actions", Wick (1951, 50) arrived at his main issue of interest, namely that the decisions behind theoretical frameworks "are social in the manner of those by which members of any community settle upon rules for the conduct of their common business." In the social determinateness of the practical decisions behind theoretical issues, Wick identified a hidden, but still quite substantive role for political and social concerns within analytic philosophy. Even though the logical empiricists (or the majority of them) did not have any direct and sophisticated political and social theory, their own theories and philosophies are highly political in a broad but still useful and traceable way.

With all these conceptions in mind, it could be argued – and admittedly, not entirely without reason – that we have simply emptied out the whole nexus of problems by claiming that all the above-mentioned works and figures can be described as having socio-ethical concerns, either explicitly or implicitly. After all, as these logical empiricists worked in a neatly specific cultural milieu (Red Vienna slowly turning into Black Vienna), it was presumably hard to escape being entangled in political matters (see Stadler 2003). Karl Menger described this experience directly and vividly in his reminiscences:

> While the political situation in Austria during the winter of 1933-34 made it extremely difficult to concentrate on pure mathematics, socio-political problems and questions of ethics imposed themselves on everyone almost every day. In my desire for a consistent, comprehensive world view I asked myself whether some answers might not come through exact thought. [...]
>
> What I had been looking for, however, was a theory of ethics – an application of exact thinking that would bear to traditional ethics a relation somewhat comparable to that of mathematical to traditional logic. (Menger 1994, 181, 184)

In other words, the social circumstances forced themselves on the logical empiricists, making it hard for them to escape (physically and intellectually) the associated political pressures. The forms of reactions differed, of course, but everyone *acted* in one way or another.

The concern to produce new and investigate old knowledge was therefore not just a simple cognitive issue of exploring the structure of the world around us, but a form of *shaping it*. These "shapings" took many forms and levels, and it would thus perhaps be unfair to start with existing frames and produce sharp definitions at the outset that would match what comes later only at the price of making cuts here and there. Even if terms such as "morality", "ethical", "social", "political" and "engagement" are flexible and seemingly lost in transition, they in fact testify to the idea that there are *many ways* to engage in philosophy as a socially sensitive enterprise. As such, only specific case studies can bring to light the real force and structure of how the *logical empiricists* envisioned the mission of *logical empiricism*. This volume is a minor but determined attempt in that direction.

1.3 The Papers in the Volume

In her chapter, Anne Siegetsleitner provides an introduction to the main topics and issues of morality in logical empiricism. One of her main goals is to disprove the received view regarding ethics and moral philosophy as too undifferentiated. In the first part of the paper, she presents central reasons for the rejection of the prevailing view. On the other hand, she puts forth for further discussion some theses about why these findings are relevant even for mathematical philosophy, or rather, for mathematical philosophers.

Alan Richardson's essay reads the structural account of objectivity in the early work of Rudolf Carnap and Hans Reichenbach through the lens of Lorraine Daston and Peter Galison's work on objectivity but differs in its account of what "dangerous subjectivity" structural objectivity was meant to overcome. He argues that Daston and Galison's "carceral self" is at best a metaphorical figure for the real issue that needed to be addressed: the social irresponsibility of disciplines like philosophy itself that, in the view of Carnap and Reichenbach, did not adequately concern itself with using a language that had properly intersubjective meanings. Carnap and Reichenbach demand that philosophers adopt the responsible approach to language routinely deployed in the exact sciences.

After Reichenbach and Carnap, Andreas Vrahimis turns to the philosophy of Moritz Schlick and how analytic philosophers viewed the role of Nietzsche's philosophy in the First World War. The first, mainly journalistic, controversy concerned the influence that Nietzsche's ideas may have had on what British propagandists portrayed as the ruthlessly amoral German foreign policy. This soon gave way to a second controversy, waged primarily among academics, concerning the purportedly vicious political outcomes of German Idealism, from Kant through to Fichte, Schelling, and Hegel. During the autumn of 1914, and at the cusp between the two controversies, Moritz Schlick was to deliver a lecture series on Nietzsche's life and work at the University of Rostock. Responding to both debates, Schlick penned an introduction in which he sought to defend philosophy against all those who would embroil it in warfare. Though Schlick's metaphilosophical views change, as this paper shows, he remains constant both in his favourable appraisal of Nietzsche, as well as his separation between politics on the one hand, and both philosophy and *Wissenschaft* on the other hand.

Schlick is the main target of GERGELY AMBRUS as well. In the paper, he discusses Schlick's *Fragen der Ethik*, at first analyzing Schlick's metaethical views, his arguments against objective values and absolute norms, and his analysis of the meaning of "good" in general and "morally good" in particular. Further, he also tests these views against some plausible objections. The second part of his paper addresses an apparent tension in Schlick's conception regarding descriptive and normative elements. Ambrus suggests that the tension may be dissolved by turning to Schlick's anthropology.

The third paper about Schlick is delivered by ADAM TAMAS TUBOLY who argues for the following three theses: (1) Schlick never overcame the problems of "traditional philosophy" and thus tried to find a way to embrace them from his own point of view. He did not dismiss them thus, but actually held them in high esteem and regarded them as serious intellectual riddles. (2) Though Schlick's broad interests never came close to the breadth of Neurath's general scholarly concerns, both of them shared a comprehensive attitude towards questions of life in general; nonetheless, Schlick's interest had a quite peculiar theoretical and intellectualist twist, contrary to Neurath's practice-oriented, explicitly value-laden and socially engaged conception. From Neurath's perspective, Schlick thus emptied out the whole field. (3) Finally, he will argue that Schlick was not the only one who tried to streamline the movement's thinking and hammer out the deviations (which manifested themselves in allegedly simplistic, formalistic tendencies): Richard von Mises's major work on positivism was written (and read by many) in a similar vein, that is, as a reinvention of classical philosophical questions in a peculiarly positivist/empiricist framework. The main point, or the narrative that he will reconstruct is that there are *different ways* of making logical positivism *less logical*, in order to demonstrate its power of embracing classical questions and problems in new settings.

In his chapter, FONS DEWULF discusses Rudolf Carnap and Leo Apostel. The former's ideas about value judgments amount to a well-known, classical account of non-cognitivism in twentieth century meta-ethics. The latter's ideas on the relation between philosophy and ethics are virtually unknown, although, as he will show, they are a direct result of his interaction with Rudolf Carnap and arguably develop Carnap's position in an interesting way. Apostel's attempt to develop Carnap's views on ethics has until now remained hidden from sight. Uncovering the attempt is not only interesting because it sheds new light on Carnap's ethical position, but also because it sheds light on the politically activist reception of logical empiricist philosophy in the post-war period. Contrary to what most commentators have argued for, logical empiricism did not entirely lose all of its political potential after the Second World War. It did not entirely die out in Europe to become a stale, politically neutral logic of science in the United States. In Belgium, through the activity of Leo Apostel, it re-acquired some of its interbellum political agenda.

Edgar Zilsel and Otto Neurath were among the few intellectuals who subjected the concept of genius to critical examination during the first half of the twentieth century, as it will be shown by GÜNTHER SANDNER. Although they lived during the same time and their intellectual and political biographies cross at numerous points, Zilsel and Neurath worked on their examinations at different times. While Zilsel dealt with the topic during the 1910s and 1920s, Neurath addressed the topic only after he emigrated (in 1934 resp. in 1940) and in a different historical context: he dealt with the figure of the genius in connection with his reflections on the causes of National Socialism and its positive ideological acceptance within wide parts of German (and Austrian) society. Despite different starting points and, at least partly, different cognitive interests, their texts and thoughts on the concept of genius have a lot in common, that is, despite the different thematic, cultural, and political contexts of their reflections, both Edgar Zilsel and Otto Neurath understood that the idea of genius was not only a cultural question but also a political one.

PHILIPPE STAMENKOVIC discusses Philipp Frank's ideas on relativity in their context. In his 1950 *Relativity: a Richer Truth*, Frank promotes a generalized "relativism", or "doctrine of the relativity of truth", requiring the specification of the meaning of the terms used in the statements one makes (their "relativization"), not only in physics but also in ethics, politics or everyday experience. This is achieved by providing the "frame of reference" relative to which the terms are expressed and on which their truth in fact depends, in analogy with Einstein's (special) theory of relativity. He grounds this application of the concept of relativity outside physics on his empiricist interpretation of Einstein's theory, together with a pragmatic and operationalist theory of meaning, which insists on the qualification, and thus relativity, of particular statements. Frank's interpretation of relativity (both within and outside physics) is fully understandable not only because of his logical empiricist stance, but also given his fight against philosophical misinterpretations of the theory of relativity, as well as against political totalitarianism. It also has particularly convincing features. Still, Frank's interpretation of relativity in physics is debatable, since Einstein's theory could just as well be read as a defense of absolutism. What is more, his application of relativity to ethics and politics is disputable, as it allows for a kind of moral relativism.

The volume concludes with HANS-JOACHIM DAHMS' chapter that was presented at the conference. That time Dahms thought it timely to ask, whether Trump's adiminstration required a response from philosophy of science too. He tries a definition and classification of fake-news and then gives some historical information about their use up to the present day. Furthermore, he enters the field of the history of philosophy: have there been any philosophers, who contributed to current irrationality and the relativization of science? Dahms concentrates on Paul Feyerabend and his – up to now – neglected treatment of witchcraft-theory. Finally, he turns to a more positive outlook, namely the scientific world-conception of the Vienna Circle: what is to be learned for todays situation from their fight against metaphysics in the 1920th and 1930th? Dahms asks that in which respects was (and is) their conception perhaps not enough in the struggle with today's irrationalism?

Acknowledgements This paper and volume were supported by the MTA BTK Lendület *Morals and Science* Research Group, by the MTA Lendület Values and Science Research Group, and by the MTA Premium Postdoctoral research scholarship. The work and discussion of the contributors and referees of the papers are gratefully acknowledged.

References

Boda, István. 1935. "A VIII. nemzetközi filozófiai kongresszus" ["The 8th international philosophy congress"]. *Athenaeum* 21: 75–79.

Carnap, Rudolf. 1934. Theoretische Fragen und praktische Entscheidungen. *Natur und Geist* 2 (9): 257–260.

———. 1963. 'Intellectual Autobiography' and 'Replies and Systematic Expositions'. In *The Philosophy of Rudolf Carnap*, ed. P.A. Schilpp, 1–84. LaSalle: Open Court.

Cartwright, Nancy, Jordi Cat, Lola Fleck, and Thomas Uebel. 1996. *Otto Neurath: Philosophy Between Science and Politics*. Cambridge: Cambridge University Press.

Carus, André W. 2007. *Carnap and Twentieth-Century Thought: Explication as Enlightenment.* Cambridge: Cambridge University Press.

Cat, Jordi, and Adam Tamas Tuboly, eds. 2019. *Neurath Reconsidered: New Sources and Perspectives*. Cham: Springer.

Damböck, Christian. 2017. *»Deutscher Empirismus«. Studien zur Philosophie im deutschsprachigen Raum 1830–1930*. Dordrecht: Springer.

Damböck, Christian, Günther Sandner, and Meike Werner, eds. 2021. *Logical Empiricism, Life Reform, and the German Youth Movement/Logischer Empirismus, Lebensreform und die deutsche Jugendbewegung*. Dordrecht: Springer.

Dewey, John. 1939. *The Theory of Valuation*. Chicago: University of Chicago Press.

Dutilh Novaes, Catarina. 2020. Carnapian Explication and Ameliorative Analysis: A Systematic Comparison. *Synthese* 197: 1011–1034.

Frank, Philipp. 2021. *The Humanistic Background of Science*. Edited by G. Reisch and A.T. Tuboly. Albany, N.Y.: SUNY Press.

Hempel, Carl G. 1969. Logical Positivism and the Social Sciences. In *The Legacy of Logical Positivism*, ed. P. Achinstein and S.F. Barker, 163–194. Baltimore: The Johns Hopkins Press.

Howard, Don. 2019. Otto Neurath: The Philosopher in the Cave. In Cat and Tuboly 2019: 45–65.

Kant, Otto. 1932. *Zur Biologie der Ethik*. Wien: Springer.

Kaufmann, Felix. 1936/2014. *Felix Kaufmann's Theory and Method in the Social Sciences*. Edited by R. S. Cohen and I. K. Helling. Dordrecht: Springer.

Köhler, Wolfgang. 1938. *The Place of Value in a World of Facts*. New York: Liveright Publishing Corporation.

Kraft, Viktor. 1937/1981. *Foundations for a Scientific Analysis of Value*. Dordrecht: D. Reidel.

Krohn, Wolfgang, and Diederick Raven. 2000. The 'Zilsel Thesis' in the Context of Edgar Zilsel's Research Programme. *Social Studies of Science* 30 (6): 925–933.

Menger, Karl. 1934/1974. *Morality, Decision and Social Organization*. Dordrecht: D. Reidel.

———. 1994. *Reminiscences of the Vienna Circle and the Mathematical Colloquium*. Dordrecht: Kluwer.

Nemeth, Elisabeth. 2003. Philosophy of Science and Democracy: Some Reflections on Philipp Frank's *Relativity—A Richer Truth*. In *Wissenschaftsphilosophie und Politik / Philosophy of Science and Politics*, ed. M. Heidelberger and F. Stadler, 119–138. Wien/New York: Springer.

———. 2011. Edgar Zilsel on Historical Laws. In *Explanation, Prediction, and Confirmation*, ed. D. Dieks et al., 521–532. Springer.

Neuber, M. 2022. Nagel on the Methodology of the Social Sciences. In *Ernest Nagel: Philosophy of Science and the Fight for* Clarity, ed. M. Neuber and A.T. Tuboly, 215–232. Cham: Springer.

Neurath, Otto. 1913/1983. The Lost Wanderers of Descartes and the Auxiliary Motive. In *Otto Neurath: Philosophical Papers 1913–1946*, ed. R.S. Cohen and M. Neurath, 1–12. Dordrecht: D. Reidel.

———. 1921/1973. Anti-Spengler. In *Empiricism and Sociology*, ed. M. Neurath and Robert S. Cohen, 158–213. Dordrecht: D. Reidel.

———. 1946/1983. The Orchestration of the Sciences by the Encyclopedism of Logical Empiricism. In *Otto Neurath: Philosophical Papers 1913–1946*, ed. R.S. Cohen and M. Neurath, 230–242. Dordrecht: D. Reidel.

Pearson, James. 2017. Carnap, Explication, and Social History. *Social Theory and Practice* 43 (4): 741–774.

Reisch, George. 2005. *How the Cold War Transformed Philosophy of Science: To the Icy Slopes of Logic*. New York: Cambridge University Press.

Richardson, Alan. 2003. Tolerance, Internationalism, and Scientific Community in Philosophy: Political Themes in Philosophy of Science, Past and Present. In *Wissenschaftsphilosophie und Politik/Philosophy of Science and Politics*, ed. M. Heidelberger and F. Stadler, 65–89. Wien/New York: Springer.

Romizi, Donata. 2012. The Vienna Circle's 'Scientific World-Conception': Philosophy of Science in the Political Arena. *HOPOS: The Journal of the International Society for the History of Philosophy of Science* 2 (2): 205–242.

Schlick, Moritz. 1908. *Lebensweisheit*. München: C.H. Beck.

———. 1930/1939. *Problems of ethics*. New York: Dover.

Siegetsleitner, Anne, ed. 2010. *Logischer Empirismus, Werte und Moral. Eine Neubewertung*. Wien, New York: Springer.

———. 2014. *Ethik und Moral im Wiener Kreis. Zur Geschichte eines engagierten Humanismus*. Böhlau: Wien.

Sommerville, John. 1936. Logical Empiricism and the Problem of Causality in Social Science. *Erkenntnis* 6: 405–412.

Stadler, Friedrich. 2003. Zur politischen Relevanz und zum kulturellen Kontext des Logischen Empirismus. In *Wissenschaftsphilosophie und Politik / Philosophy of Science and Politics*, ed. M. Heidelberger and F. Stadler, 9–24. Wien/New York: Springer.

Stadler, Friedrich, and Thomas Uebel, eds. 2012. *Wissenschaftliche Weltauffassung. Der Wiener Kreis. Hrsg. vom Verein Ernst Mach (1929)*. Vienna/New York: Springer.

Stanley, Matthew. 2008. Mysticism and Marxism: A.S. Eddington, Chapman Cohen, and Political Engagement Through Science Popularization. *Minerva* 46 (2): 181–194.

Tuboly, Adam Tamas. 2017. Philipp Frank's Decline and the Crisis of Logical Empiricism. *Studies in East European Thought* 69 (3): 257–276.

Uebel, Thomas. 2000. Logical Empiricism and the Sociology of Knowledge: The Case of Neurath and Frank. *Philosophy of Science* 67: 138–150.

———. 2007. Philosophy of Social Science in Early Logical Empiricism: The Case of Radical Physicalism. In *The Cambridge Companion to Logical Empiricism*, ed. A. Richardson and T. Uebel, 250–277. Cambridge: Cambridge University Press.

———. 2010. Opposition to *Verstehen* in Orthodox Logical Empiricism. In *Historical Perspectives on Erklären and Verstehen*, ed. U. Feest, 291–309. Dordrecht: D. Reidel.

———. 2012. Carnap, Philosophy, and 'Politics in its Broadest Sense'. In *Rudolf Carnap and the Legacy of Logical Empiricism*, ed. R. Creath, 133–148. Dordrecht: Springer.

———. 2019. Neurath on Verstehen. *European Journal of Philosophy* 27 (4): 912–938.

Wick, Warner A. 1951. The 'Political' Philosophy of Logical Empiricism. *Philosophical Studies* 2 (4): 49–57.

Chapter 2
Ethics and Morality in the Vienna Circle

Anne Siegetsleitner

Abstract The paper provides an introduction to the main topics and issues of morality in logical empiricism. One of the main goals is to disprove the received view regarding ethics and moral philosophy as too undifferentiated. In the first part of the paper, I will present central reasons for the rejection of the prevailing view. In the last part, I put forth for further discussion some theses about why these findings are relevant even for mathematical philosophy, or rather, for mathematical philosophers.

Keyword Vienna Circle · Ethics · Morality · Scientific humanism · Logical empiricism · History of philosophy

2.1 Introduction

This contribution to ethics and morality in the Vienna Circle is an extended version of an invited talk given at a workshop at the Munich Center of Mathematical Philosophy. For at least two reasons, this invitation in the context of mathematical philosophy was not to be taken for granted. The first pertains to the fact that mathematical philosophy is not closely associated with morality or ethics. The second is that the Vienna Circle itself is not closely associated with morality or ethics, either. However, I hoped that these reservations would vanish or at least be diminished by the end of my talk and my impression is that my hope was fulfilled. In keeping with the purpose of the workshop, my paper aims to provide thus some input from the past by giving a selective overview and discuss whether this is of more than antiquarian interest.

A. Siegetsleitner (✉)
Department of Philosophy, University of Innsbruck, Innsbruck, Austria
e-mail: anne.siegetsleitner@uibk.ac.at

C. Damböck, A. T. Tuboly (eds.), *The Socio-Ethical Dimension of Knowledge*,
Vienna Circle Institute Yearbook 26, https://doi.org/10.1007/978-3-030-80363-6_2

From the Vienna Circle and other European philosophical groups (especially the Berlin Group) emerged a philosophical movement that now is known as *Logical Empiricism*, *Logical Positivism* or *Neopositivism*. The characterization as well as the relationship of this movement to ethics and morality is not considered difficult without reason. Nevertheless, the questions that arose in this context are worth reconsidering. This may be surprising to someone who still holds at least one of the two main prejudices concerning the Vienna Circle that I will concentrate on. If we take the term "morality" to refer to a certain common practice of daily life you are all sufficiently familiar with and the term "ethics" to the philosophical inquiry of morality, these prejudices may be formulated in the following way.

1. Most members of the Vienna Circle were not interested in morality at all, not as philosophers and scientists and presumably not even as human beings and citizens.
2. As far as some members of the Vienna Circle regarded morality in a philosophical way, that is, practiced ethics, they supported the standard view of logical empiricist ethics.

I will explain what I mean by "the standard view of logical empiricist ethics" in Sect. 2.3.

The important issue for here is that these two theses constitute the prevailing view on the role and conception of ethics and morality in the Vienna Circle. In my monograph on this topic (Siegetsleitner 2014a), I disprove this view and reject it as too undifferentiated. Although the monograph treats the ethical main topics and positions as those developed in the respective personal and cultural contexts, I cannot address these questions here. Instead, in the first part of my paper, I will present central reasons for my rejection of the prevailing view. In the last part, I put forth for further discussion some theses about why these findings are relevant even for mathematical philosophy, or rather, for mathematical philosophers.

2.2 Scientific Humanism

As far as the first thesis of the prevailing view is concerned, it is of great interest that most members of the Vienna Circle not only were interested in morality but that core members of the circle even shared an enlightened and humanistic version of morality. This further motivated them to a certain respective practical engagement. They endorsed what Carnap called "scientific humanism" in his "Intellectual Autobiography" published in 1963 (Carnap 1963a). By the way, Feigl had already used the expression in 1949 (Feigl 1949/1981, 368).

Scientific humanism is a moral position comprised of the following four principles.

1. Human living conditions and life should be improved.
2. Whatever can be done to improve human living conditions and life is the task of human beings themselves. ("man has no supernatural protectors or enemies")

3. Many sufferings can be avoided.
4. Science is one of the most valuable instruments in the improvement of human living conditions and life (Carnap 1963a, 83).[1]

The first principle states the fundamental normative dimension of this moral stance. The other principles identify whose task this is, what can be achieved and the importance of science in this endeavor. Carnap (1963a, 83) writes, "[...] mankind is able to change the conditions of life in such a way that many of the sufferings of today may be avoided and that the external and the internal situation of life for the individual, the community, and finally for humanity will be essentially improved". Scientific humanism considers science one of the most valuable tools in achieving these aims. Hence, Carnap (1963a, 83) continues: "all deliberate action presupposes knowledge of the world, that the scientific method is the best method of acquiring knowledge and that therefore science must be regarded as one of the most valuable instruments for the improvement of life". As far as we know, this moral position was shared by most members of the Vienna Circle, as Carnap stated in his autobiography.[2]

This common view can also be found in the program pamphlet of the Verein Ernst Mach *Scientific Conception of the World: The Vienna Circle* published in the year 1929. Written by Neurath, Hahn and Carnap, it is also called "the manifesto". It clearly states:

> The attitudes toward questions of life also showed a noteworthy agreement, although these questions were not in the foreground of themes discussed within the Circle. For these attitudes are more closely related to the scientific world-conception than it might at first glance appear from a purely theoretical point of view. For instance, endeavours toward a new organization of economic and social relations, toward the unification of mankind, toward a reform of school and education, all show an inner link with the scientific world-conception; it appears that these endeavours are welcomed and regarded with sympathy by the members of the Circle, some of whom indeed actively further them. (Neurath et al. 1929/1973, 304f.)

Scientific humanism was closely linked to the scientific world-conception, and most members of the circle were united in endorsing it. Since scientific humanism is a very general attitude, it was interpreted and specified in several ways. As far as political engagement was concerned, Neurath, for example, combined it with a socialist political agenda, Schlick with a more liberal one.

In addition, the manifesto outlined the historical roots of this moral attitude. It invokes eudaimonism, liberalism, and utilitarianism. In addition to antimetaphysical research and interdisciplinarity, the manifesto generally stresses principles of this-worldliness and life-affinity (*Lebensverbundenheit*) (Stadler 1985, 117). These principles are pronounced in scientific humanism.

Because science is regarded as of prime importance to this conception of morality and logic as central to science, even Carnap's logical work is no retreat from

[1] Carnap only mentions three convictions, because he does not state principle 1 explicitly.

[2] "A number of my friends in the Vienna Circle probably shared these views in their essential features; but in detail, naturally, there were important differences" (Carnap 1963a, 83).

engagement but one form of relevant practice. The well-known sentences in the last paragraph of the manifesto hence read:

> Of course not every single adherent of the scientific world-conception will be a fighter. Some, glad of solitude, will lead a withdrawn existence on the icy slopes of logic; some may even disdain mingling with the masses and regret the "trivialized" form that these matters inevitably take on spreading. However, their achievements too will take a place among the historic developments. We witness the spirit of the scientific world-conception penetrating in growing measure the forms of personal and public life, in education, upbringing, architecture, and the shaping of economic and social life according to rational principles. *The scientific world-conception serves life, and life receives it.* (Neurath et al. 1929/1973, 317f., original emphasis)

Let me conclude: Quite contrary to the first thesis of the prevailing view, most members of the Vienna Circle at least as human beings and citizens were highly interested in morality, namely in a certain conception of morality that was called scientific humanism. Its holders are convinced that it is the task of humanity itself to improve human living conditions and life and that science is one of the most valuable means to this improvement.

Some analytic philosophers and supposedly mathematical philosophers still are – although rather implicitly – motivated by this general moral position as a background assumption. Therefore, some feel they are part of a morally good or even the best movement, which they see as inherently linked to analytic methods. Not surprisingly, an enemy to these methods is sometimes regarded as more than a philosophical enemy.

2.3 The Standard View

I will now turn to the standard view of moral philosophy generally ascribed to the Vienna Circle in the second thesis of the prevailing view. As I have mentioned, this thesis is often meant as a reproach.

The standard view of logical empiricist moral philosophy is characterized by (1) first, the acceptance of descriptive empirical research – but this research is not regarded as genuine moral philosophy – and (2) second, by the rejection of normative and substantial ethics. Metaethics remains the only legitimate way of doing moral philosophy. In addition, and third (3), within metaethics, logical empiricist moral philosophy is characterized by an extreme form of non-cognitivism (for further details, see Siegetsleitner 2014a and Damböck 2021). By "an extreme form of non-cognitivism", I mean the thesis that there is no room for knowledge and/or rationality in moral matters.

However, not all members supported the standard view of logical empiricist ethics. As I have shown earlier (Siegetsleitner 2014a), Carnap's position is nearest to the standard view, although not in all respects and not at all times. Menger's, Neurath's and Frank's positions already depart in more respects. Schlick's, Kraft's and Feigl's views, finally, deviate the most from it. In contrast to the commonly held picture of ivory tower logicians and metaethicists, some members of the Vienna Circle even practiced applied ethics. To my own surprise, Schlick even used this very expression, which was common in the Ethical Movement at the time.[3]

Let me now explain the deviations from the standard view in more detail by looking at the position of some members. Due to space limits, I cannot consider here all the members I have mentioned.

2.4 Proponents and Deviants

2.4.1 Rudolf Carnap (1891–1970)

Because Carnap is the member of the Vienna Circle whose conception of moral philosophy is closest to the standard view, I will present relevant aspects of his position first. In his well-known article "Überwindung der Metaphysik durch logische Analyse der Sprache" ("Elimination of Metaphysics through Logical Analysis of Language", Carnap 1931/1932), one finds nearly all elements of this view. In the English-speaking world, this perspective was widely spread by Ayer's *Language, Truth, and Logic* (1936) and Carnap's own *Philosophy and Logical Syntax* (1935). The latter contains the published version of three lectures Carnap gave at the University of London during his first visit to Great Britain in 1934.[4] Prior to this position, Carnap was more closely associated with phenomenological and Neo-Kantian approaches regarding questions of value (see, e.g., Mormann 2006 and Siegetsleitner 2014a). The traces of these approaches can be found in Carnap's *Aufbau* (1928, §152). I will not elaborate on this but instead on Carnap's position during and after his time as a member of the Vienna Circle.

First, in Carnap's conception of morality during his Viennese time, the content of morality is simply determined by individual decisions. This is why Hans Zeisel could report in his "Erinnerungen an Rudolf Carnap" about an incident in Vienna:

---

[3] In this respect, a short note about the Ethical Movement and its importance is adequate. The Ethical Movement was initiated in the last decades of the nineteenth century. Its aim was a secular humanist moral life, moral education, and philosophy. In 1890, the movement launched the *International Journal of Ethics*, known as *Ethics* today (see Siegetsleitner 2014b). Carnap and Schlick were members of the Ethical Community in Vienna. They took an active part in this movement. The initiator of the Ethical Movement in Austria was the Viennese philosophy professor Friedrich Jodl. Although he is mentioned in the manifesto, he is unduly nearly forgotten today. This important relationship between logical empiricists and the Ethical Movement still deserves closer consideration.

[4] It was Susan Stebbing who invited Carnap to give these lectures (see Siegetsleitner 2015). Stebbing and Carnap share the humanistic approach as can be seen, for example, through her conviction that morality does not require otherworldly sanctions or religion (see, e.g., her "Ethics and Materialism," 1939). Stebbing also gave ethical lectures at the Sunday meetings of the South Place Ethical Society (e.g., on 10 December 1939 and 19 April 1942).

> There, students asked Carnap: "Do you really mean that morality is something everyone has to decide on for himself?", and he said: "Yes". Hence, they were angry and said: "That means, if you liked to, you could murder and this would be moral?", and he answered: "Yes, in my sense, indeed, but luckily I have no inclination to murder". (Zeisel 1993, 220, my translation)

In "Elimination of Metaphysics through Logical Analysis of Language", Carnap expressed his position that all value statements were merely expressions of a *Lebensgefühl*, moral ones included. However, this was not a conclusion reached by logical analysis but a presupposition. We should take his "in my sense" very seriously.

In *Philosophy and Logical Syntax*, the little book from 1935, one can read in the same vein:

> But actually a value statement is nothing else than a command in a misleading grammatical form. It may have effects upon the actions of men, and these effects may either be in accordance with our wishes or not; but it is neither true nor false. It does not assert anything and can neither be proved nor disproved. (Carnap 1935, 24)

What value statements express are permanent emotional or volitional dispositions (Carnap 1935, 29f.).

That the crucial points are taken for granted still holds true for Carnap's later work in the logical treatment of value statements. As it is known, in his reply to Abraham Kaplan (Carnap 1963b) on value statements Carnap developed a formal language for pure optatives and introduced the optative-operator "utinam". Unfortunately, Carnap could not give an answer either to the question of whether we need such a language for our moral practice or to the question of whether this is an adequate interpretation of the moral vocabulary used in our moral practice.

Carnap's account of individualistic non-cognitivism ultimately turns out to be his interpretation of value statements right from the beginning. This is not problematic as long as one is aware of it. Carnap knew about this:

> Everyone has the right to determine the interpretation of any statement he makes; and the reader has to accept the interpretation of the author unless he finds a discrepancy between the interpretation explicitly stated by the author and that implied in the way in which the author uses the statement or argues about it. (Carnap 1963b, 1000)

One has to accept it as a possible interpretation, however, not to adopt it!

Carnap thus holds an individualistic version of moral epistemic decisionism. According to this position, individual decisions determine whether a moral statement is right or wrong. Expressing this view, he could say in 1964 that what is morally good or bad is "the individual decision of any human according to his conscience or value feeling or however one may call it" (Carnap 1964/1993, 146).

However, this stance does not mean that people cannot opt for the same values or that imperatives cannot stand in logical relation to each other. Carnap neither holds that moral agreement in the sense of shared values is impossible nor that moral or other value statements are in every respect irrational. Imperatives can be rational in their relations among each other and science can tell us which consequences follow our acts and if these consequences are consequences we want to accept (see

Damböck 2017, 190–203; Carus 2017). Furthermore, logical thinking characterized by clarity, consistency, and solidity, may improve moral life, depending on the conception of morality one has chosen. Nevertheless, to constitute, which basic imperatives define morality, so Carnap says, one can only make an individual decision without any formal or material prerequisites or restrictions.[5] The normative system in itself may be rational; the choice for it is not, not even for logical thinking itself as a basic value.

On these basic assumptions, there was little room for normative or substantial ethics. Regardless, Carnap did not use such possibilities at all, although in his later contribution, there opened up room for logic and rationality.[6]

2.4.2 Karl Menger (1902–1985)

Born into a socially and politically engaged family, it was Karl Menger who was closest to Carnap as far as his conception of morality and moral norms is concerned. Although Menger was primarily a mathematician, he was highly interested in questions of its application in economy, social organization, and morality. His main contribution in this respect was *Moral, Wille und Weltgestaltung* (1934/1997), a booklet published in 1934, despite Menger's concern that it could be understood as left radical (Leonard 1998, 19). A revised American version was not published before 1974 with the more sober title *Morality, Decision, and Social Organization. Towards a Logic of Ethics* (1974). In his *Reminiscences of the Vienna Circle and the Mathematical Colloquium* Menger explained his interest in morality and ethics as follows:

> While the political situation in Austria during the winter 1933-34 made it extremely difficult to concentrate on pure mathematics, socio-political problems and questions of ethics imposed themselves on everyone almost everyday. In my desire for a consistent, comprehensive world view I asked myself whether some answers might not come through exact thought. (Menger 1994, 181)

Menger hoped to provide such answers by the study of general conditions for the constitution of cohesive groups and the search "for alternatives to dictates and compromises" (Menger 1994, 190). In his contribution, Menger combines people into groups according to their preferences in respect of moral norms and asks e.g.: Who belongs into one group? Which groups can coexist? Most important, he states the formal conditions of cohesive groups. In this regard and in searching for pluralist and differential options of moral and social organization he makes clear that "a cohesive group may well include competing individuals, but it cannot include

[5] In his first lecture at the *Bauhaus* in Dessau Carnap presents as possible basic moral values e.g. happiness of one self (*Wohlergehen der eigenen Person*), happiness of society (*Wohlergehen der Gemeinschaft*). He himself opts for happiness of society (see Damböck 2017, 200), but this is only his individual decision.

[6] Regarding the importance of morality to Carnap's life see also Siegetsleitner (2019).

members bent on suppressing the decisions of other members" (Menger 1974, 106; see also Siegetsleitner and Leitgeb 2010).

That moral norms have to be of a one fits all kind, is not necessary in Menger's view either. There may be several solutions to moral problems, "each desired by *some people*", what leads Menger to stress the importance of the great middle: "*something* and *some people*" (Menger 1974, 106; original emphases) and to criticize Kant's categorical imperative. Followers of the categorical imperative are in his sense of the word "simple". However, he shows that simplicity is sufficient for self-compatibility, but not for other kinds of compatibility. (Menger 1934/1997, 172f.; 1974, 69) Uniform moral systems may even be an obstacle to universal harmony (1994, 187). Menger, however, thinks of the aim of a morality he would endorse to enable cohesive groups as big as possible (in the extreme: universal harmony) in order to give people the opportunity to satisfy their individual preferences. Moreover, he is not at all interested in separating morality from law, political or economic systems, nor biological theories. Those, who will do so, will even have to accept that morality won't treat the most important questions to modern people (Menger 1934/1997, 67f.; 1974, 2f.).

Like Carnap Menger takes moral norms as imperatives which can neither be true nor false. He sees himself in further accordance with Carnap's epistemological view of ethics. "But what troubled me was the negativity of this view" (Menger 1994, 181). Already in 1934, Menger held that imperatives may stand in logical relations and may be treated by logical-mathematical methods as shown in his booklet. Above all, he distanced himself from declaring moral imperatives as meaningless (Menger 1934/1997, 93; 1974, 20):

> While the Circle in 1934 was still indulging in a great deal of ill-defined talk about meaningful and meaningless, I strictly observed my maxim, *Assert or don't assert or say that you will not assert*. In particular, I explicitly stated in the booklet that I would not call any sentence about good or evil *meaningless*. (Menger 1994, 185, original emphasis)

Furthermore, Menger is similar to Carnap a non-cognitivist decisionist, holding that a moral system is always ultimately based on individual decisions (Menger 1994, 187).

Menger wanted to add something positive to morality as a human practice, conceiving of it as a pluralist system and pointing to "possibilities some of which may have been overlooked or neglected" (Menger 1974, 103). Therefore, he proposed to put an "emphasis on pluralistic organizations of society whenever and wherever such regulations are possible" (Menger 1974, 108). This was one way how science could be used as a valuable instrument in the improvement of human living conditions and life, call it ethics or otherwise. In some sense of the term, Menger was an applied ethicist and he felt that he supported the ideas of the Vienna Circle by his formal treatment of value judgments – "certainly of the Circle as conceived by Schlick" (Menger 1974, 112).

2.4.3 *Otto Neurath (1882–1945)*

Nevertheless, before I will turn to Schlick, let me continue with a brief section on Neurath. Neurath was also convinced that science, and up to a certain extent, philosophy could make a major positive difference to people's living conditions and happiness. His philosophical work has an obvious moral purpose, with scientific humanism as the relevant background.

In his 1928 publication *Lebensgestaltung und Klassenkampf* (*Personal Life and Class Struggle*), Neurath expressed his views on his moral basis. He is above all concerned with orientation towards this world. For him, this means an orientation towards people and their happiness. He believes that this this-worldliness is the best way to link up with modern logicians and physicists. (Neurath 1928/1981a, 284) Human happiness has to be the basis for evaluation and not history, the spirit, a God or whatever was offered theologizing or idealizing as an alternative source of reference and values. Just as in Epicurus:

> With him and his disciples we find no supernatural ideas, enthroned in the clouds. The gods, should they be there, do not care about people, why should people care about them? People who flee suffering and pain here on earth and want to be good to each other, they concern us, happiness, friendship, life as it is *really lived here on earth*; speculation on the other hand only as far as it helps to shape life and makes it happy: "Philosophy is activity in thoughts and speeches that creates a happy life". (Neurath 1928/1981a, 285, my translation)

Accordingly, Neurath sees Marxism and his own position as a kind of social Epicureanism. (Neurath 1928/1981a, 286).

However, although Neurath was concerned about the miserable living conditions of men and women, he restricted his academic work to providing tools for the social sciences to describe and finally improve these circumstances. He developed his famous felicitology using formal and mathematical economic methods with the following agenda (see Stuchlik 2011):

> [...] we shall coordinate the totality of a person's feeling, or that of a group, with his or its entire living condition and investigate the extent to which changes in the "state of felicity" [quality of life] in a positive or negative direction depend upon changes in these conditions. [...] we investigate [...] the conditions under which the totality of feeling becomes more or less pleasurable. Only these elements are significant for our approach to standards of living. We call that standard of living higher which produces a more pleasurable state of felicity characterised by a certain attitude or behavior. (Neurath 1937/2004, 515)

In philosophy, Neurath worked against harmful speculations. He saw Stebbing as an ally when he points out: "It is not only by accident that L. Susan Stebbing wrote on the one hand a book criticizing highly metaphysical speculations of modern physicists and on the other hand her *Thinking to some Purpose* and her *Ideals and Illusions*" (1941, 132). Here we find again a connection to Stebbing, who was an active member of the Unity of Science Movement.

For Neurath, antimetaphysics does not mean renouncing joint principles of life that he at times quite, to others rather not calls "morality". He himself represents a clear moral orientation which provides the framework for his many and varied activities, be it his work in political economy, visual education or housing. The goal of Neurath's morals is to improve the human living conditions and to increase the happiness of all mankind. A valuable instrument for this purpose is science, including basic terminological questions. Nevertheless, in his account morality is ultimately based on a joint decision that even moral science or moral philosophy cannot supersede. Neither science nor philosophy are superior to morality as a practice of life and they are not in a position to reign. Philosophers, who mean this, misjudge their role (Neurath 1932/1981b, 572f.).

Although Neurath was in favor of formal and mathematical methods, their application had to prove their usefulness to his moral and political enterprise. Therefore, he let Carnap know in a letter in 1945 a few days before his sudden death: "In our movement I sometimes have the feeling that some members avoid discussing problems of decision, action etc. and are using analysis as a kind of escape from life. THAT IS NOT MY APPROACH" (Neurath to Carnap, 22 September 1945, letter 34 in Cat and Tuboly 2019, 666).

It was certainly not the approach of Schlick, either, to whom I turn now.

2.4.4 Moritz Schlick (1882–1936)

Schlick deviates from the standard view in fundamental respects. I will concentrate on his stance on the legitimacy of normative ethics.

Basically, Schlick followed the ideal of a wise man, suggesting a worthwhile way of life to people. Accordingly he stated in "The Future of Philosophy":

> [...] clarification of meaning will be needed very badly also in a great many questions with which we are concerned in our ordinary human life. Some thinkers, and perhaps some of the strongest minds among them, may be especially gifted in this practical field. In such instances, the philosopher may not have to be a scientist – but in all cases he will have to be a man of deep understanding. In short, he will have to be a *wise* man. (Schlick 1932/2008, 389)

His first book, *Lebensweisheit* (*The Wisdom of Life* 1908), already deals with ethical questions. Its main focus is on the relation between evolution, happiness and morality. As Schlick takes the evolutionary process as a matter of fact as aiming at happiness but not justifying happiness as the moral goal, calling it an evolutionist ethics in the sense of G. E. Moore is misleading. However, *The Wisdom of Life* was an immature work.

In 1930, Schlick published *Fragen der Ethik* (*Problems of Ethics* 1930/1939), which mostly is about moral psychological issues like psychological hedonism. Nevertheless, and surprisingly, the last chapter proclaims a moral principle on a eudaimonist basis: "Be ready for happiness!" (1930/1939, 187). A close analysis of *Problems of Ethics* shows that we can find at least two concepts of morality in it. The first one is morality as morals, which he understands as a social phenomenon. Morals are not determined by individuals, but individuals work together in creating this social phenomenon. The second concept of morality is morality as a perspective of individual life. Both are taken as eudaimonistic and Schlick thinks that the former by a high chance endorses the latter. Therefore, respecting morals is a practically sound way to a happy life. Essentially, this was already his message in *The Wisdom of Life*. Although most of the parts of *Problems of Ethics* stress the value of philosophy to moral practice by clarification of meaning and introduces ethics as a scientific discipline, it culminates in proclaiming a moral principle for a normative life orientation.

On the same basis, his posthumously published work *Natur und Kultur* (*Nature and Culture* 1952) discusses questions of politics, war, economics, law and technology. It belongs in the category of applied ethics. In Schlick's manuscript "Ethik des modernen Lebens. Eine Kritik der gegenwärtigen Kultur" from the academic year 1927/28 even the term "Angewandte Ethik" [applied ethics] is used (Schlick [Ethik des modernen Lebens], Vienna Circle Archive, Haarlem (NL), Inv.-Nr. 10/A. 18, pp. 2f.).[7]

Schlick's ethical contributions show that there is more continuity than discontinuity between *The Wisdom of Life* (1908) and *Nature and Culture* (written in the mid 30s). Even *Problems of Ethics* is in some parts normative in character. Moreover, there is no strong cut with the moral-philosophical tradition. Schlick distances his account from Kant's, Brentano's and others. Nevertheless, he positions himself in a different line of tradition which is formed by Socrates, Epicurus, Hume, Comte, Mill, Feuerbach, not at least Schopenhauer, Nietzsche, and Guyau.

We have to realize that Schlick practiced normative and applied ethics on a eudaimonist basis. This demonstrates that Schlick clearly believed in the legitimacy of normative ethics, therein differing from the logical empiricist standard view of moral philosophy. He rejected non-cognitivism and emphatically defended an eudaimonist variety of cognitivism.

2.4.5 Victor Kraft (1880–1975)

The last member whose contribution to ethics I want to discuss is Victor Kraft. In contrast to many members of the circle, Kraft had not studied physics or mathematics, but rather philosophy, geography, and history. In 1914, he habilitated in theoretical philosophy. The topic of his thesis was *Weltbegriff und Erkenntnisbegriff* and Jodl was one of his supervisors (Topitsch 1960, IIIf.).

Kraft remained an independent thinker within the circle. Ernst Topitsch writes about Kraft's position in the circle:

> [H]e had already reached maturity, both as a man and as a philosopher, when that group was formed round Moritz Schlick in the 20's and he had by then brought to completion important works entirely of his own. [...], he was far too independent to start again as a disciple. (Topitsch 1976, xi)

[7] On Schlick's ethical works, see further the chapters of Gergely Ambrus and Adam Tamas Tuboly in this volume.

Kraft maintained a critical attitude, for example, towards physicalism and the idea of a unity of science, as Rutte has shown (Rutte 1973, 7).

In 1937, Kraft published the first edition of his *Grundlagen einer wissenschaftlichen Wertlehre* (*Foundations of a Scientific Value Theory*) in the series *Schriften zur wissenschaftlichen Weltauffassung*. The editors were Schlick and Frank.[8] The second, essentially revised edition, followed in 1951. This second version was published in English under the title *Foundations for a Scientific Analysis of Value* in 1981.

Already in the first edition, Kraft distinguishes between the factual content (*Sachgehalt*) and the value character (*Wertcharakter*) of value concepts used in value judgments. In Kraft's account, this value character does not express an individual statement, but an impersonal (*überpersönlich*) and general view: "*A value judgment means [...]: the command to adopt an attitude towards an object, [and to do so] generally and anonymously, not as a particular person with regard to particular persons*" (Kraft 1937, 164, my translation, original emphasis).

According to Kraft, there are pure value concepts like "good" or "bad". However, in most concepts, an additional component is present: a factual content. This component provides the criteria for impersonal judgment. In these cases, Kraft speaks of specialized value concepts. In his theory, moral concepts always belong to this category. If the criteria for these impersonal judgments are empirical, depending on the moral approach, there is plenty of room for rational argument and knowledge. This position is far beyond an extreme non-cognitivism.

In addition, Kraft made numerous attempts to come up with a rational justification of why everybody is bound to certain impersonal moral views. In one of these attempts he regards acting according morality as a necessary condition for reaching aims decreed by human nature. In another, he refers to basic requirements of culture. Unfortunately, all attempts failed (see Siegetsleitner 2014a, 371–383). Nonetheless, he kept trying up until old age. Therefore, when one talks about Kraft's rational justifications of morality in more detail, one has to state which attempt is discussed.

To cut this long story short: Based on these findings, it is wrong that the members of the Vienna Circle generally rejected normative/substantial ethics or endorsed an extreme form of non-cognitivism.

2.5 Lessons for Today?

Why is this historical retrospective important for contemporary and future philosophy? Which are the lessons to learn or aspects to consider? Let me put forth the following points for discussion. Some are more general, some especially important for mathematical philosophers.

[8] For Frank's contribution to questions of ethics see Siegetsleitner (2017).

1. Philosophers who think that following Carnap and being suspicious about normative/substantial ethics and/or cognitivism is the only option for staying close to the Vienna Circle's tradition are mistaken. Even some core members supported a wider range of options. Needless to say, even more options were considered than I could present today.
2. Shared background assumptions unite. In the Vienna Circle, these shared background assumptions did not primarily concern the conceptions of ethics or the use of formal methods, but rather morality itself. Scientific humanism was the common frame of reference. As a general frame, it is open to specific interpretations and supplements. However, we should keep in mind that it is exactly this shared scientific humanism that determines in some respects which philosophical approaches in the inquiry of morality could be considered adequate. A metaethical approach looking for good arguments of why a divine will determines morality was out of question. It would have not made sense within the frame of scientific humanism. Perhaps scientific humanism is open to taking a divine will into consideration when it comes to motivation. Apart from this general exclusion of options, additional premises led to specific metaethical positions like an individualistic non-cognitivism. I deem it important for mathematical philosophers to learn that methodological viewpoints were not constitutive in this respect. It was not the method but the account of what morality is about that united the circle in the first place.
3. A philosopher who engages in ethics cannot dismiss his or her conception of morality entirely as soon as it comes to questions of adequacy. The concentration on logic or other formal methods might not provide an escape. For mathematical philosophers, this means that in many cases, deciding on a method means at the same time deciding on a fundamental assumption about what morality is about or how moral language works. Take the translation of moral statements into pure optatives as an example. Frankly, I consider those mathematical philosophers good ones who not only know about the strength of their methods, but also about their limits and the respective background assumptions. Methods are like fishing nets: they have a determining influence on which fish you will catch. In general, you have to be a very good mathematical philosopher to be able to distinguish useful application from useless. To be sure, this is true for an applicant of any method.

I presume that the ethical contributions of the Vienna Circle would have been improved by a discussion of ethical issues in the meetings of the circle and to a certain extent by the application of formal methods, too. Unfortunately, this was not part of the members' common philosophical project which turned out to be so fruitful in other areas of philosophy.

References

Ayer, Alfred J. 1936. *Language, Truth, and Logic*. London: Victor Gollanz.
Carnap, Rudolf. 1928. *Der logische Aufbau der Welt*. Berlin: Weltkreis Verlag.
———. 1931/1932. Überwindung der Metaphysik durch logische Analyse der Sprache. *Erkenntnis* 2 (4): 219–241.
———. 1935. *Philosophy and Logical Syntax*. London: Kegan Paul, Trench, Trubner & Co.
———. 1963a. Intellectual Autobiography. In *The Philosophy of Rudolf Carnap*, ed. P.A. Schilpp, 1–84. La Salle Ill: Open Court.
———. 1963b. Reply to Abraham Kaplan on Value Judgment. In *The Philosophy of Rudolf Carnap*, ed. P.A. Schilpp, 999–1013. La Salle Ill: Open Court.
———. 1964/1993. Interview mit Rudolf Carnap. In *Rudolf Carnap: Mein Weg in die Philosophie*, 134–148. Stuttgart: Reclam.
Carus, André W. 2017. Carnapian Rationality. *Synthese* 194: 163–184.
Cat, Jordi, and Adam Tamas Tuboly, eds. 2019. *Neurath Reconsidered: New Sources and Perspectives*. Cham: Springer.
Damböck, Christian. 2017. *»Deutscher Empirismus.« Studien zur Philosophie im deutschsprachigen Raum 1830–1930*. Dordrecht: Springer.
———. 2021. Carnap, Reichenbach, Freyer. Non-cognitivist ethics and politics in the spirit of the German Youth-Movement. In Ch. Damböck, G. Sandner and M. Werner (Eds.), *Logischer Empirismus, Lebensreform und die deutsche Jugendbewegung* [Logical Empiricism, Life Reform, and the German Youth Movement]. Cham: Springer, forthcoming.
Feigl, Herbert. 1949/1981. Naturalism and Humanism. In *Herbert Feigl: Inquiries and Provocations: Selected Writings 1929–1974*, ed. R.S. Cohen, 366–377. Dordrecht: D. Reidel.
Haller, Rudolf, and Heiner Rutte, eds. 1981a. *Otto Neurath: Gesammelte philosophische und methodologische Schriften 1*. Wien: Hölder-Pichler-Tempsky.
———, eds. 1981b. *Otto Neurath: Gesammelte philosophische und methodologische Schriften 2*. Wien: Hölder-Pichler-Tempsky.
Kraft, Victor. 1937. *Die Grundlagen einer wissenschaftlichen Wertlehre*. Wien: Springer.
———. 1951. *Die Grundlagen einer wissenschaftlichen Wertlehre*. 2nd ed. Wien: Springer.
———. 1981. *Foundations for a Scientific Analysis of Value*. Dordrecht: D. Reidel.
Leonard, Robert J. 1998. Ethics and the Excluded Middle. Karl Menger and Social Science in Interwar Vienna. *Isis* 89: 1–26.
Menger, Karl. 1934/1997. *Moral, Wille und Weltgestaltung. Grundlegung zur Metaphysik der Sitten*. Wien: Reclam.
———. 1974. *Morality, Decision and Social Organization. Toward a Logic of Ethics*. Dordrecht: D. Reidel.
———. 1994. *Reminiscences of the Vienna Circle and the Mathematical Colloquium*. Dordrecht: Kluwer.
Mormann, Thomas. 2006. Werte bei Carnap. *Zeitschrift für Philosophische Forschung* 60 (2): 169–189.
Neurath, Otto. 1928/1981a. *Lebensgestaltung und Klassenkampf.* In: Haller and Rutte (1981a): 227–293.
———. [1932] 1981b. 'Die Philosophie' im Kampf gegen den Fortschritt der Wissenschaft. In: Haller and Rutte (1981b): 571–576.
———. [1937] 2004. Inventory of the Standard of Living. In *Otto Neurath: Economic Writings 1904–1945*, ed. Th. Uebel and R. S. Cohen, 513–26. Dordrecht: Kluwer.
———. 1941. Universal Jargon and Terminology. *Proceedings of the Aristotelian Society. New Series* 41: 127–148.
Neurath, Otto, Hans Hahn, and Rudolf Carnap. 1929/1973. The Scientific Conception of the World: The Vienna Circle. In *Empiricism and Sociology*, ed. M. Neurath and R.S. Cohen, 299–318. Dordrecht: D. Reidel.

Rutte, Heiner. 1973. Viktor Kraft. Eine philosophische Standortbestimmung. *Conceptus* 7 (21–22): 5–8.

Siegetsleitner, Anne. 2014a. *Ethik und Moral im Wiener Kreis. Zur Geschichte eines engagierten Humanismus*. Wien, Cologne, Weimar: Böhlau.

———. 2014b. On Friedrich Jodl's 'Morals in History'. *Ethics* 125: 211–213.

———. 2015. Susan Stebbing and the Vienna Circle on Moral Philosophy. In *Neuere Aspekte in der Philosophie: aktuelle Projekte von Philosophinnen am Forschungsstandort Österreich*, ed. B. Buchhammer, 206–212. Wien: Axia Academic Publishers.

———. 2017. Philipp Frank on Relativity in Science and Morality. *Studies in East European Thought* 69 (3): 215–225.

———. 2019. Carnaps Autobiographie als Autobiographie. *Deutsche Zeitschrift für Philosophie* 67 (2): 236–250.

Siegetsleitner, Anne, and Hannes Leitgeb. 2010. Mengers Logik für Ethik und Moral: Nichts von Sollen, nichts von Güte, nichts von Sinnlosigkeit. In *Logischer Empirismus, Werte und Moral. Eine Neubewertung*, ed. A. Siegetsleitner, 197–218. Wien, New York: Springer.

Schlick, Moritz. 1908. *Lebensweisheit. Versuch einer Glückseligkeitslehre*. Munich: Becksche Verlagsbuchhandlung.

———. 1930/1939. *Problems of Ethics*. New York: Prentice Hall.

———. 1932/2008. The Future of Philosophy. In *Moritz Schlick: Die Wiener Zeit: Aufsätze, Beiträge, Rezensionen 1926–1936*, ed. J. Friedl and H. Rutte, 371–390. Wien, New York: Springer.

———. 1952. *Natur und Kultur, ed. J. Rauscher. Vienna, Stuttgart: Humboldt-Verlag.*

Stadler, Friedrich. 1985. Popularisierungsbestrebungen im Wiener Kreis und 'Verein Ernst Mach'. In *Philosophie, Wissenschaft, Aufklärung. Beiträge zur Geschichte und Wirkung des Wiener Kreises*, ed. H.-J. Dahms, 101–128. Berlin, New York: De Gruyter.

Stebbing, Susan. 1939. Ethics and Materialisms. *Ethics* 50 (1): 35–44.

Stuchlik, Joshua. 2011. Felicitology: Neurath's Naturalization of Ethics. *HOPOS* 1 (2): 183–208.

Topitsch, Ernst. 1960. Zum Geleit. In *Probleme der Wissenschaftstheorie. Festschrift für Victor Kraft*, ed. E. Topitsch, III–VI. Wien: Springer.

———. 1976. Introduction. In: Kraft (1981): xi–xv.

Zeisel, Hans. 1993. Erinnerungen an Rudolf Carnap. In *Wien – Berlin – Prag. Der Aufstieg der wissenschaftlichen Philosophie*, ed. R. Haller and F. Stadler, 218–223. Wien: Hölder-Pichler-Tempsky.

Chapter 3
The Social Virtue of Science. Motivating Structural Objectivity in Logical Empiricism

Alan Richardson

Abstract This essay reads the structural account of objectivity in the early work of Rudolf Carnap and Hans Reichenbach through the lens of Lorraine Daston and Peter Galison's work on objectivity but differs in its account of what "dangerous subjectivity" structural objectivity was meant to overcome. I argue that Daston and Galison's "carceral self" is at best a metaphorical figure for the real issue that needed to be addressed: the social irresponsibility of disciplines like philosophy itself that, in the view of Carnap and Reichenbach, did not adequately concern itself with using a language that had properly intersubjective meanings. Carnap and Reichenbach demand that philosophers adopt the responsible approach to language routinely deployed in the exact sciences.

Keywords Structural objectivity · Subjectivity · Metaphilosophy · Rudolf Carnap · Hans Reichenbach

3.1 Setting the Scene with Daston and Galison

There must have been times when anyone paying attention to the early Anglophone reception of logical empiricism would have found the accounts very confusing.[1] For example, about half the way through her long discussion of logical positivism and

[1] Stebbing, of course, uses "logical positivism" to refer to this form of philosophy. "Logical empiricism" as a term was still a few years in the future in 1933. I use "logical empiricism" to discuss the views of Carnap and Hans Reichenbach in the 1930s not least because Reichenbach was explicit in rejecting positivism.

A. Richardson (✉)
Department of Philosophy, The University of British Columbia, Vancouver, Canada
e-mail: alan.richardson@ubc.ca

C. Damböck, A. T. Tuboly (eds.), *The Socio-Ethical Dimension of Knowledge*,
Vienna Circle Institute Yearbook 26, https://doi.org/10.1007/978-3-030-80363-6_3

analysis in 1933, Susan Stebbing discussed Rudolf Carnap's views on protocol sentences in his 1931 paper "Die physikalische Sprache als Universalsprache der Wissenschaft" (Carnap 1931). She had just discussed Carnap's concern that if the protocol language expresses internal states of experience of agents, then the sentences that are supposed to confirm physical and other scientific theories are private; they refer to experiential content that is unshared. She writes that Carnap offers the following solution:

> Carnap maintains that such difficulties disappear if we interpret protocol-language in the formal mode, i.e. as expressing structure, not as expressing content, as containing words, not as describing states of affairs. We then discover, he urges, that a protocol-language is a part of physical language, a sub-language of the language of physics. This language is inter-subjective and inter-sensory. It may be called 'the physicalistic language'. (Stebbing 1933, p. 21)

This all seems pretty odd and confused. The formal mode of speech is a mode of speech about languages, not within them. So, the adoption of the formal mode of speech will allow us to talk of the protocol language as containing words but that is a very different thing from learning that the protocol language expresses structure. Moreover, it is not clear that the status of the protocol language as a sublanguage is the physical language is best described as a discovery – or that that discovery depends on the claim about structure.

Of course, Stebbing is not entirely wrong about Carnap's article and is not entirely confused about the state of the logical empiricist dialectic of the early 1930s. But there is one important confusion that troubles this account of Carnap and a larger confusion that haunts the entire essay. The Carnap confusion is this: while the formal/material mode of speech distinction of the early 1930s is related to the content/structure distinction of the late 1920s in Carnap, they are not the same distinction. The move to the formal mode of speech re the protocol language precludes discussing the structure of experience just as much as it precludes the content of experience. The introduction of the distinction between the formal and material modes of speech and the location of logical form in language are moves meant, among other things, to solve some problems with Carnap's earlier epistemology. Carnap's empiricism ceases discussing experience entirely and concerns itself with issues of the formal relations of languages into one another, including translation and entailment relations between the system language and the protocol language.

This problem with Carnap's interpretation is an aspect of the bigger issue with Stebbing's entire approach to logical empiricism in the paper. She alights upon logical empiricism in an effort to understand Wittgenstein's account of what it is to do philosophy. Thus, her discussion of Carnap and Schlick – the two logical empiricists she spends most time on – and other logical empiricists presupposes that they are all trying to present a tidy version of Wittgenstein's views of analysis. Stebbing is driven by this hermeneutic presumption to find some important unity across the views of several logical empiricists and across several years. Moreover, this unity is meant to speak to gnomic statements in the *Tractatus*. This seems a hermeneutical project destined to fail – there is both less unity across a range of logical empiricist thinkers and less concern with Wittgenstein than she presumes.

I do not wish to make any more of Stebbing's paper here or to investigate the relations of logical empiricism to Wittgenstein. My topic is, rather, just one element in the odd amalgam of views Stebbing found herself struggling to understand: Carnap's adherence in the 1920s to a strong version of structural objectivity (for Stebbing's paper at greater length see Richardson 2017a). Stebbing's paper was only one of several in which Anglophone philosophers struggled to make sense of why the logical empiricists seemed to insist that objectivity demanded a concern only with the structure of experience. This account of objectivity seemed among the most opaque aspects of logical empiricism for its Anglophone readers.

I wish to discuss Carnap's structuralist objectivity and a related project in the early work of Reichenbach in the light of Lorraine Daston and Peter Galison's monumental work, *Objectivity* (Daston and Galison 2007), a recent central contribution to historical epistemology. Historical epistemology, in the parlance of Daston, is the history of the concepts employed in thinking about knowledge, concepts like "fact," "experience," "reason," and "objectivity." These concepts frame what seekers after knowledge take to be the goals, methods, and obstacles in their search. In their book, Daston and Galison are mainly interested in how such epistemological notions – and, of course, "objectivity" chief among them – were theorized, pursued, and operationalized within the work of scientific communities in the nineteenth century. Arguably, Daston and Galison are more interested in the regimes for securing objectivity variously conceived than they are in the various conceptions themselves. Thus, their main emphasis is on the knowledge-production regimes within the sciences, which given their interests is entirely reasonable. Philosophers, however, are (at least sometimes) interested more in concepts than they are in regimes for securing the methods by which to tell which objects fall under those concepts. So, whereas it is certainly the case that philosophers appear in Daston and Galison's story, a scholar interested in the history of epistemology or the history of philosophy of science might have wished to hear more about some aspects of the changing circumstances of philosophical reflection on science in this time period – after all, the history of epistemology itself should be a resource for historical epistemology.

In this essay, I want to employ certain aspects of Daston and Galison's account of the history of objectivity in order to illuminate some features of logical empiricism's place in the history of epistemological thought. This will shed light on the specific place of logical empiricism in the history of the conceptions of and regimes for securing objectivity in the early twentieth century. I hope that the essay will show by example that the history of epistemology can be a resource for historical epistemology. It will also show how the sort of historical framework Daston and Galison provide can help those of us interested in the history of epistemology and the history of philosophy of science to offer richer historical narratives in the history of philosophy. Thus, the essay also illustrates how historical epistemology is a resource for history of epistemology.

Such a project is not without methodological questions. The reaction to Daston and Galison's book has been mixed. Among historians of philosophy the work has sometimes seemed too broad and sweeping, not sufficiently microhistorical. Thus, Martin Kusch (2009) took Daston and Galison to task for being insufficiently interested in the historical details within the five regimes of objectivity they discuss, reserving their historical curiosity for the transitions between regimes. Other historians might suspect Daston and Galison of trying to inflate a special case – the history of objectivity in visual atlases – into a complete story of objectivity in science (despite their protestations to the contrary). Some philosophers might contend that Daston and Galison are insufficiently attentive to the details of philosophical theories of subjectivity and, thus, objectivity in the complicated theories of knowledge of the academic philosophers of the nineteenth century. Other quite general historiographical concerns have been raised – again notably by Kusch (2009) – about Daston and Galison's account of what it is to have an historical explanation of the changing conceptions of and regimes for securing objectivity – in particular, what regime of objectivity they themselves subscribe to.

This essay is too short and too circumscribed to speak to all these issues. My attitude toward these problems is practical and exploratory – Daston and Galison have given us enough historical texture that we can now investigate, for example, how the history of, say, objectivity in the social sciences complicates (or not), interacts with (or does not), and enriches (or not) the specific scientific work in visual representation of nature that their history discusses. Similarly, they provide enough of a framework for thinking of the relationship of conceptions of subjectivity and of objectivity that intensely scholarly work on, say, early Marburg neo-Kantianism or Husserlian phenomenology can find renewed importance in the history of epistemology. Moreover, attention to such work has the possibility, by altering Daston and Galison's framework, of substantially enhancing our historical understanding.[2] That is, despite its lengthy genesis, substantial physical being, and rich literature in response, *Objectivity*, so it seems to me, is more of an early substantial work on objectivity and subjectivity in nineteenth-century science than the last word on the matter – and I rather suspect Daston and Galison would agree.[3]

That said, I will approach the issue of structural objectivity in a highly circumscribed way. My topic will be the structural objectivity of the logical empiricists and that mainly in the 1920s. This will allow me to make some amendments to the tenor of some of Daston and Galison's remarks on logical empiricism and will exemplify some of Kusch's concerns with intra-regime dynamics. The approach I take to Daston and Galison's text is in line with several essays that also indicate the complications that lie ahead in the richer historical story we can build from the account Daston and Galison give: I have in mind essays such as Christopher D. Green's essay on Titchener's psychology, Dani Inkpen's essay on Helmholtz and Goethe,

---

[2] On Marburg theories of subjectivity and objectivity see Edgar (2008, 2015b); for recent work on subjectivity and the theory of science in Husserl, see Friedman (2010) and Feest (2012); for the interactions of Natorp, Husserl, and early Carnap, see Stone (2005).

[3] My own thinking about Daston and Galison has been informed by the work of and conversations with Jill Fellows on objectivity and the epistemic agent. Fellows (2011) is her most complete work; some of Fellows's characteristic concerns can be gleaned from her response (Fellows 2017) to Sandra Harding's rejection of "Mr Nowhere" in Harding's project of strong objectivity (Harding 2015).

and Paolo Savoia's essay on Charcot and Freud (Green 2010; Inkpen 2009; Savoia 2015). Green's essay follows the chronology of regimes of objectivity in Daston and Galison's work but, by taking psychology as the field of inquiry, also reveals ways in which objectivity and subjectivity are conceptually complicated. A science of psychology in the nineteenth century looks like a science of the subjective. But, if the principal epistemic virtue of science is its objectivity and subjectivity is the name for that which cannot be brought under objective control, then psychology is a specifically complicated field: it seems to aim to offer us objective knowledge of that which resists objectivity. This is not an insuperable practical difficulty – the psychologist might be able to police her own subjectivity sufficiently to allow her to theorize objectively about subjectivity in others or in general – but there are conceptual thickets here. For we are here in a position of claiming that subjectivity is an object of knowledge within one of the sciences while in the general theory of scientific knowledge subjectivity stands as the unruly threat to the possibility of objectivity.

Insofar as the objects of science become, in virtue of being known, domesticated, made transparent and subject to intervention and manipulation, a successful science of subjectivity would at least be likely to make the subjective/objective divide less a source of concern within epistemology and less, therefore, of a source of science's sense of its own achievement. That is, the success of psychology as a science of the subjective might alter the epistemological terrain upon which the nature and value of scientific knowledge is itself understood. Similarly, Savoia's essay looks at different sensory regimes for securing the objectivity of diagnosis of mental disorder – the primacy of the visual in Charcot and of the auditory in Freud. Both Charcot and Freud try to stabilize norms of objectivity to try to stabilize a subject, the psychoanalytic self. Here again the epistemic agent polices its own subjectivity while attempting objectively to know another subject.

Inkpen's essay looks at one of the key figures in Daston and Galison's story, Hermann von Helmholtz. Precisely by taking seriously Daston and Galison's claim that the virtues of objectivity were motivated by specific concerns with subjectivity, Inkpen was able to extract a different concern with subjectivity in Helmholtz than the one Daston and Galison allege. Daston and Galison's framing of the history of objectivity as the policing of dangerous subjectivity puts objectivity into a specific form of historical epistemology, an historical *virtue* epistemology. Consider Daston and Galison's own account of the sources of mechanical objectivity: here they find the motivation for regimes of mechanical objectivity in anxieties about a willful subjectivity unable to prevent itself from mingling its own presuppositions with the nature of the world. This quasi-Kantian (via Fichte and Schopenhauer and so on) account of the epistemic subject gives rise to a form of self-conscious discipline of the will played out in a willful suppression of the empirical use of the will in speculative contexts – a discipline of pure reason directed against a specific conception of the Kantian subject, so to speak. The scientific mind, in understanding its own tendencies toward error, induces a corresponding regimen of epistemic virtue, involving restraint of the personal will and its aestheticizing tendencies.

Inkpen accepts the general framework for understanding Helmholtz's own concerns with objectivity but locates his particular worries with subjectivity – especially in his remarks on Goethe's scientific work – elsewhere. Rather than unbridled willfulness, Goethe's work exhibits a sort of elitist, poetic passivity: the mind of the poet is peculiarly attuned to the objects in the world. Seen in this light, Helmholtz's rigorous account of objectivity and the experimental regimes for securing it are not attempts to discipline the will but rather to activate the understanding in a world in which the laws and objects of nature are not transparent to the mind.

These essays (and there are many others) show how detailed attention to Daston and Galison's story can provide a starting place for a yet more complete and complicated story of objectivity and subjectivity throughout this period. While not following either any of these authors very closely in topic or concern, this essay pursues the same overall strategy – my aim here is to begin from the work of Daston and Galison and complicate and enrich our understanding of early logical empiricist structural objectivity.[4]

I take the general idea of Daston and Galison's book, then, to be this: at some places and in some times the epistemic virtue of science is presented in part or primarily in the language of objectivity. However, accounts of objectivity and regimes for producing objective judgment in science following on from those accounts differ one from another – what it is paradigmatic instance of the objective for one scientist in one generation can become a clear case of the subjective in another generation. Thus, for example, the objective representations of ideal botanical types in the late eighteenth century become exactly the sort of intermingling of the subjective with the objective that mechanical objectivity is meant to root out. This explains the insistence among the promoters of mechanical objectivity on the virtues of self-denial and of specificity – the camera becoming, in consequence, the exemplary producer of objective images and, ultimately, the principal metaphor for the objective mind. Objectivity, that is, has a history.

Four more points are worth noting about the Daston and Galison story for our purposes here. First, the real history of objectivity is, they maintain, a story that begins in the nineteenth century. If I may put the point more in my own idiom than in theirs, the argument is that it is only in the nineteenth century that the mind becomes not a resource in explaining the acquisition of secure knowledge about the world, but a problem for any such explanation. It is in this sense that issues of subjectivity came more to the fore in nineteenth-century epistemology and methodology, bringing with it a greater demand for objectivity. Second, as is presumed in the first point, specific notions of objectivity are crucially developed against the background of accounts of the epistemic defects of subjects; notions of subjectivity drive notions of objectivity; regimes of objectivity are meant to police precisely those places in scientific practice where epistemically dangerous forms of subjectivity can

[4] My approach owes most to Inkpen's. As will be clear, I allege a form of virtue epistemology in logical empiricism and a concern not unlike Helmholtz's with a form of epistemic elitism, indeed a form of elitism Carnap and Reichenbach associate with literary scholars. See especially the figure of the "literary person" in the final section of Reichenbach (1931).

be expected to be lurking. Third, we can expect then a sort of dialectical interplay of notions of subjectivity inducing notions and regimes of objectivity, which, when they in theory and in practice come into problematic situations, yield both new notions of the subject and new notions of the objective and new regimes for attaining it. Fourth, this historical trajectory of the concept of objectivity and of the means to secure it is not a linear progression toward the truth about objectivity – there is no clear *telos* in Daston and Galison's story. Nor is there a wholesale replacement of conceptions of objectivity – rather there comes to be an ever more complicated and confusingly sedimented concept of objectivity with locally dominant conceptions and regimes.

3.2 Structural Objectivity and Logical Empiricism per Daston and Galison

Did anyone really ever think that the key to objective knowledge of the world was to be found in the purely structural? Yes. In fact, as is implicit in the chapter on structural objectivity in Daston and Galison's book, this idea was probably the single most important account of objectivity in early twentieth-century philosophy of science, as pursued both by scientists like Henri Poincaré and by philosophers from Ernst Cassirer to Hans Reichenbach. One can find the idea expressed with both maximal philosophical ambition and maximal clarity in perhaps the greatest work of early logical empiricism, Carnap's *Der logische Aufbau der Welt* of 1928 (Carnap 1928/1967). There Carnap introduced his epistemological project with several striking claims about scientific knowledge. Here are just a few of them from the early sections of the book that set and motivated the problem of objectivity that the book meant to solve:

> In the following we shall maintain and seek to establish the thesis that science deals only with the description of structural properties of objects. (Carnap 1928/1967, 19)

> Our thesis, namely that scientific statements relate only to structural properties, amounts to the assertion that scientific sentences speak only of forms without stating what the elements and the relations of these forms are. (Carnap 1928/1967, 23)

And, most importantly for establishing that what is at stake here is structural objectivity, we have this from Sect. 16:

> Each scientific statement can in principle be so transformed that it is nothing but a structure statement. But this transformation is not only possible, it is imperative. For science wants to speak about what is objective, and whatever does not belong to the structure but to the material is, in the final analysis, subjective. (Carnap 1928/1967, 29)

This seems a strikingly strange account of objectivity, but it is the one Carnap develops not only in the *Aufbau* but throughout his early writings, especially in the final sections of his 1926 monograph, *Physikalische Begriffsbildung* (Carnap 1926/2019). I have had much to say about this account of objectivity elsewhere – I wrote a book about it that I invited readers to read as "one chapter in the history of objectivity" (Richardson 1997; quotation at p. 4) – but Daston and Galison's work on structural objectivity, which gives Carnap's *Aufbau* a prominent place invites revisiting this account of objectivity with the question of the dangerously subjective in mind.

Before doing so, we should also note that this sort of view was not, among the early logical empiricists, uniquely Carnapian. A version of structural objectivity is given in a somewhat different idiom already in 1920 in the work of Hans Reichenbach in his book, *Relativitätstheorie und Erkenntnis A Priori* (Reichenbach 1920/1965). In that book, Reichenbach claims that the mathematical form of scientific theories is necessary for securing the objectivity of knowledge, since in the absence of mathematical form the very notion of an object of empirical knowledge is not properly defined. He speaks of co-ordinating physical reality with structures of concepts given in mathematical axiom systems, but then notes that the notion of "co-ordination" here is peculiar and sui generis. He says this:

> The coordination performed in the physical proposition is very peculiar. It differs distinctly from other kinds of coordination. For example, if two sets of points are given, we establish a correspondence between them by coordinating to every point of one set a point of the other set. For this purpose, the elements of each set must be defined. [...] Such definitions are lacking on the one side of the coordination dealing with cognition of reality. Although the equations, that is, the conceptual side of the coordination, are uniquely defined, the 'real' is not. On the contrary, the 'real' is defined by coordinations to the equations. (Reichenbach 1920/1965, 37–38)

One can express this idea as follows: something is an object of knowledge, a content of objective knowledge, only if it can find a unique place within a well-defined structure of relations. Such structures are defined in pure mathematics, and thus only through a coordination of some mathematical structure with physical reality can objects be meaningfully spoken of in the first place. Structure secures objecthood and objecthood secures objectivity.

Nor was this structuralist view of objectivity unique to the logical empiricists. As the positive review of Reichenbach's book by Ernst Cassirer made clear, this very idea was at the heart of what Cassirer called, in works such as his *Substanzbegriff und Funktionsbegriff* (Cassirer 1910/1923) his "logical idealism" or "logic of objective knowledge."[5] Carnap's dissertation director, Bruno Bauch, had a similar notion of objectivity in play right into the 1920s. Here is a characteristic passage of Bauch's *Wahrheit, Wert, und Wirklichkeit*:

> The logical must, therefore, itself be the archetypical, the objective; the structure of judgment must be the structure of objects; this [structure] cannot be merely projected into that [structure]. And because the structure of judgment is itself the structure of objects and this is, for its part, also the structure of judgment, only thereby can the sphere of possession be projected into subjectivity, can subjectivity possess objectivity at all, can, moreover, objectivity be discovered from subjectivity. (Bauch 1923, 139; my translation)

Here Bauch argues that logical structure must be the structure of objects of knowledge and promises that this will allow us to understand the relations of the subjective and

[5] On objectivity in Cassirer, see Heis (2014), Edgar (2015a), and Richardson (2017c).

the objective in ways that guarantee the possibility of objective knowledge. This is the remnant of Kantian transcendental idealism that persists in Bauch's work: the vehicles of knowledge are structured by the logic of judgment and for objective knowledge to be possible this selfsame structure must structure the objects of knowledge themselves.

As noted above, none of this is news to Daston and Galison who have a large chapter on structural objectivity in the book. In that chapter they follow the career of structural objectivity in the work of figures such as Hermann von Helmholtz, Gottlob Frege, Henri Poincaré, Bertrand Russell, Albert Einstein, and others, including the logical empiricists. They point to Carnap's *Aufbau* in particular and note its relations to other logical empiricist work, including Moritz Schlick's interpretation of the methodological lessons of Einstein's theory of relativity. Daston and Galison are fully informed about the historical phenomenon of structural objectivity and they give it a prominent place in their book – and this despite the fact that, as the note throughout the chapter, structural objectivity is not really an account of objectivity that gives pride of place to visual representation in science and thus falls outside the main trajectory of their book. What I wish to query in the account of structural objectivity for the logical empiricists is Daston and Galison's two main interpretative claims about the motivation for structural objectivity: that the sort of dangerous subjectivity at issue is at once importantly motivated by empirical understandings of the variability of human experience and importantly an issue of communicability specifically directed against a radical form of solipsism.

Consistent with their overall plan in the book, Daston and Galison need to find a dangerous form of subjectivity against which the structural account of objectivity arises and is arrayed. Their account is complicated and I cannot do sufficient justice to it here, but the two principal claims within it are these. First, structural objectivity arose in response to empirical work in psychology, psychophysiology, anthropology, and other social sciences that stressed the variability, individual and cultural, of human experience: "the self [...] renounced by structural objectivity was in part the discovery of science itself, particularly the then-young sciences of sensory physiology and experimental psychology." (Daston and Galison 2007, 258). This self, revealed largely through the regimes of mechanical objectivity, "revealed differences in perception, judgment, and even logic" (Daston and Galison 2007, 258). They provide a long example of such work in the late nineteenth-century science of color (Daston and Galison 2007, 273–283).

The radical variability of subjectivity revealed empirically drove an important epistemological worry with solipsism and the communicability of knowledge, on Daston and Galison's account. The variability across individuals and cultures of sensation, judgment, indeed reason itself, raised the specter of a radical lack of communication across people and communities. Structural objectivity sensed this danger and sought to remedy it:

> The metaphors of structural objectivity were rather of a fortress self, locked away from nature and other minds alike. Structural objectivity addressed a claustral, private self menaced by solipsism. The recommended countermeasures emphasized renunciation rather than restraint: giving up one's own sensations and ideas in favor of formal structures accessible to all thinking beings. (Daston and Galison 2007, 257)

I do not wish to be heard as saying that this is all there is to Daston and Galison's account. They emphasize in addition to these very general considerations, for example, on-the-ground efforts at the time – and sometimes led by scientists such as Helmholtz and Poincaré – to introduce international standards for measures and terminology. The communicability of scientific results had real, practical limitations in a world divided by language, politics, and expert interest. Nonetheless, given their overall argument structure in which concerns with objectivity inevitably arise from concerns with dangerous forms of subjectivity, the more general considerations must drive the story.

3.3 Structuralism and Subjectivity in Logical Empiricism

These two motivational claims are surely not simply wrong as claims to be made within the larger development of structural objectivity. I do wish to query them as the most telling interpretation of the structural objectivity of the logical empiricists in the 1920s. By the 1920s, within the philosophical trajectories within which Carnap's and Reichenbach's early work found a place, the questions of the variability of the empirical self had been, I would argue, importantly domesticated and overcome. Carnap and Reichenbach were very much within the antipsychologistic line of thought in German epistemology – the facts about the empirical self were not understood importantly to be constraints on epistemology within this line of thought by the 1920s. The sort of puzzlement induced by work in psychophysiology by scientists such as Helmholtz and Ernst Mach just does not seem to be in the field of attention that Carnap and Reichenbach have within the epistemology of science they produced in the 1920s. There are lingering issues about the place of empirical psychology in motivating the starting point of the "autopsychological constitutional system" of the *Aufbau*, but insofar as Carnap does rely on empirical accounts of the self and its experience there, he relies on Gestalt psychology. This Gestalt psychology, far from leading to anxieties for his epistemology, seems very much in line with and perhaps even motivational for his structuralism – for, on Carnap's reading in any case, the Gestalt psychologists have shown empirically that experiences are given as structured wholes, not Machian atomic sensations (for more on Carnap and Gestaltpsychology, see Feest 2007).

More generally, the epistemological vocabulary Reichenbach and Carnap employ in the 1920s seems to come from a different wing of philosophy—a neo-Kantian wing in which the empirical self is not at issue and the transcendental self had itself retreated in favour of an increasingly formal notion of transcendental logic. In 1920, Reichenbach stated his structuralist view explicitly within a quasi-Kantian idiom of the synthetic a priori – it is synthetic a priori principles that constitute the object of knowledge through the co-ordination of experience with mathematical structures, which co-ordination first grounds objectivity. This view has obvious anti-metaphysical import of a peculiarly Kantian sort. It is clearly opposed to any version of transcendental realism that demands that objective knowledge is secured as objective by referring said knowledge to objects given as such outside the realm of knowledge. Similarly, both Reichenbach's and Carnap's views are motivated against any form of presentational realism that posits an immediate relation between the mind and the objects of knowledge. On the other hand, there is very little left of the subject of knowledge, very little left to be said of the mind, in specifying the view; perhaps the mind does the co-ordinating in Reichenbach's view but we hear very little of any other activity or feature of the mind.[6] From within a generally Daston-and-Galisonian perspective this diminishment of the mind as the locus of objective knowledge renders pressing the question of what anxieties about subjectivity are motivating the adoption of structural objectivity.

Such anxieties are not obvious on first glance. After all, while it might be true, as Carnap insists in the *Aufbau,* that the content of experience is private and, thus, incommunicable and subjective, it is not clear how the private nature of the content of experience creates anxieties about dangerous subjectivity. The danger that Daston and Galison claim, as we have seen, is the danger of skepticism or solipsism – if the content of experience is the content of knowledge and if the content of experience is private and subjective, then all knowledge is private and subjective. That is, if solipsism is true, then there is nothing that divides knowledge from mere opinion. Even worse, I cannot even understand your judgments. I can speak only of my experiences; you can speak only yours; I have no access to your experiences and, thus, cannot even know what you are talking about, let alone whether it is true. And the same, mutatis mutandis, holds for you.

Now, Daston and Galison are right that this sort of solipsism was invoked by some figures in the generation before Carnap and Reichenbach to call into question views of knowledge and judgment that seem to tie them to essentially private experience. However, given that the tenor of remarks about solipsistic consequences of certain views of knowledge, those arguments really were reductions to absurdity. The solipsistic consequences of such theories are so absurd that no one could possibly believe them. The fact of communication renders such solipsism and any philosophy of knowledge that ends up endorsing it simply silly. Far from being concerned with solipsism, many proponents of structural objectivity took for granted that solipsism was simply false or meaningless.

This is certainly the tenor one finds in famous remarks against solipsism in the work of Carnap's teacher, Gottlob Frege. In a famous passage in the introduction to his *Grundgesetze der Airthmetik* of 1893, Frege inveighs against logicians who view the content of judgment to be ideas, that is, psychological objects:

> Everything is eventually dragged into the realm of psychology; the boundary between the objective and the subjective disappears more and more, and even actual objects are treated psychologically as ideas [...]. Thus everything leads into idealism and with perfect logical consistency into solipsism. [...] If we could grasp nothing but what is in ourselves, then a conflict of opinions, a reciprocity of understanding, would be impossible, since there would

[6] In 1920 Reichenbach insists that the mind is not the proper topic of epistemology – his work is not an analysis of reason but an analysis of scientific knowledge.

> be no common ground, and no idea in the psychological sense can be such a common ground. (Quoted by Beaney 1997, 206)

Throughout his career, Frege returns again and again to his objections to psychological accounts of logic, but across these passages he evinces little worry that the subjective realm of ideas might intrude upon the properly objective judgments in science. Instead, the securing of objectivity in science and, indeed, in everyday life is a presupposition, an established fact, for Frege and what is to be rejected is the psychological account of logic. Our most ordinary practices demand that we set aside as confused psychological understandings of logic, as we can see in this typical passage from his late (1918) essay, "Negation":

> It must be possible for several people who hear the same interrogative sentence to grasp the same sense and recognize the falsity of it. Trial by jury would assuredly be a silly arrangement if it could not be assumed that each of the jurors could understand the question at issue in the same sense. (Quoted by Beaney 1997, 350)

Frege's invocation here of the objective/subjective divide is not motivated by any serious concern with solipsism. Rather, Frege's concern is with poor epistemological and logical theory that has solipsism as an unintended but inevitable consequence. He presupposes that the grasp of objective thoughts is possible and seeks to provide an account of that grasp that indicates how it is possible. His own view is not obviously a version of structural objectivity but the logical empiricists and neo-Kantians who do endorse structural objectivity similarly show little genuine anxiety about solipsism and seek, rather, to provide a robust philosophical explanation of the achievement of objectivity. As the neo-Kantians were wont to say, epistemology needed to explain the fact of science; Carnap and Reichenbach follow them in this.

So, in order to find a dangerous subjectivity against which Carnap and Reichenbach's structural objectivity is arrayed, we need to dig a bit deeper into the motivations for their philosophical views. The principle virtue of structural objectivity is, as Daston and Galison have stressed, that it secures communicability: our judgments are mutually comprehensible, we speak about the same objects, and what we say is in principle true or false; we can argue about things. Now, communicability is necessary for knowledge but certainly not sufficient – if "the First Critique has 500 or more pages" is communicable then so is "the First Critique has fewer than 500 pages," but since at most one of those claims is true, at most one can be known. Communicability is a precondition for knowledge – only if we can intelligibly judge of the same objects and say things that agree or contradict one another can we be inquiring into a knowable state of affairs. Communicability shows that a judgment is a candidate for knowledge, is a suitable topic of joint inquiry.

These considerations suggest certain moves that Carnap and Reichenbach make on their way toward a generally anti-metaphysical point of view and, thus, move us in the direction of finding a dangerous subjectivity that does motivate their interest in structural objectivity. When Daston and Galison discuss dangerous subjectivity in the case, for example, of mechanical objectivity, the problem is very much individualistic: any one of us might confuse the operation of our idealizing will with the truth about nature; we must each individually discipline our will. In fact, as the problems-in-principle play themselves out within the actual practices of science, of course, it is much more likely that the operative worry is that certain specific scientists, in virtue of their authority, their persuasiveness, their social position, or what have you might introduce arbitrary, willful elements in our picture of nature. But the philosophical problem is individualistic and the disciplining, however much it is codified into social practices, is a policing of individual wills.

The situation is very different in the case of the motivating features of structural objectivity, as I see it, in Carnap and Reichenbach. The issue is not that each one of us might get caught within the wholly subjective contents of our experience and divorce ourselves from the larger community of inquirers into nature. Rather the issue is that widely circulating theories of knowledge that confuse having knowledge with being in a certain state of mind – one in which, on one version of such theories, the essence of the object revealed itself directly to the mind or that tied the content of knowledge to the concrete contents of specific personal experience – threatened to tear apart the community of inquirers. These theories did this in two ways.

First, when deployed generally, such theories allowed challenges to expert epistemic claims to be deflected with the claim that the objector did not have the right sort of or command of or attitude toward experience or the world – this amounts to a sort of elitism that asserts that only those who have certain sorts of experiences or the right interpretative attitude toward them or relation to the world can even understand certain knowledge claims. Second, when brought home to philosophy and made into an epistemology of a distinctively philosophical, phenomenological, or metaphysical knowledge, it allowed philosophers to make claims that others could neither understand nor see any evidence for and yet to present these claims as true. In a perverse way, the very unintelligibility of philosophical theorizing could then to be viewed as evidence of a sort of profundity of the subject's relation to experience on the part of the philosophical master, a vision of the true world but a world wholly outside the realm of ordinary experience of ordinary people. The dangerous subjectivity here is the dangerous subjectivity of the allegedly elite knower, armed with a false theory of knowledge, who sought to make uncontrolled theoretical interventions that were beyond the bounds of legitimate criticism. It was a problem of epistemology in the service of rationally uncontrolled power; it was the logical empiricist version of the problem of ideology, which was, for them, the problem of metaphysics.

That is, I see the principal target of logical empiricist structural objectivity to be attempts at the establishment of intellectual authority on the part of groups claiming to have access to sorts of experience or attitudes toward experience or relations to the world that others cannot have and who sought to prop up their authority on the basis of bad philosophical theorizing about the relations of knowledge to experience (or the world). In their response, the logical empiricists presupposed the ability of science to achieve objective knowledge. This allowed the proper philosophical account of such objective knowledge to be deployed in an effort to explode the claims of nonscientific intellectual authority. This is the project Carnap both explicitly develops in Part Five of the *Aufbau* and that he motivates in his informal remarks on unscientific philosophy in the introduction to the same work.

Part Five of the *Aufbau* is entitled "Clarification of some philosophical problems on the ground of constitution theory." The main business of that part of the book is to argue that various concepts employed by metaphysical philosophers – "essence," "reality" – are without objective meaning and that philosophical understandings of other concepts – "identity," causation" – introduce unsolvable pseudoproblems for otherwise clear concepts in science. The larger point of this section of the book is to extrude metaphysics from the realm of science, which is to say, to extrude it from the realm of the objectively meaningful. Of most importance to our current concerns is his explicit invocation of Henri Bergson and Bergson's account of metaphysics in providing his understanding of what he is arguing against. Citing specifically Bergson's claim that metaphysics is "that science which wants to get by without concepts," Carnap glosses the point this way:

> Other philosophers [such as Bergson – AR] use the name "metaphysics" for the result of a nonrational, purely intuitive process; this seems the more appropriate usage. [...] If the name "metaphysics" is used in this sense, then it follows immediately that metaphysics is not a science (in our sense). (Carnap 1928/1967, 295)

Indeed, it is clear that Carnap endorses Bergson's account of metaphysics here precisely because it is easy to show, on Carnap's own account of the epistemology of science and structural objectivity, that Bergsonian metaphysics is impossible – no science can be formulated without symbols, no nonrational, intuitive process can yield knowledge. More than this, however, Bergson's metaphysics is a move within a general project for securing an independent authority for philosophical, nonscientific knowledge that Carnap finds not only intellectually bankrupt but socially deleterious. Indeed, in the section just before the one in which he endorses Bergson's account of metaphysics in order to show it yields nothing, Carnap attempts to show that "nonrational knowledge" or faith is not properly understood to be a species of knowledge at all. Commenting on this, Carnap says:

> It should be more favorable *for the peaceful relations between various spheres of life*, if we do not designate two such heterogeneous spheres with the same name. It is only in this way that contradiction and strife arise, which are not even possible as long as the complete heterogeneity is clearly seen and emphasized. (Carnap 1928/1967, 295; translation altered only to note Carnap's original emphasis)

It is the reduction of social conflict that comes through avoiding this confusion of spheres that is the social project embedded within the *Aufbau*.

This seems to me a much more apposite context than a bland concern with solipsism within which to interpret the remarks from the forward to the 1928 edition to the *Aufbau* that Daston and Galison cite (Daston and Galison 2007, 291). Carnap finds this confusion of spheres of life to be primarily the responsibility of philosophers, who have been confused about the sort of intellectual work philosophy properly is. In outlining in the forward the goals of the book, Carnap emphasizes a new form of scientific philosophy that he means to promote; it is a philosophy that partakes of the seriousness and progressiveness of science:

> The basic orientation and the line of thought of this book are not property and achievement of the author alone but belong to a certain scientific atmosphere which is neither created nor maintained by any single individual. The thoughts which I have written down here are supported by a group of active or receptive collaborators. This group has in common especially a certain basic scientific orientation. That they have turned away from traditional philosophy is only a negative characteristic. The positive features are more important; it is not easy to describe them, but I shall try to give a loose characterization. The new type of philosophy has arisen in close contact with the work of the special sciences, especially mathematics and physics. Consequently they have taken the strict and responsible orientation of the scientific investigator as their guideline for philosophical work, while the attitude of the traditional philosopher is more like that of a poet. This new attitude not only changes the style of thinking but also the type of problem that is posed. The individual no longer undertakes to erect in one bold stroke an entire system of philosophy. Rather, each works at his special place within the one unified science. (Carnap 1928/1967, xvi-xvii)

We see here clearly that among the problems with philosophy has been an epistemology of philosophical knowledge that allowed an irresponsible individualism specifically in philosophy. This individualism and idolatry of philosophical genius in turn has rendered philosophy an unproductive form of intellectual activity and a source of conflict with the progressive and collaborative sciences. This view of the situation was echoed by Reichenbach in an informal newspaper essay of 1929 in which he expressed the point this way:

> The exaggerated stature attributed to the great men of the past could not but cripple initiative, could not but turn the attention of present thinkers from present possibilities and produce that unproductive epigonism the sight of which must repel anyone who works in the stimulating atmosphere of the sciences and who has even once looked at contemporary philosophy from this vantage point. (Reichenbach 1929/1978a, 250)

For Reichenbach, also, this scientific philosophy did social good. In addition simply to being more honest and accurate about the conditions of knowledge, a properly scientific philosophy, by comprehending rather than opposing the forms of knowledge production in science, scientific philosophy could lead to a greater understanding of precisely those processes. This in turn could lead to a greater understanding of the social conditions under which such knowledge production could be organized for epistemic and social good. Here's how he put the point in the introduction to his 1928 *Philosophie der Raum-Zeit-Lehre*[7]:

> The style of modern science [*Fachwissenschaft*] has gradually taken on the competition-accelerated haste of technology, the tempo of sports records; one can regret this antihumanist [*unhumanistisch*] movement, but it appears to be the necessary form of the productivity of our times. What we can set against this is not a competition via a less technical means but rather only the philosophical permeation of the knowledge process itself, the discovery of the sense and meaning of this mechanized knowledge, which indeed remains a machine in the brains of many handymen, but which reveals in its systematic whole a depth of knowledge as can only be achieved by the socialized joint work of an organized class of individual workers. (Reichenbach 1928/1958, xiii)

[7] I find the accepted English translation of this passage unsatisfactory so this is my own translation. The standard translation entirely misses the metaphor of philosophy as the permeating oil of the well-ordered machine of scientific knowledge.

3.4 Conventionalism, Responsibility, and Objectivity

Daston and Galison are certainly correct that Reichenbach's and Carnap's views are not motivated by worries about the unruly subjectivity of the willful Kantian subject that was at issue in mechanical objectivity. The will, far from being the locus of danger in the unruly subject, becomes, at least in Reichenbach's version of the story and by the late 1920s, part of the solution to the problem of knowledge. Both Carnap and Reichenbach ultimately adopt a form of voluntarist conventionalism in which the role of volition or practical decision or the will cannot be eliminated from the construction of knowledge. Indeed, as the explicit formulation of their views in Kantian terminology retreats through the 1920s, the voluntarism becomes more prominent. In 1928, Reichenbach presented his view in a way that makes specific reference to faculties of the mind, the will having pride of place:

> Rational knowledge in our sense is not tantamount to categorization within the pre-established cubby-holes of a reason that governs *a priori*, but simply amounts to unconditional faith in the power of the human capacity for knowledge–within the framework of a critique of its own goals. Thus the rational element is itself subject to change; and it emerges with increasing clarity that the basic stance of science is a faith more akin to an instinct than to rational insight, to will than to knowledge. Thus the will, the tenacious, malleable, indefatigable, and yet, eternally modifiable will is probably the basic element that truly represents the world view standing behind the scientific investigation of nature. (Reichenbach 1928/1978b, 244)[8]

Reading such sentiments with Daston and Galison's history of objectivity in mind, we can see that not only was will not the enemy of objectivity for Reichenbach, but indeed insistence upon the role of volition in knowledge was itself a demand for a responsible epistemic attitude. The dangerous subjectivity to be argued against had, correspondingly, two faces. First, there was the dangerous subjectivity of the person who did not take care to coordinate his will with the will of the community, thus rendering his language unintelligible to the community and, potentially, his ethical life outside the bounds of his community. In the work of Carnap, especially, it was clear that this was the fate of the metaphysician who offered only pseudoarguments in support of unintelligible positions. Second, there is for Reichenbach a dangerous subjectivity involved in a more robustly Kantian account of understanding and reason as having pregiven and unchangeable forms. This leads one to submit to traditional forms as an external and immutable authority and to abdicate proper responsibility for one's own epistemic choices and the structure and content of one's belief system. In many ways, it was this later view that Reichenbach was most concerned to argue against.

One faced-up to one's subjective will not by denying it or simply restraining it but by employing it explicitly and responsibly. This is the lesson embedded within

[8] I have discussed this passage and others similar to it in the work of Reichenbach in Richardson (2005) and (2006). The contrast with the concerns with the will in mechanical objectivity helps make the case for the philosophical importance of such passages.

Carnap's principle of tolerance, which is at once a license of practical freedom together with an account of what it is to use such freedom responsibly:

> *It is not our business to set up prohibitions, but to arrive at conventions.* [...] *In logic, there are no morals.* Everyone is at liberty to build up his own logic, i.e. his own form of language, as he wishes. All that is required of him is that, if wishes to discuss it, he must state his methods clearly, and give syntactical rules rather than philosophical arguments. (Carnap 1937, 51–52; original emphases)

Needless to say, on Carnap's view, philosophers routinely fail to follow these strictures. Similarly, Reichenbach's late endorsement of what he dubbed "the democratic principle" – everybody is entitled to set up his own moral imperatives and to demand that everyone follow these imperatives – in his noncognitivist metaethics is also a principle of practical freedom (Reichenbach 1954, 295; see further Richardson 2017b). The epistemic agent who has imbibed the lessons of Carnap and Reichenbach recognizes herself as having a practical epistemic freedom and then sets about constructing systems of knowledge in which that the use of such freedom is explicitly acknowledged, displayed with precision, and deployed responsibly. The epistemic problem becomes one of rooting out those who attempt to hide the use of their practical freedom behind bad epistemological doctrine or whose language does not suitably display where practical epistemic freedom is deployed. The epistemic problem is largely a problem of consilience of wills – which is why the epistemic problem is at the same time a moral and a political problem. This epistemology displays how a social responsibility to one another is as important as is a responsibility to the objects of knowledge. The discipline here is not primarily personal discipline but social discipline.

The dangerous subjectivity this view is directed against is, to repeat, the result of bad epistemology and bad metaphysics: a subject who feels subjected to immutable forms of the mind or someone who feels no responsibility to his or her larger epistemic community is likely to have been led to such bad habits by the discipline of philosophy itself. The logical empiricists do not wish to free the objective from the taint of the subjective, do not wish to deny the having of experience, but rather to free subjects of knowing from the lessons of incorrect philosophy.

Interestingly, after his linguistic turn of the early 1930s, the subject is no longer a topic of epistemology at all for Carnap, and the question of the relation of the subjective to the objective, in consequence, also disappears. The subject now is just beyond what can be spoken of meaningfully in philosophy – the practical decisions must still be made but there is no account in philosophy of the practical decision maker. It is voluntarist epistemology without a willful subject.[9] It is an interesting coda to Daston and Galison's history of objectivity that a particular formulation of structural objectivity ends with extruding both the mind and the objective/subjective distinction itself from the bounds of the theory of knowledge. This might count as a radicalized version of the discipline of the self at the heart of Daston and Galison's

[9] I explore this theme in logical empiricism in relation to Bas van Fraassen's notion of a philosophical stance in Richardson (2010).

story – the properly disciplined self disappears as a topic within epistemology entirely. The voluntarism is, still however, inherently social: Carnap now speaks the language of recommendation: for specific purposes the following sort of language would be appropriate. But these purposes are never intrinsically private. At worst they are simply not shared; they express different valuational attitudes to the world. In this sense the recommendation is itself a clarification of one's own values and an invitation to share in them.

When the guarantee of an isomorphism between reason and the world was given up and with it the guarantee that the world was, in the end, intelligible to humanity, the logical empiricists joined Deweyan pragmatists and a few other practitioners of scientific philosophy in viewing this as a liberating and hopeful thing. We need not get stuck in the ways of thinking of past but could remake our cognitive lives and bear new responsibility for our knowledge of the world. What the demand for mutual intelligibility embedded in the voluntarist version of structural objectivity at the heart of logical empiricist philosophy of science insisted upon was that we give up the idea that there were either substantial metaphysical or substantive cognitive constraints that absolved us from bearing the epistemic responsibility for our system of scientific knowledge. I have here attempted to modify Daston and Galison's account of logical empiricist structural objectivity both to make the account more accurate to the motivations of the logical empiricists and to draw out this logical empiricist theme of social responsibility in the realm of knowledge-making. That final lesson from Carnap and Reichenbach is one we ought not forget as we think about the problem of objective knowledge and the dangerous subjectivities of our own age, an age that bears all too many similarities to their own.

References

Bauch, Bruno. 1923. *Wahrheit, Wert, und Wirklichkeit*. Leipzig: Meiner.

Beaney, Michael. 1997. *The Frege Reader*. Oxford: Blackwell.

Carnap, Rudolf. 1926/2019. Physikalische Begriffsbildung/Physical Concept Formation. In *The Collected Works of Rudolf Carnap. Volume 1: Early Writings*, ed. A.W. Carus et al., 339–439. Oxford: Oxford University Press.

———. 1928/1967. *The Logical Structure of the World*. Trans. R. George. Berkeley/Los Angeles: University of California Press.

———. 1931. Die physikalische Sprache als Universalsprache der Wissenschaft. *Erkenntnis* 2: 432–465.

———. 1937. *The Logical Syntax of Language*. London: K. Paul Trench.

Cassirer, Ernst. 1910/1923. *Substance and Function*. New York: Dover.

Daston, Lorraine, and Peter Galison. 2007. *Objectivity*. New York: Zone Books.

Edgar, Scott. 2008. Paul Natorp and the Emergence of Anti-Psychologism in the Nineteenth Century. *Studies in History and Philosophy of Science* 39: 54–65.

———. 2015a. Intersubjectivity and Physical Laws in Post-Kantian Theory of Knowledge: Natorp and Cassirer. In *The Philosophy of Ernst Cassirer*, ed. J.T. Friedman and S. Luft, 141–162. Berlin: DeGruyter.

———. 2015b. The Physiology of the Sense Organs and Early Neo-Kantian Conceptions of Objectivity: Helmholtz, Lange, Liebmann. In *Objectivity in Science*, ed. F. Padovani, A. Richardson, and J.Y. Tsou, 101–122. Cham: Springer.

Feest, Uljana. 2007. Science and Experience – Science of Experience: Gestalt Psychology and the Anti-Metaphysical Project of the *Aufbau*. *Perspectives on Science* 15: 38–62.
———. 2012. Husserl's Crisis as a Crisis of Psychology. *Studies in the History and Philosophy of Biological and Biomedical Science* 43: 493–503.
Fellows, Jill. 2011. *Making Up Knowers: Objectivity and Categories of Epistemic Subjects*. Unpublished PhD thesis, Department of Philosophy, University of British Columbia.
———. 2017. The Way of Life of Mr. Nowhere: Examining Harding's 'Objectivity and Diversity'. *Philosophical Studies* 174: 1807–1818.
Friedman, Michael. 2010. Science. History, and Transcendental Subjectivity in Husserl's *Crisis*. In *Science and the Life-World*, ed. D. Hyder and H.-J. Rheinberger, 100–115. Stanford: Stanford University Press.
Green, Christopher D. 2010. Scientific Objectivity and E.B. Titchener's Experimental Psychology. *Isis* 101: 697–721.
Harding, Sandra. 2015. *Objectivity and Diversity*. Chicago: University of Chicago Press.
Heis, Jeremy. 2014. Ernst Cassirer's *Substanzbegriff und Funktionsbegriff*. *HOPOS* 4: 241–270.
Inkpen, Dani. 2009. On the Subject of Goethe: Hermann von Helmholtz and Scientific Objectivity. *Spontaneous Generations* 3: 178–194. (Originally published by the name "Hallet").
Kusch, Martin. 2009. Objectivity and Historiography. *Isis* 100: 127–131.
Reichenbach, Hans. 1920/1965. *The Theory of Relativity and A Priori Knowledge*. Berkeley/Los Angeles: University of California Press.
———. 1928a. *Philosophy of Space and Time*. New York: Dover.
———. 1928b. The World View of the Exact Sciences. In *Hans Reichenbach: Selected Writings, 1909–1953, Volume 1*, ed. M. Reichenbach and R.S. Cohen, 241–244. Dordrecht: Reidel.
———. 1929/1978. New Approaches in Science: Philosophical Research. In *Hans Reichenbach: Selected Writings, 1909–1953, Volume 1*, ed. M. Reichenbach and R.S. Cohen, 249–253. Dordrecht: Reidel.
———. 1931. *Ziele und Wege der heutigen Naturphilosophie*. Leipzig: Meiner.
———. 1954. *The Rise of Scientific Philosophy*. Berkeley/Los Angeles: University of California Press.
Richardson, Alan. 1997. *Carnap's Construction of the World*. Cambridge: Cambridge University Press.
———. 2005. The Tenacious, Malleable, Indefatigible, and Yet, Eternally Modifiable Will': Hans Reichenbach's Knowing Subject. *Proceedings of Aristotelian Society* 79: 73–87.
———. 2006. Freedom in a Scientific Society: Reading the Context of Reichenbach Contexts. In *Revisiting Discovery and Justification*, ed. J. Schickore and F. Steinle, 41–54. Dordrecht: Springer.
———. 2010. But What Then Am I, This Inexhaustible, Unfathomable Historical Self? Or, upon what ground may one commit empiricism? *Synthese* 178: 143–154.
———. 2017a. From Scientific to Analytic: Remarks on How Logical Positivism Became a Chapter of Analytic Philosophy. In *Analytic Philosophy: An Interpretive History*, ed. A. Preston, 146–159. London: Routledge.
———. 2017b. 'Neither a Confession nor an Accusation': Michael Polanyi, Hans Reichenbach, and Philosophical Modernity after World War One. *Historical Studies in the Natural Sciences* 47: 423–442.
———. 2017c. On Making Philosophy Functional: Ernst Cassirer's *Substanzbegriff und Funktionsbegriff*. In *Ten Neglected Classics of Philosophy*, ed. E. Schliesser, 177–194. New York: Oxford University Press.
Savoia, Paolo. 2015. Seeing and Hearing: Charcot, Freud, and the Objectivity of Hysteria. In *Objectivity in Science*, ed. F. Padovani, A. Richardson, and J.Y. Tsou, 123–144. Cham: Springer.
Stone, Abraham D. 2005. The Continental Origins of Verificationism: Natorp, Husserl and Carnap on the Object as Infinitely Determinable X. *Angelaki* 10: 129–143.

Chapter 4
Philosophy (and *Wissenschaft*) Without Politics? Schlick on Nietzsche, German Idealism, and Militarism

Andreas Vrahimis

Abstract With the outbreak of the First World War in 1914, there emerged two controversies related to the responsibility of philosophical ideas for the rise of German militarism. The first, mainly journalistic, controversy concerned the influence that Nietzsche's ideas may have had on what British propagandists portrayed as the ruthlessly amoral German foreign policy. This soon gave way to a second controversy, waged primarily among academics, concerning the purportedly vicious political outcomes of German Idealism, from Kant through to Fichte, Schelling, and Hegel. During the autumn of 1914, and at the cusp between the two controversies, Moritz Schlick was to deliver a lecture series on Nietzsche's life and work at the University of Rostock. Responding to both debates, Schlick penned an introduction in which he sought to defend philosophy against all those who would embroil it in warfare. Schlick offers a series of arguments defending Nietzsche against his accusers. He also argues that, though their contributions to the history of philosophy often amounted to no more than 'beautiful nonsense', the German Idealists' philosophical views cannot be held responsible for the rise of German nationalism. Finally, Schlick mounts a general defense of the search for truth, both in philosophy and in *Wissenschaft*, as a type of activity which presupposes peace. Though Schlick's metaphilosophical views change, as this paper shows, he remains constant both in his favourable appraisal of Nietzsche, as well as his separation between politics on the one hand, and both philosophy and *Wissenschaft* on the other hand.

Keywords Moritz Schlick · Vienna circle · Nietzsche · Militarism · German idealism

---

A. Vrahimis (✉)
Department of Philosophy, University of Cyprus, Nicosia, Cyprus
e-mail: vrahimis.andreas@ucy.ac.cy

C. Damböck, A. T. Tuboly (eds.), *The Socio-Ethical Dimension of Knowledge*,
Vienna Circle Institute Yearbook 26, https://doi.org/10.1007/978-3-030-80363-6_4

4.1 Introduction

The Vienna Circle's conception of the relation between their philosophical work and their politics has been the subject of an ongoing debate.[1,2] Recent work has primarily focused on the so-called 'left' faction of the Circle, often aiming to demonstrate the falsehood of an older stereotype of Logical Empiricism as apolitical.[3] But what exactly was it that motivated those within the Circle who wanted to divorce philosophy from politics? In this chapter, I will specifically direct my attention to Moritz Schlick's views on the relation between philosophy and politics.[4] In an attempt to shed further light on this topic, I will look to a rare instance, at the outset of the First World War, in which Schlick defended his attempt to demarcate between politics and philosophy.

The most important sources relevant to this issue are Schlick's notes for his lecture course on Nietzsche, which he had presented at the University of Rostock between 1912 and 1923; the notes were published for the first time in 2013. Having first presented his lecture course on Nietzsche in 1912, it was planned that Schlick would present it again during the winter semester in 1914 (see Iven 2013b, 67–68). An obstacle to this plan arose when the First World War broke out. There is inadequate evidence for determining whether the course was cancelled due to the war or not. Nonetheless, Schlick had prepared an addendum to his 1912 notes, in which he specifically responds to two controversies concerning the relation between, on the one hand, the militarist attitude prevalent in Germany during the war, and, on the other hand, the philosophical work of Nietzsche, at first, and then of the German Idealists.

This chapter will open with a very brief overview of the Nietzsche controversy in the popular press, showing how it was soon displaced by an academic controversy concerning the responsibility of German Idealism for the war.[5] This allows us to situate Schlick's 1914 addendum to his Nietzsche lectures as a response to both controversies. Having examined Schlick's defence of Nietzsche against his

---

[1] All translations from the German sources, without a previous English translation, are mine.

[2] E.g. Mulder (1968); Stadler (1982) and (1991); Cartwright, Cat, Fleck and Uebel (1996); Uebel (2005), (2008), (2012); Romizi (2012); and Tuboly (2021a).

[3] The existence of a uniform 'left' Vienna Circle was recently disputed by Richardson (2009a) and (2009b) and defended by Uebel (2010). For an account of the role subsequently played by McCarthyism in this depiction, see Reisch (2005). On an altogether different front, Horkheimer (1937/1972) had played an important role in disseminating the miscomprehension of the Vienna Circle as apolitical; see e.g. Dahms (1994); O'Neil and Uebel (2004); and Vrahimis (2020).

[4] As Mormann (2010, 263–264) notes, Schlick's early work has remained, until relatively recently, understudied. Apart from the excellent editorial work involved in the publication of Schlick's *Gesamtausgabe* (e.g. Iven 2013a, b), other recent studies that consider the early Schlick include Siegetsleitner (2010), Ferrari (2016), Textor (2018), as well as Tuboly's and Ambrus' chapters in this volume.

[5] For a more extended analysis of these controversies, see e.g. Hoeres (2004); Akehurst (2010); Vrahimis (2015) and Jeffs (2017).

interpretation as a militarist in 1914, I will turn to Schlick's development of a similar defense in *Natur und Kultur*. As we shall see, Schlick remains constant in his appreciation of Nietzsche even when he denounces fascistic attempts to employ his work as a justification for waging unrestricted war between nations. Connecting the two 1914 controversies, Schlick holds that it is contradictory to blame both Nietzsche and German Idealism for Germany's wartime aggression. Schlick has some highly critical things to say about the Idealist system-builders and their preference for 'beautiful nonsense' over scientific facts. Regardless of Schlick's estimation of their philosophical worth, he maintains that a clear line can and must be drawn between their philosophy and their political rhetoric. As I shall demonstrate, Schlick's response to both controversies relies on an overall conception of the value of *Wissenschaft* (of which philosophy is part) that runs throughout his work. This conception underlies Schlick's argument that philosophy presupposes peace, and cannot be put to use for the purposes of waging war.

4.2 The Nietzsche Controversy in the Popular Press

On June 28, 1914, Gavrilo Princip assassinated Archduke Franz Ferdinand, soon leading to the outbreak of the First World War. Scrutinizing Princip's background, the contemporary popular press discovered a secret society called the Black Hand, which they depicted as being responsible for the assassination. Allegedly, the Black Hand were inspired by the writings of a German philosopher named Friedrich Nietzsche, and they would even begin their secret meetings by reciting some of his poetry (Martin 2006, 147–148). This set the wheels in motion for a subsequent controversy concerning the connection between Nietzsche's philosophy and the war.

With the United Kingdom's entry into the war effort on August 4, Nietzsche's philosophical work would become one of the chief targets of the British press.[6] Nietzsche was declared to be the mastermind behind amoral German militarism and imperialism (see, e.g., Martin 2006). The propagandists would caricature Nietzsche as directly responsible for German aggression and the brutal tactics employed in the burning of Louvain. They identified a transmission of his thought through the influence of the historian Heinrich von Treitschke,[7] as well as Friedrich von Bernhardi, a general and writer on military matters.

Nietzsche was presented as having taught the 'mandarins' of the German state the lesson that politics is an expression of the will to power, a game in which morality (and especially Christian slave-morality) constitutes a mere hindrance to success.[8] Nietzsche's influence was by no means portrayed as limited to policy-makers:

---

[6] On the parallel Franco-German controversy over Nietzsche within Catholic theology, see Köster (1998, Ch. V).

[7] For a discussion of Treitschke's debates with Dilthey (in the broader context of the development of Logical Empiricism) see Damböck (2017, 99–106).

[8] I borrow the term 'mandarin' here from the title of Ringer's (1969) famous work.

it was rumored that German soldiers went to war with a copy of *Zarathustra* in their knapsacks (as opposed to British soldiers with their Bibles).[9] The author of the *Antichrist* was even sometimes portrayed as the devil (Martin 2006, 153).[10]

In his academic estimation of Nietzsche's influence on the war published in 1915, H. L. Stewart, Professor of Philosophy at Dalhousie University, would describe the debate in the popular press as follows:

> Newspapers and magazines have been filled with estimates which may well bewilder the public. One critic declares that chapter and verse may be quoted from *Zarathustra* to justify every step which Germany has taken; another finds that no philosopher has spoken out more boldly against the spirit of racial aggressiveness that, in short, Nietzsche is the champion of the German *Geist*, not as embodied in, but as contrasted with, the German *Reich*. In the heated language of the moment there is, of course, much exaggeration and much misunderstanding. (Stewart 1915, v)

Interestingly, as Stewart notes, the controversy had included various attempts to answer the British propagandists' claims about Nietzsche's responsibility for the war. Stewart nonetheless dismisses such views as serving the interests of "German apologists" (1915, 150), who would rather "disown" Nietzsche than have their real beliefs made public internationally.

Despite purporting to give a disinterested scholarly take on the debate that would avoid the fallacies of the wider public, which "loves to fix odium upon an individual" (1915, 153), Stewart soon falls into analogous journalistic tropes. Though he admits that philosophy rarely has the public influence that is attributed to Nietzsche, he expressly rejects the claim that Nietzsche was not influential in Germany. As evidence of Nietzsche's influence, he cites his verbal conversations with "a British critic, very familiar with Continental philosophy" (1915, 150) and a German professor, both unnamed, who told him that Nietzsche is constantly talked about "in educated circles" (1915, 151). Stewart (1915, 155) acknowledges the view that Nietzsche was in fact an individualist, but claims that this leads him towards idolizing ruthless warriors (1915, 159). He sees that Nietzsche criticized "the rising military spirit" (1915, 155) which threatened to undermine German civilization, but he insists that Nietzsche had no great love for civilization and was delighted by its destruction (1915, 160–161).[11] Stewart sees that Nietzsche valued cosmopolitanism

---

[9] As Martin (2006, 154–155) points out, publishers at the time did issue pocket-sized versions (intended for soldiers) of the bible and various literary and philosophical works, including Nietzsche's *Zarathustra*, as well as anthologies of his sayings about war. It should be noted that the existence of this industry partly explains the proliferation of wartime writings by philosophers, since many such works were intended to be printed and disseminated in this form. (For this insight, I am grateful to Bernie Linsky who mentioned to me that Wundt's *Die Nationen und ihre Philosophie* was printed in this way.)

[10] Martin (2006, 153) shows that this description was used in Britain as a marketing strategy to sell Nietzsche's books.

[11] Interestingly, as we shall see, Schlick's defenses of Nietzsche (both in 1914 and during the 1930s) raise the three points Stewart would reject: he rejects the idea that Nietzsche was influential in Germany, he insists on interpreting Nietzsche as an individualist, and on this basis he rejects the view that Nietzsche valued war among states.

and despised nationalism (1915, 156–157); yet behind his cosmopolitanism, according to Stewart, is an imperialistic effort to establish an internationally 'dominant "exploiting" class' (1915, 161).[12] He detects a Nietzschean "immoralism" as the underlying principle guiding the foundation and running of the German state:

> What the statesmen of Berlin had for generations been whispering into one another's ears in secret Nietzsche has proclaimed upon the housetops; he inspired the thought that the unscrupulous selfishness which Prussia had plainly practised [*sic*], and the ruthlessness which had marked the campaigns of her troops, were not something to be ashamed of, but something to be gloried in; he cast the halo of an intellectual vindication round the methods of aggression which the bureaucracy had followed, but which they had for merely thought it desirable to mask before the public opinion of Europe. (Stewart 1915, 171)

The above passage illustrates the floundering of Professor Stewart's supposedly 'scholarly' attempt to avoid the excesses of the popular press that he decries. It is perhaps unsurprising that Stewart's effort to give the popular controversy over Nietzsche a kind of academic veneer was not further discussed during the war.[13]

4.3 The Academic Controversy About German Idealism

Apart from Stewart's unwarranted rhetorical flourishes, there is a better explanation as to why the Nietzsche controversy was not widely addressed by academics. The controversy, which took place in newspapers at the early stages of the war in 1914, soon gave way to a second, academic controversy that would continue throughout the war.

Contrary to Stewart's claims, Nietzsche had been a relatively minor figure in the contemporary academic philosophical establishment, whether Germanophone or not.[14] Rather than Nietzsche, contemporary international academic philosophical debate was primarily concerned with various brands of Idealism.[15] In the wake of Kant, this had been a product that Germanophone academia had successfully exported throughout the world. At the time, most of the Philosophy Departments of the Universities of the key players involved in the war (including Britain, France, the United States and Germany) had been dominated by Idealists of one sort or

[12] In support of this view, Stewart (1915, 163) mentions the specific interpretations of Nietzsche as a militarist that Schlick would vehemently argue against, namely the idea that States are locked in a perpetual struggle for dominance.

[13] Voegelin (1944, 177) briefly refers to Stewart in his discussion of Nietzsche's relation to the next war.

[14] Nietzsche did begin to appeal to popular audiences before the war, and he had considerable influence over various social movements; see e.g. Thomas (1983) and Akehurst (2010, 20).

[15] Nevertheless, as Damböck (2020) highlights, the influence of post-Kantian Idealism in nineteenth German philosophy was exaggerated by later figures (such as Heidegger) who underemphasized other developments in German philosophy after Hegel's death.

other.[16] After the end of the war, most of these Idealist schools would be in decline. In large part this was due to their participation in the wartime controversy, as the Professors' enthusiasm for the war in 1914 would be hard to justify after 1918.[17]

The controversy can be traced back to Bergson's wartime lectures in France, as well as to a declaration by a number of British authors who, in response to the German offensive on Belgium, defended England's entry into the war. On October 4, 1914, there came an answer signed by 93 German Professors attempting to justify Germany's participation in the war.[18] Their defense closes with a promise that

> we shall carry on this war to the end as a civilized nation, to whom the legacy of a Goethe, a Beethoven, and a Kant is just as sacred as its own hearths and homes. ("Professors of Germany", 1919, 285)

The reference to Kant led to a subsequent international debate about whether he and the post-Kantian Idealists were somehow responsible for the development of German militarism. Illustrating the shift that took place from one controversy to the next, a *New York Times* review of Dewey's 1915 *German Philosophy and Politics* states:

> Not Nietzsche, but Immanuel Kant is responsible for the spirit of twentieth century Germany. Not belief in the superman but belief in the categorical imperative and the thing-in-itself has sent Germany to war with the world. Not Thus Spoke Zarathustra but the Critique of Pure Reason explains the amazing utterances of Bernhardi, of Treitschke, of Wilhelm himself. (Anonymous 1915)

Dewey, Bergson, Boutroux, L.T. Hobhouse, F.C.S. Schiller, and other philosophers would become embroiled in a debate as to whether the war justifies a severing of all ties with Germanophone philosophy, whose malevolent influence it purportedly exemplifies. Amongst the German Neo-Kantians, we find figures like Hermann Cohen, Paul Natorp, and Alois Riehl attempting to justify Germany's involvement in the war on the grounds of the superiority of German *Kultur* (a concept which they trace back to Kant) over the mere civilizations of the rest of Europe.[19]

---

[16] As Akehurst (2010, 18–19) notes, for example, despite the fatal blow they had earlier received from Moore's and Russell's philosophical objections, in 1914 the British Idealists were still a leading force in British academic and political institutions. (The impact of Idealism on Anglophone academia would long continue to be felt even outside philosophy, for example in the philosophical writings of physicists like Eddington, and later Jeans.) In fact, German Universities were at the forefront of international academia before the war, and it was only Germany's defeat that tipped the scales towards Anglophone Universities; see e.g. Windsor (2014).

[17] On the aftermath of these debates, see e.g. Ringer (1969, esp. Ch. 4); Kusch (1995, 224–271); Vrahimis (2015, 2021c) and Reynolds (2017).

[18] It should be noted that, though Schlick was not among the ninety-three Professors, he did sign the declaration of October 16, 1914, produced by Ulrich von Wilamowitz-Moellendorf and Dietrich Schäfer, and signed by more than three thousand German teachers and professors (which included all of the teachers in his University); see Iven (2008, 2013a, 30–32); Wolters (2017, 11–13; 2021).

[19] See Vrahimis (2015, 2021c); Beiser (2018, 300–304). Like many German academics at the time, Schlick opposed culture to civilization in his early work; see Mormann (2010, 268–269). Outside the Neo-Kantian mainstream, other German philosophers prominently engaged in wartime propa-

I will avoid providing further details concerning these philosophers' perplexing and complex attempts to fight the war by means of their arguments and theses, limiting myself instead to the mention of two facts about this controversy insofar as they are relevant to Schlick's response. First of all, for the purposes of this study, it is important simply to note, as I already have, that this controversy followed after the Nietzsche controversy in 1914. Schlick, as we will go on to see, responded to the Idealism controversy at its very early stages, before the publication of extended analyses of Idealism developed by some of the abovementioned philosophers. He thus does not consider it in its details; primarily, Schlick is concerned with rejecting the central claim that the philosophies of Kant, Fichte, Schelling, or Hegel are somehow responsible for militarism. Secondly, the international debate did not only pit Idealists against their critics, but also led to internal tensions among the various schools that claimed to be continuations of the German Idealist legacy. If they were to avoid the accusation of betraying their nations, Idealists on either side of the war needed to defend opposite views. While the Idealist schools in Germany tended to portray the legacy of idealism as supporting German militarism, the defenders of Idealism in the rest of the world tended to demarcate between philosophy and its application to wartime politics (Vrahimis 2015, 91–92). Within the Germanophone context, the kind of divorcing of philosophy from militarist politics that Schlick advocated was in line with his overall anti-Idealistic stance. Yet outside the Germanophone context, Schlick's position is aligned with defenders of Idealism who proclaimed the political neutrality of its strictly speaking philosophical content, and thereby its innocence as far as the war is concerned.

4.4 Nietzsche's Influence on Schlick

Schlick had encountered Nietzsche's work both quite early on in relation to his own intellectual development, and also at a very early stage in the history of the reception of Nietzsche's thought (Iven 2013a, b). Nietzsche had not yet died in 1898 when the 16-year-old Schlick discovered his works. Like many adolescents after him, Schlick enthusiastically read *Zarathustra,* having read very little philosophy before that (Iven 2013b, 55). He would later note in his unpublished autobiography that during his lifetime no other book would "so shake and enrapture [him] as much as *Zarathustra*" (quoted in Iven 2013a, 18), while elsewhere he thanks Nietzsche for causing in him "so many tears of high enthusiasm" (quoted in Iven 2013a, 18). Schlick's unpublished (and undated) writings even include his attempt at emulating Nietzsche's language in a prose poem consisting of a dialogue with Zarathustra (Iven 2013b, 63). Traces of Nietzsche's style can be found throughout Schlick's less

ganda included Wilhelm Wundt and Max Scheler; see Kusch (1995, 211–224); see also de Warren and Vongehr (2018).

technical work, from the 1908 *Lebensweisheit* to his posthumously published *Aphorismen* (1962).

Before the 1912 lectures, the influence of Nietzsche is chiefly felt in *Lebensweisheit* (Ferrari 2016). I shall return towards the end of this chapter to Schlick's account of the way in which, in Schlick's reworking of Nietzsche's 'will to power', scientific knowledge [*Wissenschaft*] is connected to the 'will to pleasure' [*Wille zur Lust*] and the 'will to truth' [*Wille zur Wahrheit*]. As I will outline in Sects. 5.8 and 5.9, this Nietzschean account will later constitute a central aspect of Schlick's conceptions of epistemology and metaphilosophy, and thus of his account of the relation between philosophy and politics.

Given the above, it should not be surprising that Schlick chose Nietzsche as the topic of his lectures at the University of Rostock in 1912. Before moving on to analyze Schlick's introduction to the Nietzsche lectures penned in 1914, two things must be noted concerning the content of Schlick's 1912 notes.

First, a note should be made on the methodology of Schlick's lectures. Interestingly, Schlick takes a kind of Nietzschean approach to his presentation of Nietzsche, insofar as he intertwines his account of the thinker's life with an account of his philosophy. In order to make sense of Nietzsche's *oeuvre*, Schlick's lectures explore biographical details, e.g. concerning Nietzsche's health, his relations to his family and friends, his various travels. Schlick (2013a, 102–103) in fact argues that such an account is indispensable to understanding Nietzsche's work. In one of many examples of Schlick's approach, he explains how Nietzsche's choice of an aphoristic style was dictated by the state of his health, which only allowed him to concentrate on writing, aided by an amanuensis, for limited spans of time (Schlick 2013a, 240–242).[20]

Secondly, perhaps the most important aspect of Schlick's lectures is the way in which it presents Nietzsche's relation to positivism. Schlick divides Nietzsche's work into three phases: (i) an early metaphysical phase under the influence of Schopenhauer and Wagner, (ii) an anti-metaphysical positivist phase, and (iii) a phase in which he is concerned with the development of an approach to values *within* positivist strictures. Schlick is careful to argue that the third phase in Nietzsche's development is still compatible with positivism, and that the late Nietzsche is *not* a metaphysician. Thus, for example, Schlick (2013a, 293–294)

[20] A tension may be here detected between Schlick's intertwining of Nietzsche's life and thought on the one hand, and on the other hand his views concerning the autonomy of the search for truth from practical concerns (which I examine in Sect. 5.9). If the search for truth is truly autonomous from practical concerns, in the way Schlick describes, then why should an account of Nietzsche's life be relevant to his philosophical ideas? Schlick's lectures are clearly not only concerned with Nietzsche's writing styles and ideas, but also with accounting for their genesis. Nietzsche's life is interesting in connection with the latter purpose, and Schlick uses it to helps his audience better understand the context in which Nietzsche's style and ideas developed, as well as to clarify some of his intentions. Nonetheless, though Schlick does not state this in the Nietzsche lectures, he would probably agree that the genetic account has little to do with the assignment of truth-values to Nietzsche's propositions.

interprets Nietzsche's view of eternal recurrence as scientific, and not metaphysical.

Intriguingly, Schlick's 1912–1923 lectures overlap with a time in his intellectual development in which he upheld what has been characterized as critical or structural realism, and *not* positivism. In fact, during this time, Schlick was critical of specific types of positivism, and had raised a series of objections against the Machian positivist (or 'empiriocriticist') view of science (see Lewis 1988; Textor 2018). His portrayal of Nietzsche as a positivist does not, however, diminish the degree of influence he had on Schlick's early realist phase. In fact, as we shall see in the discussion of Schlick's conception of *Wissenschaft* in Sect. 5.8, Schlick explicitly appeals to Nietzsche and his notion of a 'joyful science' in opposing Mach's and Avenarius's explanations of scientific knowledge in terms of a drive to pleasure (see also Mormann 2010, 268–269; Ferrari 2016, 9; Textor 2018, 114–115). In other words, Schlick's early critique of positivism focuses precisely on the Machian conception of knowledge, from which he seems to exempt Nietzsche, whom he interprets primarily as an ally to his view.

Schlick's early realist phase ended with his conversion to positivism in 1922, which would bring him even closer to Nietzsche as he portrays him in his lectures. Though Schlick does not directly address this, other Vienna Circle members follow Schlick in portraying Nietzsche as a predecessor of the positivist overcoming of metaphysics. Whether directly through Schlick's influence or not, Frank (1917/1970, 232–234),[21] Neurath (1921/1973, 196; 1936/1981, 692, 687; 1938/1955, 10–11, 18; 1932/1987, 10–11; 1938/1981, 826), and even Carnap (1928/2003, 105, 108–109, 271; 1931/1959a, 80; 1932/1959b, 167–168) will, in various passages, interpret Nietzsche as a positivist critic of metaphysics, and thus as a predecessor of the Circle.[22]

---

[21] See also Von Mises (1938/1987, 169, 171).

[22] See also Fischer (1982); Tuusvuori (2000, 159–162, 233, 290–292, 678–680); Mormann (2015).

A possible objection about this claim might be raised concerning its compatibility with what Carnap (1931/1959a) states about Nietzsche, given that he explicitly classifies him as 'the metaphysician who perhaps had artistic talent to the highest degree' (80). In what may appear at first glance confusing, Carnap goes on to argue that Nietzsche divided his work into empirical studies, on the one hand, and on the other hand poetry, through which 'he expresses most strongly that which others express through metaphysics' (p. 80). In other words, Carnap at first classifies Nietzsche as a metaphysician, and then proceeds to say that he avoided metaphysics, expressing his *Lebensgefühl* through artistic means. Is Carnap claiming that Nietzsche is a metaphysician who avoids doing metaphysics? One possible way of resolving the tension here would be to see Carnap as endorsing the tripartite understanding of Nietzsche's work that Schlick also presented in his Rostock lectures. According to this interpretation, Nietzsche was initially a metaphysician, and then came to overcome his metaphysical inclinations, becoming instead a positivist. Carnap can thus be understood to claim that Nietzsche, who was once a representative of Schopenhauerian metaphysics, abandoned its errors and confusions, becoming instead a positivist by focusing his work on either empirical studies or poetry. Interpreting Carnap thus allows us to align his position with that upheld the rest of the Circle's members, except for Waismann.

By contrast to his Vienna Circle colleagues, Waismann (1994a, b) wrote critically about Nietzschean ethics. Furthermore, perhaps unaware of his teacher's views on the matter, Feigl

The above summary illustrates what is at stake in Schlick's 1914 defense of Nietzsche. Schlick is concerned with much more than simply a disinterested dispelling of misinterpretations of some random figure in the History of Philosophy. Schlick's motivation for criticizing the militaristic reading of Nietzsche is, to some extent, politically motivated, insofar as he seeks to answer the British propagandists' claims. But Schlick is also motivated philosophically, insofar as he wants to defend the reputation not only of a writer who had urged his teenage self to take up philosophy, but also of an ally in upholding a *Wissenschaftliche* conception of philosophy. A further, connected, motivating factor is the urge to clarify the proper relation between philosophy and war, on the one hand, and to politics more broadly on the other.

4.5 Schlick's Defense of Nietzsche

As already established above, Schlick had written his lecture notes on Nietzsche in 1912. The lecture course was not offered in 1913, and, as Iven (2013b, 67–68) notes, Schlick had prepared to lecture on the topic again during the winter semester of 1914/1915. It is unclear, as already mentioned, whether the plan had to be aborted due to the outbreak of the war.[23]

In October 1914, Schlick wrote a special introduction to his lectures which would address a topic that was publicly debated at the time, namely the relation between Nietzsche's ideas and the outbreak of the war. At a time, as we have seen, when most of Schlick's colleagues were seeking for ways in which philosophy could contribute to the war effort, the main impetus for Schlick's response to both controversies is that of calling for moderation and opposing the contemporary militaristic frenzy (see also Wolters 2017, 2021). As we have seen, Schlick wrote at a time when attention was shifting away from the Nietzsche controversy, and towards the nascent controversy about German Idealism.[24] Thus, though his primary concern was with Nietzsche, Schlick was well placed to respond to both controversies. In fact, as we shall see, Schlick will set one controversy against the other, showing that

would later join other Anglophone analytic philosophers in talking of "such views as those of Nietzsche, Hitler, and Mussolini on the 'greatness' of war" (1952/1981, 383).

[23] The lectures were paused again in 1917–1918. Upon resuming the lectures in 1919 up to 1923, Schlick added a preliminary section on Schopenhauer, whose thought he presented by way of an introduction to Nietzsche. See Schlick (2013a, 34–35).

[24] Schlick's position can be opposed to one of the few subsequent Germanophone academic responses to the Nietzsche controversy, namely that by the sociologist Werner Sombart. By contrast to Schlick, Sombart (1915, 53) conceded the propagandists' claim that this is indeed 'Nietzsche's war', only to qualify this in a manner which connects to his overall conception of the war as a battle between German Culture and the Entente powers' mere Civilisations. (Coincidentally, Schlick also rejects contemporary caricatures of English civilization's reliance on a type of mercantile utilitarianism, which later find their way into Sombart's work; see Vrahimis 2015, 89–90; Wolters 2017, 14).

one cannot, without contradiction, blame the war both on Nietzsche and on the Idealism which he opposed.

In the introduction to his lectures, Schlick objected not only to the use of Nietzsche's work as a justification for militarism, but also to any urge to seek justifications for militarism in philosophy. Schlick proposes a psychological hypothesis to explain this type of urge. War, according to Schlick (2013a, 77), is so overwhelming that it can bring about certain cognitive and affective distortions and biases. Schlick thinks that the association of ideas can explain such biases, and claims that through repetition, the idea of war comes gradually to be associated with almost every thought and feeling in every realm of human activity.[25] As a result of association, there emerges a kind of involuntary drive or "inner compulsion" (Schlick 2013a, 78) towards connecting all facets of life to the war. Schlick thinks his hypothesis explains why even his contemporary philosophers are obsessed with the war. In Schlick's account, the desire to embroil philosophy in debates about the war is diagnosable as a symptom of the overwhelming nature of war as he has outlined it. War seeks to infiltrate all human "life, feeling, and thought" (Schlick 2013a, 77), even in philosophy which is the most abstract form of thought.

Schlick concedes that his own entry into the debate concerning philosophy's relation to the war is also explained by his psychological hypothesis. In other words, like his opponents, he is also subject to the same law of association which makes connections between war and all kinds of other phenomena. Schlick acknowledges that his hypothesis does not necessarily amount to an argument against debating the relation between philosophy and war, or other such overwhelming contemporary events. Thus his hypothesis does not bar him from entry into the debate. Entering the debate, Schlick will argue for the dissociation between war and philosophy. In fact, as we shall see in Sect. 5.9, Schlick's claim is based on the view that philosophy, qua *Wissenschaft*, presupposes an ability to stand back from all drives associated with practical pursuits (among which we can include the involuntary drive to connect everything with war).

Schlick further seems to imply that his hypothesis concerning the all-encompassing nature of war could account for the types of aggression that were involved in the disputes of his contemporary philosophers and journalists. The aim on each side is to claim for itself the 'correct' philosophical legacy, while blaming its opponents' participation of the war on their choice of the wrong philosophical legacy. If it is true, as Schlick will show, that the very notion of a philosophical justification for war is absurd, then it also follows that the attribution of blame for war to some philosophical tradition or other is misguided.

Having presented his psychological hypothesis, Schlick proceeds to examine specific examples which draw this connection. His main focus is the aforementioned British propaganda and its connection to the Nietzsche controversy. In Schlick's account, the propagandists make extravagant claims about the role of

---

[25] On the role associationist psychology plays in the early Schlick's response to Mach, Avenarius, and Schopenhauer, see Textor (2018, 111–112, 116, 119).

power in German *Realpolitik*. Schlick summarizes the British press' depiction as follows:

> Power, brutal power, was proclaimed to be the highest ideal by us, our spiritual leaders openly stand in opposition to morality, a new master morality [*Herrenmoral*] was preached by us (viz. a morality of authoritarian strongmen, by contrast to a Christian morality of humility); we were striving for a revaluation of all values, we wanted to impose the new anti-moral values on Europe and the whole world. (Schlick 2013a, 79)

Schlick goes on to note that the propagandists depict Nietzsche as the mastermind behind this type of mentality, with newspaper headlines using slogans like "Germany at Nietzsche's feet" (Schlick 2013a, 79).[26]

Schlick's notes do not go into further detail about the nature of the link that British propagandists had drawn between Nietzsche and German politics. This may partly be explained by the fact that the journalists' polemical exaggerations did not really offer much by way of argument for their position. As we have seen above, it was simply asserted that Nietzsche was an amoral atheist whose power-hungry vision had served to shape German politics. However, it is also likely that Schlick could expect from his audience in 1914 to have been familiar with the relevant accusations, and so he does not need to repeat them at great length. Thus after his brief summary, Schlick immediately leaps to a defense of Nietzsche.

Schlick's first response is to note that to anyone familiar with Nietzsche's life and work – as his students will (ideally) be by the end of his course – such accusations will appear strange, and their couching in Nietzschean vocabulary (using terms such as 'power' or 'revaluation of all values') is misleading. Schlick's defense of Nietzsche focuses on two points, which the propagandists confusedly combine. On the one hand, the propagandists miscalculated Nietzsche's influence on German culture and politics. On the other hand, they misunderstood Nietzsche's work.

Schlick summarily rejects the claim that Nietzsche's thought had heavily influenced the development of German culture and politics before 1914. It suffices for Schlick (2013a, 79) to note that few political or military leaders in Germany had even heard of Nietzsche before the British press informed them of his influence on them. Even if some German policy-makers did hear of Nietzsche, they were by no means enthusiastic about his work.[27] Furthermore, Schlick connects this first claim to the second one, insofar as he doubts that Nietzsche's work has been widely understood in Germany.

---

[26] Schlick does not cite a specific source for this quote, and I have been unable to find a specific mention of this phrase in a contemporary newspaper. It is most likely that Schlick here is simply mimicking the exaggeratedly polemical tone of the popular press (in a manner that is relatively mild by comparison to the claims about Nietzsche's diabolical status that we examined above).

[27] But see e.g. Thomas' (1983) overview, which demonstrates that Nietzsche had an significant influence on German politics, including Socialism, Anarchism, Feminism, the Youth Movement to which various Vienna Circle members were connected, and also Social Darwinist tendencies. Whether this implies Schlick was wrong to note that Nietzsche did not influence state policy depends on determining the extent to which these tendencies had such an effect, and the degree to which such an effect connected to their influence by Nietzsche; these are historical matters that I cannot resolve here.

Here, Schlick seems to have missed an opportunity to more adequately address the specific claim at stake in the controversy, i.e. Nietzsche's purported influence on state policy through Treitschke and Bernhardi. In fact, in his original sketch for his lecture notes, Schlick (quoted in Iven 2013b, 68) noted down both these names, but seems to have later given up the effort to directly address the topic in any more detail. It would have been easy to show (as, for example, we have seen Stewart did later on) that the journalistic claim concerning Treitschke and Bernhardi commits both mistakes that Schlick detects in the Nietzsche controversy. On the one hand, Schlick could question whether there is any evidence that either Treitschke or Bernhardi had a significant influence on the events of the First World War. On the other hand, Schlick could argue that the militaristic reading of Nietzsche that the journalistic controversy attributed to Treitschke and Bernhardi was a mistake, due to an overall misunderstanding of his philosophy.

Schlick thinks that even if the caricature of Nietzsche as an anti-moral, power-hungry philosopher were to be accepted, this could not entail that

> the Germans [...] thereby could have become wild beasts that attack their neighboring peoples so as to satisfy their own will to power, at the expense of all moral feelings. (Schlick 2013a, 80)

It would, in other words, take a few further steps to show that, even if the militaristic interpretation of Nietzsche (whether mistaken or not) was widely accepted in Germany, this did *in fact* result in the kinds of amoral actions attributed by the British propagandists to the German state.

It should be noted that a degree of vagueness surrounds the second of Schlick's claims, i.e. that the propagandists misunderstood Nietzsche. Schlick seems to promise that he will offer a more robust philosophical defense of this claim, and hopes to convince his students that his interpretation of Nietzsche clears him of the accusation that he is a militarist. Sadly, Schlick does not specify in which way the subsequent lectures correct the propagandists' misunderstandings of Nietzsche, and simply notes that it will be made clear to his audience that they are.

At first glance, it should be noted that Schlick's interpretation of Nietzsche is quite controversial. As already noted, in Schlick's lectures he interpreted Nietzsche as becoming, after his early metaphysical Schopenhauerian phase, a positivist. Insofar as he holds that this is the correct interpretation of Nietzsche, he would be justified to assert that almost all of his interpreters at the time had fundamentally misunderstood him.[28] If this is to be the basis of Schlick's claim that Nietzsche is not a militarist, then it may not be very convincing to anyone who resists the view that he was a positivist.

There are two ways in which Schlick does explicitly substantiate his claim that Nietzsche cannot be correctly interpreted as a militarist. Both are much more specific than the vague contention found in his introduction to the Nietzsche lectures. It should be remembered that the 1914 text consists merely of lecture notes, a fact

[28] A significant exception was Kleinpeter's (1913) book (published later than Schlick's 1912 lectures), in which he argued that Nietzsche's epistemology is comparable to Mach's.

which may explain their incompleteness – so perhaps it would be justifiable to look elsewhere for a more extended account. The first of Schlick's explicit defenses of the claim that Nietzsche was not a militarist is developed, within a different context, in *Natur und Kultur*, which I examine in Sect. 5.6. A second defense of his claim, to which I will return in Sect. 5.7, relates to an inconsistency between the Nietzsche controversy and the German Idealism controversy, since, as Schlick detects, they involve incompatible claims.

There is a further way in which Nietzsche's influence on Schlick connects to his separation between philosophy and militarism. In the early Schlick, this is a result of his conception of philosophy as the pursuit of truth. Schlick pits this metaphilosophical view against the possibility of any involvement of philosophy in militaristic discourse. Schlick's early metaphilosophy, as I will endeavor to show in Sect. 5.8, is an outgrowth of his interpretation of Nietzsche (as already intimated by my account of Schlick's influence by Nietzsche in Sect. 5.4). The relevant separation between philosophy and politics also continues, as I will show, in Schlick's later view of philosophy as the pursuit of meaning.

4.6 Schlick's Rejection of the Interwar Far-right's Interpretation of Nietzsche

Schlick's return to the question of Nietzsche's connection to militarism in *Natur und Kultur* is no longer a response to the British propagandists of 1914, but rather primarily an answer to the far right's attempt to appropriate of Nietzsche's during the interwar period. Schlick's brief commentary on Nietzsche in this unfinished book from the 1930s is presented within a broader context of criticizing authoritarian conceptions of the state.[29] Schlick mentions the "prophets" (1952, 77) who misinterpret Nietzsche as endorsing a kind of supra-moral status for states. According to such misinterpretations, which are almost identical to those presented in the journalistic controversy of 1914, Nietzsche's critique of Christian morality applies to states, and entails that they may exercise their power without restraint, for example by waging war through any means possible.

Schlick acknowledges that it is easy to misunderstand Nietzsche in this militaristic manner, given that his writings do in fact appear to glorify war. However, as Schlick points out, like anyone who had not witnessed the First World War, Nietzsche could not possibly conceive of modern technological warfare. Though traumatizing, Nietzsche's experience during the Franco-Prussian war, which Schlick details in his 1912 lectures (2013a, 151–154), was far removed from the horrors of trench warfare:

> The word 'war' has, today and in politics, a very different reference than it had in Nietzsche's mouth: it refers today to a horrible, nationwide organization of annihilation, by which all

[29] Schlick also includes a criticism of Hitler in the same chapter (1952, 86). For an analysis of Schlick's account of the state, see Schleichert (2003).

> sorts of machines hurl, from as far away as possible, exploding masses of steel and poisonous gases [...] it is an indiscriminating rampage of the will to destroy; not a competition, not a selection, but a natural catastrophe, a mass phenomenon [*Massenerscheinung*] – all quite incompatible with that which Nietzsche understands as war. (Schlick 1952, 77–78)

Schlick does not think that Nietzsche would have any praise for such technologically induced horrors, and the herd mentality that underlies them.

Nietzsche does not, in Schlick's account, condone unrestricted technological warfare between states, i.e. herds fighting herds. Instead, according to Schlick's individualist interpretation, Nietzsche's praise of war should be better understood as praise for altogether different types of warfare. The main examples Nietzsche has in mind, according to Schlick, are the kinds of warfare that take place among individuals within society, or of "the struggle of the individual against the masses" which "it would be ridiculous to imagine as a battle with riffles and rapid-fire guns" (Schlick 1952, 77). What best instantiates, according to Schlick, the kind of warfare that is a prerequisite for the Nietzschean *Übermensch* is humanity's struggle against nature (notably also William James' proposal of "an army enlisted against Nature" (1911, 290) that should come to replace wars among humans). As evidence for this, Schlick cites part of a relevant passage from Nietzsche's plan for the third book of his unfinished *The Will to Power*, where he sketches his conception of the will to power as follows:

> The will to power.
>
> How the men undertaking this revaluation of themselves would have to be constituted.
>
> The order of rank as an order of power: war and danger required for a rank to maintain its conditions. The grandiose prototype: man in nature; the weakest, cleverest being making itself master, subjugating the more stupid forces. (Nietzsche 2003, 87).

In this passage, while Nietzsche seems to take war to be unavoidable, he does indeed designate the type of war that he has in mind as that between humanity and nature. The broader context for this is Nietzsche's aristocratic conception of a master-morality. Though Schlick notes that he disagrees with this idea, he states that a complete attempt to refute it cannot be undertaken within the context of his discussion of the state.[30] Against Nietzsche's aristocratic critique, he mentions that a morality of kindness [*die Moral der Güte*] is not in any way exclusive to Christianity, or to any "individual peoples or states" (Schlick 1952, 78), but exists throughout all human cultures.

Contrary to a lot of subsequent scholarship that had focused on this element in Nietzsche, Schlick rejects the idea that his conception of power and its relation to morality, which Schlick thinks to be highly problematic, is a central part of Nietzsche's work. Instead, Schlick presents such elements in Nietzsche (which would perhaps undermine his own account) as the weakest, and most misguided, aspect of his overall worldview. Schlick closes his commentary on Nietzsche as follows:

[30] Waismann (1994) would later undertake a more extensive critique of Nietzsche's ethics.

> I hold him to be the richest, most ardent mind of the nineteenth century, but concerning the nature of power I seek from him no guidance, just as I would not concerning a mathematical question. It is the bitterest fate for a Great man, when he works more through his falsehoods than through his truths. (Schlick 1952, 78–79)

Schlick's contention about Nietzsche's falsehoods in the context of *Natur und Kultur* is a direct attack against the claim to Nietzsche's legacy made by his contemporary far-right. Contrary to many of their Anglophone analytic colleagues, many of whom would have been directly or indirectly aware only of the British side in the Nietzsche controversy,[31] Schlick and other Vienna Circle members resisted the far-right appropriation of Nietzsche. In Schlick's view, there may have been elements in Nietzsche, at his worst, that could be misunderstood and put to political use by militarists and fascists.[32] However, at his best, Nietzsche was a positivist predecessor of the overcoming of metaphysics, a project that was later to be undertaken by Logical Empiricism.

4.7 'Beautiful Nonsense': Schlick on German Idealism

As mentioned above, in 1914 Schlick has a further reason for rejecting the blame put on Nietzsche's work as a cause for German militarism. This concerns what he sees as a contradiction between the two prominent views in the two controversies during the war, one about Nietzsche, the other about German Idealism. What is debated in the former is whether Nietzsche's thought is purportedly responsible for German militarism. What is at stake in the latter is the relation between German wartime aggression and German Idealism, in a line of influence that goes back to Kant. In response to the two controversies, Schlick notes that Nietzsche's and Kant's views in normative ethics are polar opposites. Kant conceives of morality as

[31] See Akehurst (2010, e.g., 18–25, 55–58, 69–70, 96, 101–104).

[32] An interesting parallel case to be noted here is that of Neurath's and Lauwerys' (1944a, 1944b, 1945) contention that Plato's *Republic* should be banned from being taught in Germany after the Second World War. Their main argument was that Plato's work presents some of the basic racist and militaristic ideas that were deployed by the Nazis, and so their adherents could go on preaching such ideas after the war, referring to Plato instead of Hitler; see also Soulez (2019), Tuboly (2021b). Interestingly, Nietzsche's name only makes a brief appearance in their polemic against Plato – he is not mentioned as an inspiration for the far right, but merely admonished for his admiration for Frederick II (Neurath and Lauwerys 1944b, 575). In other words, Neurath and Lauwerys resist the urge to blame Nietzsche for the rise of Nazism (as would many others in the aftermath of the Plato controversy).

Interestingly, Schlick's defence of Nietzsche and Neurath's and Lauwerys' indictment of Plato, though parallel in many ways, move in opposite directions. Schlick pronounces Nietzsche's innocence by attempting to dispel any possible militaristic interpretations of Nietzsche as misguided. By contrast, Neurath and Lauwerys indict Plato by warning against the danger of a possible racist and militarist interpretation of the *Republic*.

pertaining to duty, and not to natural inclination.[33] Nietzsche holds the reverse view, and according to Schlick, Nietzsche and Kant are "antipodes" (2013a, 80).[34] Thus, Schlick argues, the propagandists will have to choose: do they believe that the war was caused by too strict a Kantian adherence to moral duty, or by Nietzsche's devilish undermining of all moral standards?

Schlick rejects both claims: neither Nietzsche nor the German Idealists, *qua* philosophers, are responsible for the war. In rejecting the latter claim, Schlick offers a historical argument concerning the origins of the association between German Idealism and German nationalism. According to Schlick (2013a, 81), such an association was common in the early nineteenth century, when philosophers in the Idealist tradition, most notably Fichte, happened also to present their defenses of nationalist ideas in the public sphere. Interestingly, Schlick addresses the case of Fichte by arguing for a separation between his philosophical and his political ideas. He states that the popular appeal of the latter is primarily due to Fichte's "admirable moral character and personality" (Schlick 2013a, 81), and has little to do with his highly problematic philosophical outlook. A strict line must be drawn, in Schlick's account, between Fichte's work as a political orator and his philosophical ideas. For Schlick, Fichte's political works are explicitly *not philosophy*.[35]

In his 1914 lecture notes, it becomes clear that Schlick does not think very highly of the philosophical works of the German Idealists. He claims, for example, that the Idealists' promises of 'Wissenschaft' (e.g. in the title of Fichte's *Wissenschaftslehre*) are deceitful. Instead, their work often contains statements that fly in the face of scientific fact. Schlick gives the example of Hegel's denial, two centuries after Galileo, that the stars are celestial bodies, in preference to the view that they are "only abstract points of light" (Hegel 2004, 297). Contrary to those who would see them as the epitome of German thought, Schlick (2013a, 82) argues that their work constitutes a "most lamentable regression" in the history of philosophy. By contrast, he considers the preceding phase in the history of German philosophy, namely the late eighteenth century, to have been "a *true* heyday" (2013a, 82 original emphasis). The systems of Fichte, Hegel, and Schelling fail to live up to the rigorous standards of their predecessors:

> in place of the strict discipline of thought that we find in Kant, we find in Fichte, Schelling [and] Hegel airy speculation, leading to a momentary dazzling, but not to lasting truth. (Schlick 2013a, 82)

As Schlick (2013a, 81) clarifies, his rejection of the idealist systems relies on a conception of philosophy as the pursuit of truth, which he designates as the primary

---

[33] Kantian morality was among Schlick's targets already in *Lebensweisheit*, and would continue to be so in later work; see Ferrari (2016).

[34] Neurath (1938/1955, 11) would also later present Nietzsche as a predecessor to Logical Empiricism in part due to his critique of Kant.

[35] Schlick's account seems to presuppose that Fichte's political works are generally misunderstood to be works of philosophy, a view which he is contesting. More problematically, Schlick does not address the fact that, however one categorises them, Fichte's political oratory draws its authority from the status of its author as a canonical Great Philosopher.

task of philosophy. Notably, though he thinks they don't live up to the task, Schlick would later explicitly attribute a version of this conception of philosophical methodology to Hegel and Schelling. According to Schlick, they conceive of philosophy as a special kind of path to truth:

> In Germany at the beginning of the nineteenth century Schelling, Fichte, and Hegel believed that there was some kind of royal path leading to truth which was reserved for the philosopher, whereas the scientist walked the pathway of the vulgar and very tedious experimental method, which required so much merely mechanical technique. They thought that they could attain the same truth that the scientist was trying to find but could discover it in a much easier way by taking a short cut that was reserved for the very highest minds, only for the philosophical genius. (Schlick 1932/2008a, 376–377)

The early Schlick was in agreement with the Idealists, insofar as they both conceived of philosophy as a pursuit of truth. As I will discuss in detail in what follows, he would subsequently come to reject this metaphilosophical thesis, replacing it with a conception of philosophy as the pursuit of meaning. It is, in fact, in this context that the passage quoted above was written, as part of Schlick's critique of the conception of philosophy as a pursuit of meaning. Under both conceptions of philosophy, however, Schlick thinks Fichte, Schelling, and Hegel fare pretty badly for the same reasons, namely due to their ignorance of scientific facts.

According to Schlick in 1914, it follows that the Idealists miss the mark by the standards set both by his and by their own conceptions of philosophy as the pursuit of truth. In Schlick's view, deprived of any value as *Wissenschaft* (i.e. as the pursuit of truth), the Idealists' work consists primarily of emotional expressions. This culminates in Schlick's conclusion, in reference to Fichte, Hegel, and Schelling, that

> If one understands philosophy as nothing more than belletristic talk about things, as throwing around provocative thoughts and as reveling in high feelings unrelated to reality, then possibly these structures of thought have their legitimacy. (Schlick 2013a, 81)

It is notable that we find here, already in this underdeveloped form from 1914, a line of thought that will come to play a central role in subsequent criticisms weighed by the Vienna Circle against metaphysical thinkers. Schlick will repeat some very similar claims about the Idealists in later works, such as his 1931 draft of "Philosophy as the Pursuit of Meaning":

> We know what happened when, in the days of Schelling and Hegel, philosophy undertook to bridge the gulf then existing between the various branches of science before patient experimental research had reached the general principles which made such an undertaking possible: the philosophers furnished principles of their own, which they believed to be the expression of some deep insight into the nature of things, but which in reality were even less than hasty generalisations of superficial truths. They were nothing but vague metaphors of no cognitional value.
>
> The result of such a "philosophy" was at best some beautiful nonsense. (Schlick 2013b, 129–130).

Notice that Schlick here uses almost the same vocabulary as found in the Nietzsche lectures, e.g. in his reference to the primarily aesthetic appeal of the Idealists' writings. Like the Nietzsche lectures, Schlick's 1931 article goes on to list various false scientific theories upheld by the idealists (2013b, 130). These now include not only Schelling's and Hegel's misunderstandings of electricity, but also the same basic misunderstandings of astronomy that we find in Hegel, which Schlick in 1931, perhaps slightly more charitably, further explains as Hegel's attempt to connect astronomy with optics.[36]

There are, however, two important differences between Schlick's verdicts in 1914 and 1931. Firstly, as already noted, in 1931 Schlick is explicitly rejecting the notion that philosophy is in any way engaged in the pursuit of truth. Having moved away from his early realist phase, Schlick would deny that philosophy is identical with *Wissenschaft*. As we have seen, he even cites Hegel and Schelling in rejecting the idea that philosophy is a kind of path to truth that is nobler than science. Instead, as we shall see in more detail soon, Schlick argues that philosophy is an activity of clarifying meaning. The shift towards a conception of philosophy as the pursuit of meaning accounts for a second noticeable difference between the 1914 and 1931 texts, namely the addition in 1931 of the term 'nonsense' to the charges laid against the Idealist system-builders. This signifies its connection to verificationism, which underlies Schlick's understanding of the pursuit of meaning.

Despite these differences, the overall polemical attitude displayed by Schlick towards the Idealists remains, as we have seen, at its core unchanged between 1914 and 1931. Like later anti-metaphysical work by the Vienna Circle, Schlick's 1914 lecture criticizes the Idealists for preferring aesthetically beautiful poetic modes of emotional expression over the philosophical pursuit for truth.[37] We can find the same attitude in Neurath's 1921 accusation of Spengler as a pseudo-rationalist, by which he means a poet or a prophet disguised as a philosopher (1921/1973, 160; cf. Vrahimis, 2021a). We can also see an analogous attitude in Carnap's 1931 contention that metaphysicians (e.g. Heidegger) are like failed poets or 'musicians without musical ability' (1931/1959a, 80), while Nietzsche's overcoming of metaphysics implied that he divided his work between undisguised poetry (e.g. in *Zarathustra*) on the one hand, and purely empirical studies on the other (see Vrahimis 2013, 31–86).

4.8 The Game of Value-Free *Wissenschaft*

A further notable aspect of the 1914 passage by Schlick is that it prefigures his subsequent apolitical rejection of metaphysics (whereby *Wissenschaft* is strictly separated from any political stance), and by extension his disagreement with the 'left' Vienna Circle's conception of the political role to be played by the overcoming of

---

[36] In his draft for the article, Schlick (2013a, 141) explicitly cites Hegel's account of the stars as abstract points of light, and also names Hegel's theory of heat (which he does not further explore, but only cites the relevant passage).

[37] Bertrand Russell had weighed similar accusations against Bergson in 1912; see Vrahimis (2011, 2019, 2021b).

metaphysics. As already noted above, Schlick's 1914 criticism of the Idealist blurring of philosophy and emotional expression is based on his conception of a purely theoretical search for truth. But what exactly was it that Schlick had in mind in 1914 when he identified philosophy with a search for truth?

In order to determine this, one must look to Schlick's evolutionary account of the value of knowledge, an account that will remain central throughout his work (see e.g. Mormann 2010, 267; Ferrari 2016). In his earliest writings, this is expressed as the view that *Wissenschaft* is the product of a "will to truth" (Schlick 1908/2006, 86; see Ferrari 2016; Textor 2018, 117–118). This view, as already mentioned, is the result Schlick's reworking of Nietzsche's 'will to power' (see e.g. Iven 2013b, 58; Ferrari 2016, 7). In *Lebensweisheit*, Schlick (1908/2006, 170–181) chooses to present it under a motto from Nietzsche's *Zarathustra*: "Then, however, was Life dearer unto me than all my Wisdom had ever been" (Nietzsche 1885/1905, 232). A central element in Schlick's evolutionary account of *Wissenschaft* is a distinction he draws between two possible attitudes that can be taken towards objects in the world (Schlick 1908/2006, 86): a theoretical one, which concerns facts (or in Schlick's terminology, the nature ("Wesen", 1908/2006, 86) of objects), and a practical one concerned with values. According to Schlick, all practical evaluations are guided by a fundamental drive that he calls the "will to pleasure". *Lebensweisheit* is primarily concerned with working out the consequences of this conception of values within a broad evolutionary framework, in which the will to pleasure guides the interests of all living things, from animals to humans. Within this framework, Schlick is faced with the problem of explaining how it is possible that a theoretical attitude can arise at all. In other words, Schlick asks: how is it possible that the human animal, primarily concerned with its survival and driven by its will to pleasure, may develop a 'will to truth', a pursuit of knowledge for the sake of knowledge rather than for some specific other practical end?

Schlick's 1908 conception of the evolutionary emergence of *Wissenschaft* in the human animal can be broken down into four elements:

1. *Pure will to pleasure*: in this first stage, *qua* animal, the human being is guided in its activities purely by considerations concerning the fulfillment of some practical end, through the 'will to pleasure'. Everything in this stage is seen "from the point of view of value" (Schlick 1908/2006, 170).
2. *Technological thinking*: in this intermediary stage, we have a kind of technological thinking guided by practical ends, namely an attempt to make predictions based on past experiences. According to Schlick, the human brain evolved so as to be able to associate past experiences with conclusions, which in turn can be used as predictions for future actions, and is thus conducive to the human animal's survival.
3. *(Purposeless) games*: In another intermediary stage, Schlick proposes that we imagine a human animal that, for whatever reason, has temporarily had all other drives satiated. Out of curiosity, the human can come to pose some question or puzzle that is irrelevant to its practical interests. It can pursue such a puzzle in the form of a game, i.e. an activity primarily undertaken for its own sake, and which,

after a process of habituation, can even become pleasurable without being guided by practical interests.[38]

4. *Games guided by the 'will to truth'*: The fourth stage is a combination of the basic elements of the previous two stages. Guided by practical concerns, humans find that posing questions about predictions works better when set in the context of a game, rather than when pursued for this or that specific practical end. This, in Schlick's account, explains the development of the 'will to truth', and thus of *Wissenschaft*. On the one hand the search for truth is undertaken for its own sake, while on the other hand it is instrumental in predicting future events, and thus conducive to survival. By contrast to one intermediary stage (2), it is guided by theoretical, not specific practical interests. By contrast to another intermediary stage (3), which may involve any playful stream of thought, *Wissenschaft* is limited insofar as it is determined by the 'will to truth'.

Schlick thinks of *Wissenschaft* as, ideally, a kind of game, in so far as it is an activity primarily undertaken for its own sake, and which can be pleasurable without being guided by practical interests (see also Mormann 2010, 269–272). This is how Schlick interprets Nietzsche's notion of "joyful wisdom", the joy involved in the game of pursuing truth. He opposes this to the pursuit of knowledge as work, i.e. directed towards specific practical goals other than itself. In fact, Schlick takes a very critical stance towards the idea of *Wissenschaft* undertaken as a kind of job, named "research" (1908/2006, 180), whose sole aim is academic advancement (or 'fame and glory' (1908/2006, 180)), and which is associated with a certain seriousness and gravity. Instead, Schlick waxes lyrically about

> [t]he sensitive Nietzsche, who felt this disharmony perhaps especially in himself, therefore praises, in contrast to this unhappy practice of erudition, the 'joyful *Wissenschaft*' of the wise.
>
> Wise is he [*sic*] to whom the contemplation of the world and the search for truth have bestowed all their blessings [...] which the Play of the Spirit effuses on all who give themselves to it (Schlick 1908/2006, 181)

Schlick recruits Nietzsche as a proponent of his notion of *Wissenschaft* as 'joyful', in that it involves a pleasurable game of pursuing truth.

Like the technological thinking that gave rise to it, *Wissenschaft* does of course serve practical purposes, insofar as it makes successful predictions. Nonetheless it is of crucial importance for the production of successful predictions that truth itself, and no other end, is what is pursued! Successful predictions presuppose the ability to take a purely theoretical attitude, which aspires only to truth and is thus divorced from practical evaluations. As Schlick puts it,

> [e]ven if science does not draw its conclusions for the purpose of action, the truth of its propositions can of course (this follows from the above definition of truth) only be tested by the success of actions; the latter are then called experiments. (Schlick 1908/2006, 173)

[38] Schlick designates this moment as the birth of *Wissenschaft* (1908/2006, 171).

Schlick returns to this theme in later work, most strikingly perhaps in the midst of his contribution to the Protocol-Sentence debate,[39] in his 1934 "The Foundation of Knowledge". There we find Schlick making the following claim:

> Cognition is originally a means in the service of life. In order to find his way about in his environment and to adjust his actions to events, man must be able to foresee these events to a certain extent. For this he makes use of universal statements, cognitions, and he can make use of them only in so far as what has been predicted actually occurs. Now in science this character of cognition remains wholly unaltered; the only difference is that it no longer serves the purposes of life, is not sought because of its utility. With the confirmation of prediction the scientific goal is achieved: the joy in cognition is the joy of verification, the triumphant feeling of having guessed correctly. And it is this that the observation statements bring about. (Schlick 1934/1959, 222–223)

Though the addition of Schlick's conception of *Konstatierungen* is a new, perhaps more technical, element here, the view of *Wissenschaft* developed in *Lebensweisheit* is clearly repeated in the above passage more or less unaltered (similar claims are found e.g. in Schlick 2013b, 209–210). The same approach to the value of knowledge, i.e. in terms of a drive to pleasure which is explained in evolutionary biological terms, is earlier found in his 1918/1925 *General Theory of Knowledge* (Schlick 1918/1925/1974, 94–101). Schlick discusses critically the idea that science "serves only the needs of practical foresight" (1918/1925/1974, 96), clarifying that

> the scientist, in pursuing his cognitive goals, fares better if he does not think of practical applications and does not take as the aim of his endeavors the discovery of only practically applicable truths; he should therefore set to work *as if* truth itself were the final goal. As experience teaches, this is the only way to make the great advances that subsequently prove so fruitful; if when we began we had in mind only what was useful to man, we would never have arrived at such knowledge. But even though it is thus useful for humanity to pretend that truth and pure knowledge constitute the ultimate aim of science, the real goal of cognition nevertheless is actually practical use, which alone provides the struggle for truth with its reason for being. The striving for knowledge 'for its own sake' is merely a game. (Schlick 1918/1925/1974, 96–97, original emphasis)

Like earlier in *Lebensweisheit*, Schlick goes on to note that the type of play that he has in mind is not just a "waste of time" (1918/1925/1974, 97), but becomes itself a source of pleasure. This takes place after a process of transformation in which, through habituation, what was originally undertaken as a means becomes an end. Play, the product of this transformation, is the pinnacle of human activity[40]:

> Speech, at first a means of communication, becomes song; walking, originally a means of getting about, grows into dance; seeing turns into looking- at, hearing into listening-to, work into play. At the apex are those activities connected with play. They alone satisfy immediately, while all behavior that serves simply as means directed to ends - work - acquires value only in relation to the results obtained. (Schlick 1918/1925/1974, 97)

[39] For the classic discussion of the debate and its significance for the development of Logical Empiricism, see Uebel (2007); see also Uebel (2020).

[40] See also Tuboly's chapter in this volume for a discussion of the concepts of work and play in Schlick.

In his 1927 "On the Meaning of Life", Schlick further develops the idea of play as exemplifying the meaning of life. There, Schlick explicitly connects his account of play with an insight that he attributes to Nietzsche. In order to do so, Schlick first presents a very compressed summary of his classification of three phases in Nietzsche's work, which he had more extensively developed in his 1912 lectures (see Iven 2013a, 43–46; Ferrari 2016). In 1927, the three stages are presented in terms of Nietzsche's answer to Schopenhauerian pessimism, i.e. as Nietzsche's three replies to the contention that life is meaningless:

> We know how Nietzsche, for example, sought to conquer this pessimism. First by the flight into art: consider the world, he says, as an aesthetic phenomenon, and it is eternally vindicated! Then by the flight into knowledge: look upon life as an experiment of the knower, and the world will be to you the finest of laboratories! But Nietzsche again turned away from these standpoints; [...] it is hard to reduce to a brief formula what the wisest Nietzsche, the Nietzsche of *Zarathustra*, saw as the meaning of life. For if it be said that henceforth the ultimate value of life, to him, was *life itself* that obviously says nothing clear and does not find the right expression for the deep truth which he then perceived or at least suspected. For he saw that life has no meaning, so long as it stands wholly under the domination of purposes [...] In truth, we shall never find an ultimate meaning in existence, if we view it only under the aspect of purpose. (Schlick 1927/1979, 113, original emphasis)

In the above passage, Schlick underlines the way in which the way of play traverses all periods in Nietzsche's thinking, remaining a concern both in his positivist valuation of scientific knowledge as the ultimate *telos*, and his later reconfiguration in terms of the value of life itself (which Schlick thinks is still within the bounds of his earlier anti-metaphysical positivism). Schlick goes on to outline his own positive understanding that follows from such a critique of the exclusive focus on purposive action, namely his idea that the meaning of life is to be found in play, defined as activity undertaken for its own sake, rather than for the sake of the attainment of some other end. Schlick's conception of play as the meaning of life, and not solely connected to the value of knowledge, clearly indicates that he follows Nietzsche into his third period. He remains in agreement with this third phase in Nietzsche even when he presents his conception of play as epitomized by *Wissenschaft* as 'joyful' (in the Nietzschean sense that we have already seen presented in *Lebensweisheit*, and which continues to permeate later writings). Thus, as I have demonstrated above, Schlick's Nietzschean conception of play remains interconnected with his conception of value-free *Wissenschaft* throughout his philosophical *oeuvre*.[41]

[41] Similar claims about the continuation of these Nietzschean themes throughout Schlick's work have also been made by Mormann (2010) and Ferrari (2016).

4.9 Philosophy, *Wissenschaft*, and Politics

Having clarified Schlick's conceptions of play and value-free *Wissenschaft*, the question remains: where exactly does Schlick situate philosophy within this picture? Schlick, as we have seen, changes his metaphilosophical views, explicitly resisting the identification of philosophy with *Wissenschaft* and the pursuit of truth. As e.g. Tuboly's chapter in this volume shows, Schlick had a broader (because more "traditional") conception of the scope of philosophy than some of his fellow Vienna Circle members. Would such an expanded conception of philosophy perhaps allow for some values to creep in?

To answer this, we need to look back to Schlick's analysis of the four stages in the evolution of *Wissenschaft*, i.e., as already outlined above: (1) pure 'will to pleasure', (2) technological thinking, (3) ('purposeless') games, and 4) games guided by the 'will to truth'. It is clear that the early Schlick has a precise conception of science, which he identifies with (4). But what about philosophy?

As we have already seen, the early Schlick identifies philosophy with the pursuit of truth, and thus with *Wissenschaft*. Yet this appears to be a prescription rather than a descriptive account of the history of philosophy. In fact, in various passages, the early Schlick contends that not all work that has historically gone under the banner 'philosophy' has attained the level of *Wissenschaft* in its 'joyful wisdom'. Indeed, sometimes philosophers seem to regress back to the stage preceding the evolution of *Wissenschaft*, engaging in mere intellectual games instead of the pursuit of truth. When regressing to this stage, according to Schlick, "one cannot deny even the philosophemes of Hegel or Schelling any justification, for in the game it is not so much a question of whether it is meaningful or meaningless" (Schlick 1908/2006, 173).

When people like Hegel or Schelling are playing the game, they are sometimes not guided by the pursuit of truth (even though they designate this as their motivation). Instead, they are simply indulging in 'beautiful nonsense'. Though it is likely that no knowledge will thereby be attained, one could plausibly find pleasure in playing the game of trying to piece together such puzzle, e.g. asking what exactly Hegel might have meant by some combination of words. But even if the game is played for its own sake, the possibility that it might (even accidentally) give rise to *Wissenschaft* is not *a priori* excluded. For example, Schlick (1932/2008b, 400–401) discusses a number of questions that were at first thought by philosophers to be insoluble (and so, in the language of *Lebensweisheit*, they would remain within the domain of an aimless game), which upon further refinement became scientific questions. Some of Schlick's own work could possibly be thought of in this way, e.g. his views on the meaning of life or the philosophy of youth, which he talks about as undertaken for leisure (Schlick 1927/1979, 123).[42] That the game does not have any practical outcome is not in itself a reason to abolish it – but one should, at the same

[42] Mormann (2010, 265) talks of ethics as the early Schlick's "hobby".

time, not have any expectations of useful outcomes from it, especially in terms of some power to make predictions.

As already outlined above, Schlick changes his metaphilosophical views, rejecting the idea that philosophy is engaged in a scientific pursuit of truth. This is set out most clearly in "The Future of Philosophy", which discusses four conceptions of philosophy, namely:

(i) philosophy as science (i.e. the pursuit of truth),
(ii) philosophy as the 'wisdom of life' (here identified as 'the pursuit of meaning'),
(iii) philosophy as epistemology (i.e. a kind of pursuit of truth about the pursuit of truth), and
(iv) philosophy as metaphysics (i.e. as a pursuit of extra-scientific truth).

Schlick explicitly rejects all three conceptions of philosophy as different kinds of pursuit of truth, in preference for what he holds to be a Socratic (Schlick 1932/2008a, 376, 379–380, 383; see also Schlick 1936/1938, 395–397) and Wittgensteinian (Schlick 1932/2008a, 387) understanding of philosophy as the pursuit of meaning.

Interestingly, the metaphilosophical picture presented here is a kind of reconfiguration of the one we already saw developed in *Lebensweisheit*, and it is perhaps no accident that Schlick retains the title of 'wisdom of life' for his later conception of philosophy. According to Schlick, the pursuit of meaning is a mental *activity* of clarification that is a prerequisite for the pursuit of truth.[43] In order for the scientist to seek for the truth of propositions, the philosopher needs to have played the game in which the meanings of its terms are clarified.

Schlick's conception of the pursuit of meaning is in a sense a regimented version of the playful activity that in *Lebensweisheit* was presented as coming one stage before the evolution of *Wissenschaft*. In *Lebensweisheit*, the game of asking questions which have no practical relevance is not regimented. Since it is a kind of free activity, primarily characterized by its being undertaken for its own sake, there is no reason for which to exclude nonsense from it. By contrast, Schlick's regimented version of the game of the pursuit of *meaning* does include the exclusion of nonsense in its rules. Thus armed with the Vienna Circle's critique of metaphysics, Schlick (1932/2008a, 386–387; 1931/2008c, 301–302) would consign a significant part of the history of philosophy to meaninglessness. Philosophy as the pursuit of meaning is neither the pursuit of truth, nor just any unregimented, aimless intellectual game, but falls somewhere in between the two.[44] Though Schlick's definition of philosophy as the pursuit of meaning does set a boundary, it is quite a broad one at

[43] 'Prerequisite' does not here necessarily imply a temporal precedence – sometimes the truth-value of a statement happens to have been determined before its meaning has been clarified.

[44] A large part of the history of what has traditionally been designated as 'philosophy', in Schlick's (1932/2008a, 386) view, was simply disguised science. Indeed, as Schlick (1932/2008a, 388) clarifies, a distinction between philosophy and science comes relatively late in history, and before that, people called 'philosophers' would engage both in the pursuit of truth and the pursuit of meaning. The scientific pursuit of truth and the philosophical pursuit of meaning are two entirely separate games that interact in many important ways.

that (see Tuboly's chapter in this volume). It can thus encompass, for example, the various different projects Schlick was engaged in until the end of his life, including his concern with culture in *Nature und Kultur*.

The primary change in Schlick's metaphilosophy between 1908 and 1931 is his rigorous differentiation between philosophy and *Wissenschaft*. What remains the same, however, throughout Schlick's different conceptions of philosophy, is the element of abstention from practical purposes.[45] Though both may arise from practical concerns, neither the pursuit of truth, nor that of meaning, can be determined by practical ends. Both must be undertaken, in Schlick's conception, for their own sake, as a kind of game.[46]

This explains why Schlick, in 1914, draws a strict line that separates the Idealists' politics from their philosophy. It is also the reason why, in the Nietzsche lectures, Schlick extends this account to make the following claim: once it is conceived in terms of a game undertaken for its own sake (and in 1914 this game is the pursuit of truth, while later it is the pursuit of meaning), philosophy cannot play *any* political role, let alone the type of political role that it is thought to play in the exaggerated debates about which philosophical idea caused the war. Philosophy, whether as a type of *Wissenschaft* or as another type of game, presupposes the ability to take a step back from all practical purposes.

Notice here that the prerequisite disinterested state of mind is at odds with the kind of psychological phenomenon which Schlick had described at the beginning of his lectures. A mind preoccupied with war and the accompanying struggle for survival will have difficulty to undertake the ascent above such practical preoccupations and into a state which plays a game only for its own sake, let alone a state in which it wills only truth. If anything, Schlick (2013a, 85–86) contends, it is an *absence* of philosophical thinking which allows for the kind of falling back into baser instincts that drives people to war. In Schlick's defense of philosophy against the charge of militarism, he claims that the war is not explained by appeals to the fact

> that individual nations filled their minds with some philosophy, but at most one could put the blame on the *lack* of philosophy. All wars, perhaps all conflicts, arise from instincts that are much baser, but more powerful, than the drive to philosophy. And if it were powerful enough to influence people's actions and nations' destinies, then there would perhaps be no wars at all. True philosophy is always peaceful, the spirit of philosophy, which is the spirit of truth, goes hand in hand with the spirit of peace. (Schlick 2013a, 85–86)

Thus, in the midst the jingoist turmoil of 1914, we find Schlick arguing that there is a kind of performative contradiction in the appropriation of philosophy for the purposes of warmongering. Philosophy, whether as the pursuit of truth or as the pursuit of meaning, is only possible in peace, insofar as peace makes the pursuit possible.

---

[45] Elsewhere, for example, Schlick clarifies that, in his conception, philosophy "cannot interfere with science or with human values" (1932/2008b, 414).

[46] While the earlier separation between *Wissenschaft* and *praxis* already sets Schlick apart from Neurath, the designation of a special role for philosophy stands in even more stark opposition to Neurath's conception of Unified Science.

4.10 Conclusion

Many of the leading academic philosophers of Schlick's generation entered into the militaristic frenzy of the first years during the Great War, and at least some of them would later be shunned for this. After the war, Idealism would eventually lose the place it occupied at the forefront of pre-war academic philosophy (see Vrahimis 2015). Since it is unknown if Schlick presented his lectures, it is unclear whether anyone ever actually witnessed Schlick's attempt to intervene in the war controversies. There were, to the best of my knowledge, no public responses to Schlick (as there were to the various other philosophers who intervened in the Idealism controversy). Furthermore, Schlick advocated an apolitical view of philosophy, and did not endorse the squabbling that other academics were involved in.[47]

Schlick was still in many ways part of the pre-war philosophical establishment (what Ringer calls the 'German mandarins') when he moved to Vienna after the war.[48] However, due the two factors mentioned above, Schlick would be well-positioned to communicate with scholars who had remained outside this philosophical establishment.[49] Unlike other philosophers in this establishment, who would have plenty of reasons to regret their wartime defenses of militarism, the outcome of the war gave Schlick no incentive to revise his opinions concerning the apolitical nature of philosophy. On the contrary, he may have felt vindicated in at least thinking about, if not publicly calling for, a philosophical abstinence from the militarist frenzy. This may explain why Schlick would insist on his apolitical stance during the interwar (though e.g., as we have seen, his last unpublished writings did contain criticisms of philosophical positions upheld by the far-right). By contrast with Schlick, the very different war-time experiences of the Vienna Circle's 'left wing' mostly gave them reasons to become more politically active,[50] and even to be less wary of explicitly connecting their political with their philosophical outlooks. This difference at least partly explains their subsequent disagreements concerning the

---

[47] As Mormann (2010, 272) points out, it was this 'political naivety' that led Schlick to the failed attempt to present the Vienna Circle as apolitical so as to accommodate the Austrofascist regime. Schlick even wrote to Dollfuß to express his congratulations and sympathy (Stadler 1982, 200); see also Stadler (1982, 196); Stadler (1991); Hegselmann (1987, 273–274).

[48] Mormann (2010, 284–285) claims that Schlick, as opposed to other Vienna Circle members, belonged to this mandarin tradition.

[49] For example, his work on the meaning of life and on the philosophy of youth had some relation to Wyneken's youth movement (itself significantly influenced by Nietzsche); see e.g. Ferrari (2016, 8). It is notable that both Carnap and Reichenbach were involved in the youth movement, the latter being active (alongside his friend Walter Benjamin) in Wyneken's group; see e.g. Wolters (2021).

[50] E.g. in Carnap's autobiography, he says "I have always had an intense interest in moral problems, both those concerning the life of individuals and, since the First World War, those of politics" (Carnap 1963, 82); on Carnap's change of heart from his initial militaristic enthusiasm for the war, see Wolters (2017, 18–22; 2021); on Carnap's 'activism', see e.g. Uebel (2012).

ethical, social, and political consequences of upholding a *Wissenschaftliche Weltauffassung*.[51]

Acknowledgments In the summer of 2019, I presented some of the ideas that eventually developed into this paper at the annual conference of the Society for the Study of the History of Analytical Philosophy, held at Boston University, and at the TiLPS History of Analytic Philosophy workshop, held at Tilburg University; I am thankful to all those present for their many useful comments, suggestions, and encouragement. I owe thanks to Mathias Iven for reading and commenting on a draft of this chapter. For their many helpful suggestions and overall editorial contributions, I am especially grateful to Christian Damböck and Adam Tamas Tuboly.

References

Akehurst, Thomas. 2010. *The Cultural Politics of Analytic Philosophy: Britishness and the Spectre of Europe*. London: Continuum.

Anonymous. 1915. German Spirit due to Kant, not Nietzsche: Professor Dewey Traces German Militarism Back to the Famous Philosopher of the Eighteenth Century and his Categorical Imperative. *The New York Times*, July 18, 1915.

Beiser, Frederick C. 2018. *Hermann Cohen: An Intellectual Biography*. Oxford: Oxford University Press.

Carnap, Rudolf. 1928/2003. *The Logical Structure of the World and Pseudoproblems in Philosophy*. Chicago: Open Court.

———. 1931/1959a. The Elimination of Metaphysics Through Logical Analysis of Language. In *Logical Positivism*, ed. A.J. Ayer, 60–81. Glencoe: The Free Press.

———. 1932/1959b. Psychology in Physical Language. In *Logical Positivism*, ed. A.J. Ayer, 165–198. Glencoe: The Free Press.

———. 1963. Intellectual Autobiography. In *The Philosophy of Rudolf Carnap*, ed. P.A. Schilpp, 3–83. LaSalle: Open Court.

Cartwright, Nancy, Jordi Cat, Lola Fleck, and Thomas Uebel. 1996. *Otto Neurath: Philosophy Between Science and Politics*. New York: Cambridge University Press.

Dahms, Hans-Joachim. 1994. *Positivismusstreit, Die Auseinandersetzung der Frankfurter Schule mit dem logischen Positivismus, dem amerikanischen Pragmatismus und dem kritischen Rationalismus*. Frankfurt: Suhrkamp.

Damböck, Christian. 2017. *Deutscher Empirismus Studien zur Philosophie im deutschsprachigen Raum 1830–1930*. Dordrecht: Springer.

———. 2020. (Dis-)similarities: Remarks on 'Austrian' and 'German' Philosophy in the 19th Century. In *Franz Brentano and Austrian philosophy*, ed. D. Fisette, G. Fréchette, and F. Stadler, 169–180. Cham: Springer.

de Warren, Nicolas, and Thomas Vongehr, eds. 2018. *Philosophers at the Front: Phenomenology and the First World War*. Leuven: Leuven University Press.

Feigl, Herbert. 1952/1981. Validation and Vindication: An Analysis of the Nature and the Limits of Ethical Arguments. In *Inquiries and Provocations: Selected Writings 1929–1974*, 378–392. Dordrecht: D. Reidel.

---

[51] How exactly to draw the line between the Vienna Circle's 'left' and 'conservative' wings remains debatable, and I will not attempt to settle the debate here. However one construes the difference, a central issue that is pertinent to it concerns the different conceptions of philosophy's relation to politics that I have explored above. There are, nonetheless, certainly also things that divide the two that are irrelevant to their political disagreements.

Ferrari, Massimo. 2016. Wiener Gefühle. Fragen der Ethik zwischen Schlick und Carnap. In *Wissenschaft und Praxis. Zur Wissenschaftsphilosophie in Frankreich und Österreich in der ersten Hälfte des 20. Jahrhunderts*, ed. Ch. Bonnet and E. Nemeth, 5–28. Dordrecht: Springer.
Fischer, K. R. 1982. Nietzsche und der Wiener Kreis. *Grazer Philosophische Studien* 16 (1): 255–269.
Frank, Philipp. 1917/1970. The Importance of Ernst Mach's Philosophy of Science for our times. In *Ernst Mach: Physicist and Philosopher*, ed. R.S. Cohen, 219–234. Dordrecht: Springer.
Hegel, Georg W. F. 2004. *Philosophy of Nature: Being Part Two of the Encyclopedia of the Philosophical Sciences*. Translated by Anton V. Miller. Oxford: Clarendon Press.
Hegselmann, Rainer. 1987. Introduction. In: *Unified Science*, ed. B. McGuinness, ixx–xxi, 273–74. Dordrecht: Reidel.
Hoeres, Peter. 2004. *Der Krieg der Philosophen: Die Deutsche und Britische Philosophie im Ersten Weltkrieg*. Paderborn: Schöningh.
Horkheimer, Max. 1937/1972. The Latest Attack on Metaphysics. In *Critical Theory: Selected Essays*, 132–187. New York: Continuum.
Iven, Mathias. 2008. Moritz Schlick im Ersten Weltkrieg. Adlershof 1917/18. In *Moritz Schlick. Leben, Werk und Wirkung*, ed. F.O. Engler and M. Iven, 59–90. Berlin: Parerga.
———. 2013a. Einleitung. In Schlick 2013: 17–52.
———. 2013b. Editorischer Bericht. In Schlick 2013: 55–76.
James, William. 1911. The Moral Equivalent of War. In *Memories and Studies*, 265–296. London: Longmans, Green.
Jeffs, Rory. 2017. Hegel in Dark Times: The Resurrections of Geist from the Ashes of War. In *100 Years of European Philosophy Since the Great War: Crisis and Reconfigurations*, ed. M. Sharpe, R. Jeffs, and J.A. Reynolds, 161–182. Dordrecht: Springer.
Kleinpeter, Hans. 1913. *Der Phenomenalismus: Eine naturwissenschaftliche Weltanschauung*. Leipzig: Barth.
Köster, Peter. 1998. *Der verbotene Philosoph: Studien zu den Anfängen der katholischen Nietzsche-Rezeption in Deutschland (1890–1918)*. Berlin/New York: de Gruyter.
Kusch, Martin. 1995. *Psychologism: A Case Study in the Sociology of Philosophical Knowledge*. New York: Routledge.
Lewis, Joia. 1988. Schlick's Critique of Positivism. In *PSA: Proceedings of the Biennial Meeting of the Philosophy of Science Association*, 1, 110–17.
Martin, Nicholas. 2006. Nietzsche as Hate-Figure in Britain's Great War: 'The Execrable Neech'. In *The First World War as a Clash of Cultures*, ed. F. Bridgham, 147–166. New York: Camden House.
Mormann, Thomas. 2010. Zwischen Weisheit und Wissenschaft. Schlicks weites philosophisches Spektrum. *Grazer Philosophische Studien* 80: 263–285.
———. 2015. Review of *Moritz Schlick: Nietzsche und Schopenhauer (Vorlesungen)*. *Journal for General Philosophy of Science* 46 (2): 419–423.
Mulder, Henk. 1968. Wissenschaftliche Weltauffassung. Der Wiener Kreis. *Journal of the History of Philosophy* 6: 386–390.
Neurath, Otto. 1921/1973. Anti-Spengler. In *Empiricism and Sociology*, ed. M. Neurath and R.S. Cohen, 159–213. Dordrecht: Reidel.
———. 1932/1987. Unified Science and Psychology. In *Unified Science*, ed. B. McGuinness, 1–23. Dordrecht: Reidel.
———. 1936/1981. Die Entwicklung des Wiener Kreises und die Zukunft des Logischen Empirismus. In *Gesammelte philosophische und methodologische Schriften, Volume 2*, ed. R. Haller and H. Rutte, 675–703. Vienna: Hölder-Pichler-Tempsky.
———. 1938/1955. Unified Science as Encyclopedic Integration. In *International Encyclopedia of Unified Science. Volume I. Part 1*, ed. O. Neurath, R. Carnap, and Ch. Morris, 1–27. Chicago: University of Chicago Press.
———. 1938/1981. Einheitswissenschaft als Empiristische Synthese. In *Gesammelte philosophische und methodologische Schriften, Volume 2*, ed. R. Haller and H. Rutte, 826. Vienna: Hölder-Pichler-Tempsky.

Neurath, Otto, and Joseph A. Lauwerys. 1944a. Nazi Textbooks and the Future: Part 1. *The Journal of Education* 76 (904): 521–522.
———. 1944b. Nazi Textbooks and the Future: Part 2. *The Journal of Education* 76 (905): 574–576.
———. 1945. Plato's Republic and German Education. *The Journal of Education* 77 (907): 57–59.
Nietzsche, Friedrich W. (1885/1905). *Thus Spoke Zarathustra.* Translated by Thomas Common. New York: The Modern Library.
———. (2003). *Writings from the Late Notebooks.* Edited by R. Bittner, translated by Kate Sturge. Cambridge: Cambridge University Press.
O'Neil, John, and Thomas Uebel. 2004. Horkheimer and Neurath: Restarting a Disrupted Debate. *European Journal of Philosophy* 12 (1): 75–105.
Professors of Germany. 1919. To the Civilized World. *The North American Review* 210: 765.
Reisch, George. 2005. *How the Cold War Transformed Philosophy of Science: To the Icy Slopes of Logic.* Cambridge: Cambridge University Press.
Reynolds, Jack. 2017. Philosophy and/or Politics? Two Trajectories of Philosophy After the Great War and Their Contamination. In *100 years of European Philosophy Since the Great War. Philosophical Studies in Contemporary Culture*, ed. M. Sharpe, R. Jeffs, and J. Reynolds, 215–232. Cham: Springer.
Richardson, S.S. 2009a. The Left Vienna Circle. Pt. 1: Carnap, Neurath, and the Left Vienna Circle Thesis. *Studies in History and Philosophy of Science* 40: 14–24.
———. 2009b. The Left Vienna Circle. Pt. 2: The Left Vienna Circle, Disciplinary History, and Feminist Philosophy of Science. *Studies in History and Philosophy of Science* 40: 167–174.
Ringer, Fritz K. 1969. *The Decline of the German Mandarins: The German Academic Community, 1890-1933.* Hanover/London: Wesleyan University Press.
Romizi, Donata. 2012. The Vienna Circle's 'Scientific World-Conception': Philosophy of Science in the Political Arena. *HOPOS: The Journal of the International Society for the History of Philosophy of Science* 2 (2): 205–242.
Schleichert, Hubert. 2003. Moritz Schlick's Idea of Non-Territorial States. In *The Vienna Circle and Logical Empiricism: Re-evaluation and Future Perspectives*, ed. F. Stadler, 49–61. Dordrecht: Kluwer.
Schlick, Moritz. 1908/2006. *Lebensweisheit: Versuch einer Glückseligkeitslehre* u. *Fragen der Ethik.* Edited by M. Iven. Dordrecht: Springer.
———. 1918/1925/1974. *General Theory of Knowledge.* Translated by Albert E. Blumberg. New York/Vienna: Springer.
———. 1931/2008c. The Future of Philosophy [1931]. In *Die Wiener Zeit: Aufsätze, Beiträge, Rezensionen 1926–1936*, ed. J. Friedl and H. Rutte, 297–303. Dordrecht: Springer.
———. 1932/2008a. The Future of Philosophy [1932]. In *Die Wiener Zeit: Aufsätze, Beiträge, Rezensionen 1926–1936*, ed. J. Friedl and H. Rutte, 371–390. Dordrecht: Springer.
———. 1932/2008b. A New Philosophy of Experience. In *Die Wiener Zeit: Aufsätze, Beiträge, Rezensionen 1926–1936*, ed. J. Friedl and H. Rutte, 397–414. Dordrecht: Springer.
———. 1934/1959. The Foundation of Knowledge. In *Logical Positivism*, ed. A.J. Ayer, 209–227. Glencoe, Illinois: The Free Press.
———. 1936/1938. L'École de Vienne et la Philosophie traditionnelle. In *Gesammelte Aufsätze 1926–1936*, 389–398. Vienna: Gerold.
———. 1952. *Natur und Kultur.* Wien-Stuttgard: Humbold.
———. 1962. *Aphorismen*, ed. Blanche Hardy Schlick. Vienna: Carl Ueberreuter.
———. [1927] 1979). On the meaning of life. Moritz Schlick: Philosophical Papers: Volume Two (1925-1936), ed. H. L. Mulder; B. F. B. van de Velde-Schlik, 112–129. Dordrecht: D. Reidel.
———. 2013a. *Nietzsche und Schopenhauer*, ed. M. Iven. Dordrecht: Springer.
———. 2013b. *Erkentnisstheoretische Schriften, 1926–1936*, ed. Johannes Friedl and Heiner Rutte. Dordrecht: Springer.

Siegetsleitner, Anne. 2010. Schlicks *Fragen der Ethik* und die vorherrschende Sicht logisch-empiristischer Ethik. In *Logischer Empirismus, Werte und Moral*, ed. A. Siegetsleitner, 9–19. Vienna: Springer.

Sombart, Werner. 1915. *Händler und Helden: Patriotische Besinnungen.* Munich/Leipzig: Duncker & Humblot.

Soulez, Antonia. 2019. Does Understanding Mean Forgiveness? Otto Neurath and Plato's 'Republic' in 1944–45. In *Neurath Reconsidered: New Sources and Perspectives*, ed. J. Cat and A.T. Tuboly, 435–449. Cham: Springer.

Stadler, Friedrich. 1982. *Vom Positivismus zur "wissenschaftlichen Weltauffassung".* Vienna: Löcker.

———. 1991. Otto Neurath-Moritz Schlick: On the Philosophical and Political Antagonisms in the Vienna Circle. In *Rediscovering the Forgotten Vienna Circle: Austrian Studies on Otto Neurath and the Vienna Circle*, ed. Th. Uebel, 159–168. Dordrecht: Kluwer.

Stewart, Herbert L. 1915. *Nietzsche and the Ideals of Modern Germany.* London: Edward Arnold.

Textor, Mark. 2018. Schlick on the Source of the 'Great Errors in Philosophy'. *Journal of the American Philosophical Association* 4 (1): 105–125.

Thomas, R. Hinton. 1983. *Nietzsche in German Politics and Society, 1890-1918.* Manchester: Manchester University Press.

Tuboly, Adam Tamas. 2021a. A Cricket Game, a Train Ticket and a Vacuum to be Filled: Ayer's Logical Positivism as a Focal Point for Post-War British Cultural Struggles. *British Journal for the History of Philosophy* 28 (6): 1134–1150.

———. 2021b. Neurath on 'Plato-Hitler' and the British Scene of Irritation. In *Scientific World-Conceptions: Rudolf Carnap and Otto Neurath*, ed. Ch. Damböck, J. Friedl, and U. Höfer. Rodopi/Brill. Forthcoming.

Tuusvuori, Jarkko S. 2000. *Nietzsche & Nihilism: Exploring a Revolutionary Conception of Philosophical Conceptuality.* PhD Dissertation, University of Helsinki.

Uebel, Thomas. 2005. Political Philosophy of Science in Logical Empiricism: the Left Vienna Circle. *Studies in History and Philosophy of Science* 36: 754–763.

———. 2007. *Empiricism at the Crossroads. The Vienna Circle's Protocol Sentence Debate.* Chicago: Open Court.

———. 2008. Writing a Revolution: On the Production and Early Reception of the Vienna Circle's Manifesto. *Perspectives on Science* 16 (1): 70–102.

———. 2010. What's Right about Carnap, Neurath and the Left Vienna Circle Thesis: A Refutation. *Studies in History and Philosophy of Science* 41: 214–221.

———. 2012. Carnap, Philosophy, and 'Politics in its Broadest Sense. In *Rudolf Carnap and the Legacy of Logical Empiricism*, ed. R. Creath, 133–148. Dordrecht: Springer.

———. 2020. Schlick and Wittgenstein: The Theory of Affirmations Revisited. *Journal of the History of Philosophy* 58 (1): 141–166.

Voegelin, Eric. 1944. Nietzsche, the Crisis and the War. *The Journal of Politics* 6 (2): 177–212.

Von Mises, Richard. 1938/1987. Ernst Mach and the Scientific Conception of the World. In *Unified Science*, ed. B. McGuinness, 166–190. Dordrecht: Reidel.

Vrahimis, Andreas. 2011. Russell's critique of Bergson and the divide between 'Analytic' and 'Continental' Philosophy. *Balkan Journal of Philosophy* 3 (1): 123–134.

———. 2013. *Encounters between Analytic and Continental Philosophy.* Basingstoke: Palgrave Macmillan.

———. 2015. Legacies of German Idealism: From the Great War to the Analytic-Continental divide. *Parrhesia* 24: 83–106.

———. 2019. The Analytic-Continental divide and the question of Philosophy's relation to Literature. *Philosophy and Literature* 43 (1): 253–269.

———. 2020. Scientism, Social Praxis, and overcoming Metaphysics: A debate between Logical Empiricism and the Frankfurt School. *HOPOS: The Journal of the International Society for the History of Philosophy of Science* 10 (2): 562–597.

———. 2021a. The Vienna Circle's responses to Lebensphilosophie. *Logique et Analyse* 253: 43–66.
———. 2021b. Russell reads Bergson. In *The Bergsonian Mind*, ed. M. Sinclair and Y. Wolf. Oxon: Routledge. Forthcoming.
———. 2021c. *Bergsonism and the History of Analytic Philosophy*. Hampshire: Palgrave. Forthcoming.
Waismann, Friedrich. 1994a. Ethics and Science. In *Ethics and the Will*, ed. B. McGuinness and J. Schulte, 33–52. Dordrecht: Springer.
———. 1994b. Will and Motive. In *Ethics and the Will*, ed. B. McGuinness and J. Schulte, 53–137. Dordrecht: Springer.
Windsor, Tara. 2014. Rekindling Contact: Anglo-German Academic Exchange after the First World War. In *Anglo-German Scholarly Networks in the Long Nineteenth Century*, ed. H. Ellis and U. Kirchberger, 212–231. Leiden/Boston: Brill.
Wolters, Gereon. 2017. Clash of Cultures? – German Philosophers of Science and the Great War (1914-1918). *Bollettino della Società Filosofica Italiana* 220: 7–28.
———. 2021. Philosophenkrieger? – Wie Carnap & Co den Ersten Weltkrieg sahen. In *Logical Empirism, Life Reform, and the German Youth Movement*, ed. Ch. Damböck, G. Sandner, and M. Werner. Cham: Springer. Forthcoming.

Chapter 5
Schlick on the Meaning of "Good"

Gergely Ambrus

Abstract The paper discusses Schlick's *Problems of Ethics* (1939; *Fragen der Ethik, 1930)*. It comprises of two parts. The first part analyses Schlick's metaethical views, his arguments against objective values and absolute norms, and his analysis of the meaning of "good" in general and "morally good" in particular. Further, it also tests these views against some plausible objections. The second part addresses an apparent tension in Schlick's conception. The underlying general thesis of the *Problems of Ethics* is that philosophical ethics is an exclusively descriptive, non-normative endeavor: philosophy ought not aim at proving what the good is, which norms are objectively valid. Justification of values or norms may only consist in determining whether they are endorsed, judged as values or norms by a particular society. This is, however, the task of scientific ethics, i.e. empirical scientific research. In contrast with this proclaimed view, Schlick in the last chapter of the book also puts forth his own normative ethical views, i.e. his Principle of Happiness – apparently leaving the role of the moral philosopher and moral scientist behind and acting as a moralist. Moreover, it seems, that he relies on this normative principle to ground his analysis of the meaning of "morally good". I suggest that this tension may be dissolved by turning to Schlick's anthropology. One important characteristic of this anthropology is that humans by nature have altruistic inclinations, and this grounds that individuals may genuinely desire the welfare of society. This is supported by empirical observations, according to Schlick, hence a legitimate scientific ethical claim. But there is also another, more interesting and unique characteristic: Schlick seems to preserve elements of his earlier ethical views, he put forth in *Lebensweisheit* (1908) and in "Vom Sinn des Lebens" (1927), i.e. his Schillerian anthropology, according to which the essence of humanity is playing. For according to Schlick's proposed "ethics of kindness" the ethics of desire and the ethics of demand may be synthesized by the

G. Ambrus (✉)
Institute of Philosophy, Eötvös Loránd University, Budapest, Hungary
e-mail: ambrus.gergely@btk.elte.hu

C. Damböck, A. T. Tuboly (eds.), *The Socio-Ethical Dimension of Knowledge*,
Vienna Circle Institute Yearbook 26, https://doi.org/10.1007/978-3-030-80363-6_5

ethics of love: acting morally means acting freely, without coercion, that is, acting playfully, in accordance with the essence of human nature. So our altruistic inclinations may be nurtured by Schillerian aesthetic-moral education.

Keywords Moritz Schlick · Metaethics · Happiness · Altruism

5.1 Introduction

Moritz Schlick's book, the *Problems of Ethics* shares the fate of G.E. Moore's *Principia Ethica* in a certain respect: only its first, introductory chapter became known for a wider public (by being republished in the influential collection of Ayer (1959) on Logical Positivism), and, the content of the rest of book fell into oblivion, except perhaps for a few historians of philosophy specialized in Schlick or in twentieth century ethics (see e.g. Siegetsleitner 2014, 265–331; Mormann 2010; Iven 2006a, b).[1] However, a reconstruction of Schlick's ethical views grounded solely on this first chapter may only be partial, and potentially misleading, as I aim to show.

In the introductory chapter, "What is the Aim of Ethics?", Schlick briefly summarizes the main topics of the book, namely: the material and formal characteristics of value judgments, the ontology and epistemology of values, and the theory of moral motivations. Schlick's respective theses are: the good is always desired and demanded; goodness is not an objective property, it is purely dispositional; the only cognitively meaningful questions concerning values and motivations are: what individuals or communities *judge* as valuable and as moral motivations. In line with such an approach to the subject-matter of ethics, Schlick declared that answers to the question what values and moral motivations *are*, can only be provided by *empirical* psychological and sociological research; there is no autonomous moral knowledge, based on moral intuition or moral sense, ethics has no special epistemological methods.

Thus, *scientific* ethics is solely a subdiscipline of psychology and sociology, and *philosophical* ethics ought to aim at investigating metaethical problems exclusively, i.e. at analyzing moral language, the meaning of moral terms, determining the general characteristics of moral judgments, exploring questions concerning the nature and epistemology of values. Philosophical ethics cannot and must not give answers to substantial moral questions like what the good is in itself, how we ought to act,

[1] This is perhaps a bit of an exaggeration, since in the German-speaking philosophical community Schlick's work is better-known and has been discussed; see e.g. Hegselmann's introduction to the German reprint of the book (Hegselmann 1984), Heiner Rutte's "Ethics and the Problem of Value in the Vienna Circle" (Rutte 1991), Matthias Iven's "Editorischer Bericht" to the *Fragen der Ethik* (Iven 2006b), and Anna Siegetsleitner's works on Schlick's ethics (Siegetsleitner 2010, 2014, 265–331). Also, a fairly recent development in the English reception is that Scott Soames discussed extensively Schlick's ethics in his *The Analytic Tradition in Philosophy*, vol. 2: *A New Vision* (Soames 2018, Ch. 12). This is still not to say, however, that the "Problems of Ethics" is widely known in contemporary Anglo-Saxon moral philosophy.

what the duties are. Taking a stand on these issues belongs to the authority of each individual person. Moral philosophers are not moral experts or moral peers of other people – although, of course, they also have their own views about these matters. As this brief introduction shows, Schlick's book appears to represent one characteristic attitude within the logical empiricist community concerning what philosophy is or ought to be, applied to the field of ethics.[2]

The first chapter does summarize the essence of Schlick's metaethical theses. However, it only provides an incomplete picture of Schlick's ethical views and his overall attitude towards ethics. In the ensuing chapters Schlick elaborates his account of the characteristic features of moral judgments, the nature of morally good as opposed to good in a general sense, and also his theory of moral motivations. And in the last chapter (VIII. "What Paths Lead to Value?") he also puts forth his own substantive, normative ethical view. At first glance, this is rather surprising, as he declares in the introduction that the only legitimate aim of the moral philosopher, as opposed to the moralist, is to investigate metaethical problems – and he keeps to his word in the first seven chapters. But on closer inspection we will find that there is a natural connection between Schlick's meta-ethical and substantial views, and that the latter is grounded in his earlier ideas formulated in "Vom Sinn des Lebens" ("The Meaning of Life", 1927/1979) and also already in his early *Lebensweisheit* (*Wisdom of Life*, 1908).

The structure of the paper is as follows. In the first part, I present Schlick's analysis of the general nature of the good, viz. his views on the material and formal characteristics of value judgments and on the general nature of the property good. In the second part, I discuss Schlick's account of the specifically *moral* goodness. Then I consider possible objections to Schlick's views and replies to them. Lastly, I investigate Schlick's normative ethics presented in the last chapter of the book, and argue that it may be seen as a necessary amendment to the earlier chapters, since certain important objections against Schlick's metaethics may only be answered relying on these substantial views.

5.2 The General Outlines of Schlick's Conception of the Good and Ethics

5.2.1 A General Analysis of the Concept of "Good"

Schlick begins his investigations with an analysis of the concept of good in general: this is divided into an analysis of the meaning of the term "good" and a characterization of the general nature of the property *good*.

As for the first: Schlick asserts that the meaning of the sentence "X is good" is constituted by the empirical facts that obtain when "X is good" is *judged* to be true;

[2] Menger, von Mises, Neurath or Kraft presented different attitudes though. About these differences see e.g. Tuboly's chapter in this volume.

and, he adds, our knowledge about these facts is based solely on *sense perception*. According to his proposal, the value-judgment "X is good" is similar in kind to judgments like "X is living" or "X is green" in that the meaning of these terms/sentences can be characterized by the situations in which they are asserted, i.e. those situations in which people would ordinarily judge about an object that it is living (e.g. that it is capable of locomotion, has metabolism, reproductory capacities etc.), or that an object is green (has the same color as grass, pine trees, watermelons etc.). Schlick's suggestion should be understood as an expression of the radical verificationist theory of meaning ("the meaning of a sentence is the method of its verification") he advocated in this period, applied to the term "good".

Schlick's justification for this suggestion is partly negative. He claims, first, that the reference of "good" *ought to* be determined by the very *same methods* as the reference of any other kind of terms, for example of "life" or "green", simply because there is *no other* legitimate method: neither *moral sense* nor *rational intuition* theories are eligible. Second, against "emotionist moral intuitionism" Schlick objects that *moral* pleasure (as opposed to other sorts of pleasures) has no distinctive phenomenal character[3] and neither has a specific organ – he explicitly opposes Brentano's (1889/1969) views about the existence of a special value-feeling (cf. Schlick 1930/1939, 105–106). Further, he contends, even if there existed a special "moral pleasure", moral judgments widely differ across humans, hence a common reference of the term "good" could not be based on such moral feelings. Third, against rationalist moral intuitionism (advocated for example by Moore (1903/1993), Prichard (1912) or Ross (1927, 1930/2002)), Schlick argued that there is no rational intuition whatsoever; not only moral knowledge is not based on intuition, there is no intuitive knowledge in mathematics or in metaphysics either.

This latter claim is based on Schlick's general views about the nature of the knowledge, which he advocated from the beginnings of his philosophical career. An early formulation of this doctrine may be found in "Gibt es intuitive Erkenntnis?" ("Is There Intuitive Knowledge?" 1913/1979); a detailed explication of it in his *Allgemeine Erkenntnislehre* (*General Theory of Knowledge* 1918/1925/1985); and again, emphatic pronouncements in "Erleben, Erkenntnis, Metaphysik" ("Experience, Cognition, and Metaphysics" 1926/1979). Accordingly, knowledge is always mediated by concepts, it is based on comparisons, it is always a fitting into a system; thus intuitive, *direct* knowledge is a contradiction in terms, an oxymoron as it were.

Also, there is a further reason, why the judgment "X is good" cannot be based on value-feeling or rational intuition, namely that such views explain value judgments as determining the absolute value of a certain object by such epistemic means. According to Schlick (1930/1939, 100), "[Absolute] value is something wholly independent of our feelings, something pertaining to valuable objects, in a definite amount and degree, quite independently of the way in which we react emotionally

[3] Cf. Schlick (1930/1939, 106): "I do not succeed in distinguishing between the feeling of pleasure I have when I hear 'Don Juan' or see a noble face or read about the personality of Abraham Lincoln and an elementary value-experience which, according to that view, must first assure me that what gives me joy is also a *value*" (original emphasis).

to them, and to whether anyone acknowledges the value or not." He, however, argues that there are no such values (See chapter V. *Are there Absolute Values?*). This point leads to the topic of the next section.

5.2.2 The Nature of the Good in General: Goodness Is Not an Objective Property of the World, It Is Purely Dispositional

As for the general characteristics of goodness: Schlick proposes that goodness is not an objective property of the world; it is non-objective, and in two different senses. First, goodness is subjective, i.e. *mind-dependent*: it is not a property of mind-independent objects. For, Schlick argues, it makes no sense that values would subsist even if no humans, no sensing and desiring subjects existed. Second, the good is also non-objective in the sense of being relative, i.e. *non-intersubjective*. For, as observations attest, there is no uniformity across individual value judgments. It is worth noting that this sort of non-objectivism is more radical then, for example, Hume's view, according to which moral value judgments may be characterized as non-objective in the first sense, but not in the second. (The Humean view is quite plausible concerning judgments about sense perception: perceptual judgments are subjective, i.e. non-objective in the first sense, as they are about phenomenal objects, which exist only for the human perceivers; but they are intersubjective, i.e. objective in the second sense, as there is a wide-ranging (though, of course, not exceptionless) uniformity across individual perceptual judgments such as "This is green". Whether the Humean view is also plausible about moral judgments is, of course, more controversial.)

As regards the justification of non-objectivity as mind-dependence, Schlick argues against the alternative accounts asserting that goodness is a mind-independent (judgment-independent) property. He considers the "theory of objective values" that "proclaims the existence of a system of values, which, like the Platonic ideas, constitutes a realm independent of actuality". The general objection is that the existence of such platonic entities is not verifiable. Neither the feeling of pleasure can serve as a verification criterion, since there are different types of objects that may cause pleasure, hence some further criteria are needed to determine an object's having value (Schlick 1930/1939, 102–103); nor there is the possibility of finding some objective criterion, for all such attempts are circular or lead to an arbitrary determination of what is valuable (Schlick 1930/1939, 103–104). Neither is there a subjective criteria, i.e. a specific value-feeling, as according to Brentano (1889/1969), for example, the occurrence of which would serve as verification for the value judgment (Schlick 1930/1939, 105–107).

As a positive characterization, Schlick proposes that goodness is a *purely dispositional property*. The material characteristic of value judgments is that goodness consists entirely in *being desired* by a subject. Accordingly, "X is good" means: X or its idea is capable to evoke pleasure in subject S, and X is desired by S. Similarly, "X is wrong" means: X or its idea is capable to evoke pain in subject S, and X is repulsive to S. As for the formal characteristics of value judgments: The form of "X is good" is always a command: X is demanded.

It might seem that this proposal is at odds with the plausible interpretation of Schlick as advocating a version of non-cognitivism, based on his proposal that a subject's judgement that "X is good" means that X is desired by the person. Since it may seem that Schlick's proposal here is that the meaning of "X is good" is that X it is capable of causing pleasure, which clearly is a proposition having a truth-value, hence it supports a cognitivist reading of Schlick's view about value-judgments. I do not think, however, the there is an irreconcilable conflict here. For note that Schlick does not assert that goodness is identical with pleasure, or with the dispositional property of being able to cause pleasure. Good (for a person) is whatever that is desired (by that person). It is an *additional* feature of those entities that are being desired that they are capable of causing pleasure in the desiring subject. But this results from the meaning of the concept of *desire*: i.e. that to desire X involves that attaining X is accompanied – usually – by having pleasure; and *not* from the meaning of "*good*".[4]

As to the justification of the dispositionality of good: Schlick provides the following two arguments:

- *The argument from motivation.* If goodness were an objective property such that it had *no pleasure-evoking capacity*, then it would *not have an influence on life*, since it *could not motivate* action. (According to Schlick's "law of motivation",[5] the will, the primary cause of action, is the desire with which the idea of the highest level of pleasure is associated.) So, if good objects or their ideas caused no pleasure in us, viz. we were emotionally neutral towards goodness, we would have no motivation for good conduct.
- *The argument from the correspondence between good and action.* If, on the one hand, goodness were an objective property which *also* had the dispositional property of being capable to cause pleasure in subjects, but, on the other, this capacity were only *contingent*, then it would be *possible that the good evokes pain* not pleasure in subjects, hence humans would be motivated to *avoid* good conduct; and this is an absurd consequence.

[4] As for the meaning of "morally good" in particular, the supposition that Schlick provided a cognitivist account, is rather straightforward, as we will see shortly in Sect. 5.2.3. This, however, is still not sufficient in itself to reconcile the fact that Schlick also proposes a normative ethics, i.e. a moral command of his own, with the purely descriptive nature of his ethical project, ethics as a science. This needs further elucidation – what I attempt in the last part of the paper, in Sect. 5.3.4.

[5] Cf. Schlick (1930/1939, 26–55), Chapter II. *What are the Motives of Human Conduct*? Especially sections 3–10.

5.2.3 The Analysis of the Morally Good

Thus far, we have only addressed issues about the characteristics of goodness in general, and the meaning of "good" understood in a general sense, not the features of specifically *moral* goodness. The discussions in the first, introductory chapter of the *Problems of Ethics* do not distinguish between moral and other, for example aesthetic or culinary values. Later in the book, however, Schlick provides a characterization of the general features of moral goodness that may be reconstructed as follows:

- Moral goodness is concerned with *human conduct*.
- The judgment that a conduct is morally good expresses that a conduct is *approved by the society*.[6]
- "To be approved by the society" means that the majority of those influenced by the conduct *desire* it. (Schlick 1930/1939, 82–83)

The meaning of "morally good" in contrast with "good" in general is the following. "*A* is morally good" if and only if.

1. The (majority) of society desires *A*.
2. The (majority) of society judges that *A* is beneficial for the society.

Schlick supports his view by several arguments, which may be dubbed as *the argument from ordinary language use, the argument from co-extensionality*, and *the argument from psychological law*.

According to *the argument from ordinary language use,* (1) and (2) are consonant with the meaning how the term "morally good" is used in everyday parlance. This is shown by the following facts: lawmakers always appeal to this understanding; every society takes it to be an identifying feature of moral precepts that they promote the welfare of their own society not of "strangers'"; further, that "egoism" understood as ignoring the welfare/interest of others is generally taken to be morally wrong conduct (cf. Schlick 1930/1939, 86.ff.).

The argument from co-extensivity starts from the assumption that the descriptions expressed by (1) and (2) are co-extensive: "*A* is morally good" viz. "*A* is approved by (the majority of) society", viz. "*A* is desired by (the majority of) society", is co-extensive with "*A* promotes the welfare of society". Then it is argued that the extensional equivalence of *empirical terms* is in itself a strong reason for

[6] It is important to note that Ayer, for example, was firmly against such views that identified good with general acceptance. See e.g. Ayer (1936/1971, 138–139.): "We reject the subjectivist view that to call an action right, or a thing good, is to say that it is generally approved of, because it is not self-contradictory to assert that some actions which are generally approved of are not right, or that some things which are generally approved of are not good. And we reject the alternative subjectivist view that a man who asserts that a certain action is right, or that a certain thing is good, is saying that he himself approves of it, on the ground that a man who confessed that he sometimes approved of what was bad or wrong would not be contradicting himself."

assuming their intensional equivalence as well. Hence the conclusion: the *meaning* of "morally good" is that which promotes the welfare of society.

According to *the argument from psychological law* a law may exist, according to which a conduct, which is judged by the members of society as promoting the welfare of society, is a conduct, which is also *desired* by the members of society. Such a law may be grounded in human nature: it may be the case that humans living together in a community naturally develop *altruistic desires*; accordingly, a conduct or its idea which promotes the welfare of society causes pleasure in the members of the society. If there really is such a law then:

3. The (majority of) society desires *A*, *because* it judges that *A* is beneficial for the society.

(3) is independent from (1) and (2), since it may well be that what society desires increases the welfare of society, but the members of society desire it for a different reason; for example, because they hold that it is commanded by God, or because it complies with the categorical imperative etc. So, if there exists such a psychological law, then the connection between (1) and (2) is grounded in some *natural feature* of the world.[7]

5.3 Problems for Schlick's Conception of Ethics and the Good

After outlining Schlick's conception about the nature of the good in general and the morally good in particular, as well as about the meaning "good" and "morally good", I shall address some problems these views seem to face.

[7] I slightly changed the terminology Schlick uses, but (1), (2) and (3) express the content of his claims, I believe. To make out that this is the case, consider the following quote.

> The determination of the complete extensional equality of the concepts „morally good" and „what advances the pleasure of society" leads any unbiased person to believe also in their intensional equality; and special opposing reasons would be necessary to make this belief appear to be unjustified. Without such opposing reasons the inference to identity of both concepts is simply obvious in terms of the method of empirical knowledge. If, in addition, we should succeed in deducing from psychological laws that behavior which is favorable to the genesis of pleasure in the human community must be approved by it, while what increases sorrow necessarily is subjected to its disapproval (as we have seen in the case of egoism), then no one will be able to upset our conviction that his approbation and disapprobation is nothing but 'moral' approbation and disapprobation. (Schlick 1930/1939, 86)

> Substitute "advances the pleasure of society" with "promotes the welfare of society", and note that Schlick also asserts that approval or approbation of *A* by the members of society implies that they desire *A*; then, accordingly, we get that a piece of behavior that may cause pleasure for the members of society is desired by them. Cf. also Schlick (1930/1939, 96–99).

5.3.1 Problem 1: Accounting for Moral Disagreements[8]

It is generally accepted that moral disagreements are *genuine* (or at least, some of them), so any viable theory concerning the meaning of "good" must account for this feature of moral discourse. Not all interest theories are able to do this; simple interest theories, such as Hobbes', for example, cannot. For, according to the Hobbesian view, "X is good" means: *I* desire X. On this understanding, however, moral disagreements would not be genuine. For consider the case when S_1 judges that "X is good" while S_2 judges, contrariwise, that "X is not good". According to Hobbes, this is tantamount to saying that S_1 desires X while S_2 does not. However, these are just two independent facts about the psychological characteristics of two different persons, hence the judgments of S_1 and S_2 do not contradict each other; the conflict is only illusory. Another view in the neighborhood is Hume's proposal, according to which "X is good" means: the *majority* of the community desire X. This suggestion may establish that moral disagreements within one society are genuine; but still, it cannot explain how moral disagreements between different societies may be real.

And how does Schlick's view handle this problem? As we have seen, according to Schlick's analysis, the *material characteristic* of the judgment "*A* is morally good" is: (1) the (majority of) society desires *A*, (2) the (majority of) society judges that *A* is beneficial for the society, and (3) the (majority of) society desires *A*, *because* it judges that *A* is beneficial for the society. This proposal, however, seems to face similar problems as Hume's view. But Schlick also proposes that the *formal characteristic* of the value judgment is that it is always a command: X is demanded. This latter feature may be used for explaining the genuineness of moral disagreements. First, however, we should clarify whether this formal characteristics applies to all or only to moral value judgments.

If they apply to all, then it follows that all value-disagreements are of the same nature; there is no difference in kind between moral, aesthetic or gustatory disagreements. Consider, for example, what disputes over the taste of oysters or about the character trait of frugality amount to on such an understanding:

"S_1 judges that oysters are good" *means*: S_1 desires oysters; and demands: Eat oysters!
"S_2 judges that oysters are bad" *means*: S_2 is averse to oysters; and demands: Do not eat oysters!

"S_1 judges that frugality is good" *means*: S_1 desires frugality; and demands: Be frugal!
"S_2 judges that frugality is bad" *means*: S_2 is averse to frugality; and demands: Do not be frugal!

[8]This problem is not addressed directly by Schlick in the *Problems of Ethics*. However, I find it worthy to investigate whether it is possible to provide such an account, based on Schlick's theory, for it is a widely held assumption that moral disagreements are genuine, and differ from culinary or aesthetic disagreements in this respect (though, of course, the latter is more controversial).

In this case, Schlick can account for the intelligibility of value disagreements (in contrast with Hobbes and Hume), but he does not distinguish between moral and other value disagreements. And this goes against the widely held view, that there *is* a difference.

Alternatively, if *only moral value judgments* have a command character, then the difference between moral and other disagreements can be traced back to this difference: moral disputes are genuine as they express contrary commands, while aesthetic and culinary disputes are only pseudo-disputes, since they do not involve commands. Hence, on this interpretation of the "morally good" it is also implied that moral disagreements are genuine. Consider, for example:

If S_1 judges that frugality is morally good, then he believes that the conjunction (1) (2) and (3) is satisfied concerning frugality, and that "Be frugal!" is demanded.
If S_2 judges that frugality is not morally good, then he believes that the conjunction (1), (2) and (3), is not satisfied concerning frugality, and that "Be frugal!" is not demanded.

Thus the latter interpretation of "morally good" can account both for the difference between moral and non-moral disputes and for the genuineness of moral disputes. However, such an account still has *an implausible consequence*, namely: it implies that *moral disagreements are purely factual*, hence they can be settled solely by reference to facts, i.e. by ascertaining whether X satisfies (1)–(3) or not.[9] And this poses another problem for Schlick's view.

5.3.2 *Problem 2: Moore's Open Question Argument and Schlick's View*

The view that it is a purely empirical, factual question whether the morally good is X or not, and that moral disputes can be settled exclusively by *empirical-scientific methods* contradicts what Moore's *Open Question Argument* purports to prove, namely that moral goodness cannot be defined in terms of any natural properties, that is, in terms of such properties that are identifiable by scientific methods. According to the argument when we assert that "The good is X", it is always sensible to ask: "Is X *really* the good?". And when we ask this, what we want to know *is not* whether X is what (1) the (majority of) society desires (2) judges beneficial for the society, and (3) desires it because judges it to be beneficial for the society. What we want to know is: "Is X *really* the good?" – independently of who and why desires it.

[9] That conflicting moral value judgments have a command character does not by itself imply that moral disputes cannot be settled solely by relying on facts. For when it turns out which standpoint satisfies (1)–(3), then the dispute is settled; implying also that the demand character of the refuted moral claim is illegitimate. Thus in order to reject that moral conflicts can be settled solely by appealing to facts, it is not enough to show that moral judgments have a command character.

We may wonder how Schlick would counter Moore's argument. Schlick (1930/1939, 6) refers explicitly to Moore only once in the *Problems of Ethics* but is seems obvious that he is attacking Moore's view, for example, when he argues that "it is very dangerous to withdraw from this task (i.e. of defining 'good' – G.A.) under the pretext that the word 'good' is unanalyzable, of which therefore a definition, a statement of connotation, is impossible" (1930/1939, 8). Here Schlick asserts that although it may be true that "good" is not definable, in the strictest sense of the word, but this does not imply that the meaning of "good" cannot be determined, by investigating how the term "good" is *used* (cf. Schlick 1930/1939, 8–10). He treats "good" in analogy with color words, just like Moore, and suggests that although it is true that the meaning of "green" is not definable, we can nevertheless determine its meaning by investigating how the term "green" is used (i.e. as the color of grass, pine trees etc.).

It seems that the objection formulated above, based apparently on Moore's argument, can be blocked easily by showing that it misses the point against Schlick. For Moore's *starting point* is that the good is an objective property; and the *Open Question Argument* and his other supplementary arguments aim to prove his *further* claims, namely that this objective property is simple, and not reducible to any other – natural or metaphysical[10] – properties. In contrast, Schlick's *starting point* is that the good is *not* an objective property. So interpreting Moore's question "Is X really the good?" as if it purported to show that the good cannot be subjective (non-objective in the sense of mind-dependent) misses Moore's original intention; he did not argue by the open question for the objectivity of goodness – he took this for granted – but for the claim that the good is an irreducible, *sui generis* property. Hence Moore's argument seems not to affect Schlick's account at all, since, again, he starts from the assumption that goodness is not an objective property.

Further counter-arguments against Moore along the lines of Schlick's account of the meaning of "morally good" may be run as follows.

First, Schlick's assertion that the definition[11] of "morally good", [i.e. (1) the (majority of) society desires *A*, (2) the (majority of) society judges that *A* is beneficial for the society, and (3) the (majority of) society desires *A*, because it judges that *A* is beneficial for the society], is meant to be an empirical claim (of course). It is supported by the empirical arguments, i.e. the arguments from ordinary language use, co-extensionality and psychological law. None of these prove that the meaning of "morally good" could not be something else. What is claimed is that this is in fact its meaning. But against such an empirical statement about meaning it is no objection that it is conceivable that it could be otherwise.

Second, consider the following reconstruction of the *Open Question Argument*:

4. Suppose the morally good is X.

[10] Metaphysical in the sense used in the *Principia Ethica*. Cf. Moore (1903/1993), Chapter IV. *Metaphysical Ethics*.

[11] In other words, a statement about its meaning: cf. Schlick (1930/1939, xiii).

5. If the morally good is X, then it is not conceivable that the morally good is not X (but Y).
6. It is conceivable that the morally good is not X (but Y).[12]

    Therefore

    C. The morally good is not X.

Moore takes this argument to be valid. This is supported by Moore's account, according to which (in Kripkean terminology) "good" is a rigid designator directly referring to goodness (which is a *sui generis*, non-natural,[13] simple property), since (5) is justified on this understanding of the term "good" (as accordingly, "good" refers to the same property in all conceivable worlds).[14] Hence, it seems, goodness cannot be property X, where X is what satisfies Schlick's (1)–(3) conditions (for then both (5) and (6) are true on this interpretation of "X"). Schlick, however, could reply that on his account, "morally good" refers *non-rigidly*, fixing its reference through a description, consisting of conditions (1)–(3). So the property of goodness is *whatever* that satisfies (1)–(3): hence it may well be that in fact (1) the (majority of) a society desires X, and (2) judges that X beneficial for the society, and (3) desires X because judges it to be beneficial for the society; but at the same time it is conceivable that the (majority of) society desires Y (not X), and (2) judges that Y beneficial for the society, and (3) desires Y because judges it to be beneficial for the society. Hence (5) is false, therefore the conclusion does not follow.

Of course, there are further arguments both on Moore's and Schlick's side. Schlick, for example, argued that we *must* give a scientific-empirical definition of the good, since any other kind would be illegitimate (as they are either based on illegitimate conceptions of moral knowledge or committed to illegitimate sorts of entities). On the other hand, Moore in his *Ethics* (1912) took into consideration that goodness is not an objective property, but expresses the desires of the judger. But he rejected this idea on the ground that it cannot explain the genuineness of moral disagreements.[15] Schlick, of course, could not accept Moore's explanation of moral

---

[12] Let's take "it is conceivable" as synonymous with "it can be meaningfully (i.e. non-vacuously) asserted" or with "it can be meaningfully questioned".

[13] And also a "non-metaphysical" property, in the sense Moore uses the term in the *Principia Ethica*.

[14] Although, of course, Moore did not use the terminology of rigid designation, it is plausible to interpret his view in this sense, i.e. that the term "good" refers to the same (simple, non-natural, *sui generis*) property in all possible worlds (moreover, it supervenes on the same natural properties in all possible worlds).

[15] Cf. Moore (1912, 101): "If, when one man says, 'This action is right,' and another answers, 'No, it is not right,' each of them is always merely making an assertion about *his own* feelings, it plainly follows that there is never really any difference of opinion between them: the one of them is never really contradicting what the other is asserting. They are no more contradicting one another than if, when one had said, 'I like sugar,' the other had answered, '*I don't* like sugar.' In such a case, there is, of course, no conflict of opinion, no contradiction of one by the other: for it may perfectly well be the case that what each asserts is equally true; it may quite well be the case that the one man really does like sugar, and the other really does *not* like it. The one, therefore, is *never* denying what the other is asserting. And what the view we are considering involves is that when one man holds an action to be right, and another holds it to be wrong or not right, here also the one is *never* denying what the other is asserting."

disagreements, but his own view also has problems, namely, that it implies that they are of a purely factual character. And although we have just discussed why Moore's argument need not be taken to establish that this is false, there are also other accounts spelling out the idea that moral disagreements concern not only factual matters, which, however, do not rely on Moore's assumption of the objectivity or goodness, and that moral judgments express objective facts (see e.g. Stevenson 1937).

5.3.3 Problem 3: The Command Character of "Morally Good"

According to section 1.5. of *The Problems of Ethics*, the form of "X is good" is always a command: X is demanded. As I have argued above, this formal characterization applies only to the morally good; moreover, there is textual evidence, at least implicit, that this was in fact Schlick's view. For example, Schlick refers to Kant as a philosopher who put the whole weight of his moral philosophy on this formal character. Moreover Schlick (1930/1939, sec. 1.6.) also writes that a command always presupposes a commander, who is, according to his conception, the society; since neither God, nor "No-one", as in Kant's ethics (in Schlick's wording), would qualify. This seems to imply that "X is good" has the character of command only if its source is the society. But on this reading we are led to the following problem. How is it possible that the judgment "*A* is good" acquires a command character?

Simply that (1) the majority of the society desires X, (2) judges that promoting X is beneficial for the society and (3) desires X because it judges that promoting X is beneficial for the society, does not ground that "Act as to promote X" or something similar is a command that is demanded. Plausibly, a minimal requirement for an imperative sentence to express a command is that:

(i) It is *obligatory* for the target group.
(ii) It is *effective*: (A significant number of the) members of the target group obey it.

As for (i): few philosopher would hold that simply being the majority view in a community makes a view obligatory to accept for all members of the community, and it is not very likely that Schlick intended to demand such identification with the "general will". As for (ii): the effectiveness of such demands are also doubtful: the fact that the majority of society desires X, and judges it to be beneficial for the society, *does not (and usually cannot) compel* someone to desire X and to judge it beneficial for the society.

Further, suppose that someone, in fact, *desires to identify with*, i.e. desires to accept honestly, the majority view, because she takes it to be constitutive of being a member of a society to approve of its moral values. However, her desire may not be satisfied. If, for example, her original view about values is in strong opposition to the majority view, then she *may not be able* to commit herself to the latter. (If someone is strongly convinced that abortion is wrong in all circumstances, or that certain deeds deserve capital punishment, then she may not be able to commit herself to the contrary views.) And, even if she manages, against her earlier contrary commitments, to identify with X, a value society praises, in the sense that she brings herself to act in accordance with X; this does not imply that she also desires X. It may well be that she does not desire X, but her acting in accordance with X is based on some other motivation (e.g. the desire to be a true member of society). But then it follows that according to Schlick's account of the good and the moral good, *the moral good may not be good* (as it is not desired) – and this sounds somewhat paradoxical.[16]

So, it seems, we found that Schlick's view of the morally good can account for the difference between moral and non-moral value disputes in terms of moral value judgments having a command character, but it cannot account for the command character itself and neither for the attractiveness of the moral goodness.

5.3.4 Schlick's Proposed Solution

In "What Paths Lead to Value?", Chapter VIII of *The Problems of Ethics,* Schlick – rather surprisingly – put forward a *substantial* account about what the morally good *is*. In contrast with his previous preoccupations and proclaimed view, here he addresses *not* what certain societies *believe to be* the good but the question what the good *is* – hence leaving the role of the moral philosopher and moral scientist behind and acting as a moralist.

The fundamental moral principle Schlick proposes: Be ready for happiness! (Sei Glücksbereit!). In other words, be a person who is able to receive the greatest happiness possible for him or her. In this last chapter Schlick sets out to justify this Happiness Principle, by purporting to show that the effectiveness of moral value judgments and moral commands can be based on it. The justification starts from the supposition that the idea of the happiness of society can only motivate action, if it causes pleasure in the subjects, hence only if subjects have altruistic desires. Schlick's argument may be reconstructed as follows:

1. The moral principle commands: Be ready for happiness.
2. Altruists can meet this demand *better*, since they can experience more pleasure than egoists, as they are capable of receiving pleasure from perceiving of or acting for other persons' happiness/welfare (unlike the egoists).
3. From (1) and (2) the effectiveness of moral value judgments can be explained: for altruists the content of the desire promising the maximum pleasure (according to the law of motivation) coincides with the content of the moral demand, i.e. to act for the welfare of society.

[16] Though such a view is not unprecedented. Kant's account may be interpreted as such, because of its strict anti-hedonism (i.e. that pleasure-evoking capacity is emphatically not a characteristic of moral values). However, for a pleasure-based value theory such as Schlick's this implication would be rather odd. It may be worth mentioning that even Kant's view is more subtle than being simply anti-hedonist, for he does hold that reverence for the moral law is a motivational force of morally good conduct, and reverence is a passion, which although ambivalent in character, *does* have an the attractive aspect, namely love (besides the aspect of fear).

One may object that a justification of the Happiness Principle along these lines can only work if human beings really *are* altruistic. Schlick's reply would be that it is in fact so. Schlick had a positive anthropology, similar to Hume's and Mill's, according to which we have *innate* altruistic impulses, which, moreover, can be further developed and strengthened. These motives can ground that we desire the pleasure or welfare of others and act for it without coercion. Further, Schlick ventures that it may be a *natural/psychological law* that in humans living together in a community, altruistic impulses, the capacity to feel pleasure over others' well-being, naturally evolve.

Thus Schlick's view seems to be that:

4. We have *innate* altruistic inclinations.
5. These potential altruistic desires *naturally develop* when living in a community.

Moreover, he seems to suggest that theses altruistic inclinations, the disposition to act in favour of the benefit of society may also be developed by moral education. Here there is a connection with Schlick's ethical views advocated earlier, in particular with his commitment to Schiller's conception of playing and the idea of aesthetic education. In Chapter VIII, especially in the last two sections (12. "Personality and Kindness" and 13. "Ethics of Duty and Ethics of Kindness"; Schlick 1930/1939, 202–209) Schlick echoes his earlier "moralist" views, which he put forth, for example, in "Vom Sinn des Lebens" (1927/1979), and also in his early *Lebensweisheit* (1908). In these works a guiding idea is that the nature of moral conduct is analogous with the nature of playing: in playful activities subjects act "disinterestedly", i.e. freely, without coercion, but at the same time in accordance with the general rules of the play. Thus playing is an end-in-itself, it is acting in accordance with the rules of conduct for its own, for no external purpose, which also brings pleasure with it.

As Schlick writes in "Vom Sinn des Lebens":

> Let us recall Schiller's remark, that *the principle of play* (emphasis added) as the true vocation of man will attain its deepest significance if we apply it to the seriousness of duty and destiny. What does this mean? It was Schiller who rebelled against the doctrine of Kant, whereby, of course, the moral is primarily to be found where man acts as conquering himself. For in Kant's view, an action is moral only when it springs from reverence for the law of duty as its sole motive; and since in the actual man conflicting inclinations are always present, moral action means a struggle against one's own inclinations, it means laborious work. Schiller was utterly and entirely right, for this account of the good is infinitely remote from the meaning that everyone is otherwise naturally accustomed to associate with the word. We do not call *him* the best man, who is obliged unceasingly to resist his own impulses, and is constantly at war with his own desires; we say this, rather, of the man whose inclinations are kindly and benevolent from the start, so that he simply does not fall into doubt and self-conflict. The man who struggles with and conquers himself is perhaps the type of the *great* man, but not of the *good* one. A being whose pure will flows from his natural dispositions, without hesitation or wavering is what we call an innocent person, and innocence is always the state of greatest moral perfection. This innocence is by no means a kind of ignorance, but rather a kind of freedom. (Schlick 1927/1979, 125–126)

The virtuous character, according to Schlick, is considerate, viz. takes into account others' interests and desires, but also firm, has a will of his own, autonomously determines his conduct (Schlick 1930/1939, 202–205). This harmonizes with Schiller's description in his *Letters upon the Aesthetic Education of Man* (1795/1965) according to which the man of good character is maximally open to others, he is under the "material impulse"; but at the same time not guided from without, his will is determined by himself, from the inside, in other words, he is also under the guidance of the "form impulse"; and these two impulses are synthesized in the "play impulse". The kind and non-egoistic behavior, "the ethics of kindness" Schlick praises, is clearly reminiscent of the conduct characteristic of the citizens of Schiller's aesthetic state.

> The dynamic state can only make society simply possible by subduing nature through nature; the moral (ethical) state can only make it morally necessary by submitting the will of the individual to the general will. The aesthetic state alone can make it real, because it carries out the will of all through the nature of the individual. If necessity alone forces man to enter into society, and if this reason engraves on his soul social principles, it is beauty only that can give him a social character; taste alone brings harmony into society, because it creates harmony in the individual. (Schiller 1795/1965, 293. XXVII. Letter)

So as a result of (aesthetic) moral education the demand of acting as to promote the welfare of society will become the desire of the kind, virtuous character, hence it is not forced upon him from without, and will have genuine attractive force, and thus will be effective.

5.4 Summary and Conclusion

Schlick's views on the morally good divide into two parts. One part belongs to the research program of philosophical-cum-scientific ethics. As an outcome of this project Schlick proposes a meaning for the "morally good", which, he claims, is gained from the investigation of the ordinary use of the term. Further, he also sets up the scientific task of empirically determining what moral values different societies actually endorse. The other part consists in Schlick's normative ethics, featuring his Happiness Principle.

However, the relation between these two parts is not obvious. First of all, it is not clear, why Schlick provided a normative ethics in the first place? It is a strange move, considering that he begins the book with the mocking of those philosophers who fail to realize that what they do is not philosophy but moralizing. For consider:

> Thus ethics is a system of knowledge, and nothing else; its only goal is the truth. Every science is, as such, purely theoretical; it seeks to understand; hence the questions of ethics, too, are purely theoretical problems. As philosophers we try to find their correct solutions, but their practical application, if such is possible, does not fall within the sphere of ethics. If anyone studies these questions in order to apply the results to life and action, his dealing with ethics has, it is true, a practical end; but ethics itself never has any other goal than the truth. So long as the philosopher is concerned with his purely theoretical questions, he must forget that he has a human interest as well as a cognitive interest in the object of his inves-

> tigation. For him there is no greater danger than to change from a philosopher into a moralist, from an investigator into a preacher. (Schlick 1930/1939, 1–2)

And also

> For if ethics furnishes a justification it does so only in the sense just explained, namely, in a relative-hypothetical way, not absolutely. It "justifies" a certain judgment only to the extent that it shows that the judgment corresponds to a certain norm; that this norm itself is "right", or justified, it can neither show nor, by itself, determine. Ethics must simply recognize this as a fact of human nature. Even as a normative science, a science can do no more than explain; it can never set up or establish a norm (which alone would be equivalent to an absolute justification). It is never able to do more than to discover the rules of the judgment, to read them from the facts before it; the origin of norms always lies outside and before science and knowledge. This means that their origin can only be apprehended by the science, and does not lie within it. In other words: if, or in so far as, the philosopher answers the question "*'What is good?' by an exhibition of norms, this means only that he tells us what 'good' actually means; he can never tell us what good must or should mean.* (Schlick 1930/1939, 17–18, emphasis added)

So how come that at the end of the book Schlick himself could not help becoming a moralist leaving the attitude of the philosopher behind?

We may think of Chapter VIII as a sort of appendix – after the scientific and philosophical work is done, Schlick may have felt that now he was allowed to express his personal view about what good was, as a sort of public statement of his own commitments.[17] This much is allowed even for philosophers and may serve a legitimate aim, even by Schlick's standards, namely spreading his views and helping to "convert" others to his conviction. This attitude is legitimate as long as it is not accompanied by the mistaken idea that moral philosophers are moral peers and their substantial moral views have greater authority than of other people.

I suggest, however, that the role of Chapter VIII is more than that. It is in fact *not* an appendix that could be left out. For it appears, the contradictions or tensions inherent in the philosophical and scientific part cannot be resolved without reference to the ideas proposed in chapter VIII, most importantly the Happiness Principle. For consider: Schlick claims that his formulation of the meaning of the "morally good" – i.e. what the majority of society desires and judges to be useful for society – is in line with, in fact distilled from, common sense, and how we ordinarily use the term.

This definition however does not meet some plausible requirements, namely that the concept of good must be such that:

A. It accounts for the attractiveness of the morally good.
B. It accounts for the command character of moral value judgments.
C. It implies that the question of what is morally good is not a purely factual matter, and moral disagreements are not solely disagreements about facts.

These requirements seem to be part of the commonsense conception of the morally good (cf. Stevenson 1937). Moreover, Schlick explicitly endorsed A. and B. Further,

[17] For an interpretation along these lines, see Tuboly, this volume.

although he seems not to endorse C. as he states that scientific ethics is a subfield of empirical psychology and sociology, it is not the full story. For this is only Schlick's view about scientific ethics; but as we have seen in Chapter VIII, he does formulate the question: what *is* really good? – as opposed to what people *hold to be* good.

A. and B. cannot be met solely on the basis of the scientific definition of morally good (since being the majority view cannot in itself ground attractiveness and command character). But relying on Schlick's normative conception based on the Happiness Principle may help here. We may put it this way: if Schlick is right and the commonsense/scientific notion of moral goodness is what he suggests (i.e. what the majority of society desires, and what the majority of society judges to be useful for society, and what also has a command character) then the commonsense view is inherently problematic. He, however, wants also to *justify* the commonsense view, not only to identify what it is.

And here comes his normative ethics to the rescue. For consider again: if men are truly altruistic, then the tensions are resolved. If members of the society desire what is beneficial for the society because the idea of the welfare of others cause pleasure in them, then moral goodness is attractive for them.

Nonetheless a problem still prevails, namely: what guarantees that all members agree about what is beneficial for the society? So, even if by natural law and the help of (aesthetic) moral education all members of the society develop altruistic desires, the concrete object of their desire may differ. I think, Schlick probably would have answered that although there are *no general rules* about what the good for the society is, but the virtuous, i.e. altruistic, kind, considerate, non-egoistic person finds out with high probability which conduct is the good one in the particular cases.

A last question: is then Schlick inconsistent by proposing a normative ethics, his Principle of Happiness which moreover seem to be indispensable for making his philosophical and scientific ethical views consistent?

Yes and No. Clearly, the justification of the Happiness Principle is not provided by the *science of ethics*, i.e. it is not verified by *observations about the actual norms of societies*. Nonetheless, the principle is "naturalistic" in the sense that it is based on the empirically observed features of human nature, according to Schlick, namely that humans in general seek happiness and have altruistic impulses. So the principle is in accordance with the general empiricist and anti-metaphysical spirit of scientific philosophy and in sharp contrast with such an "ethics of demand" that contradicts human nature (as, for example, Kant's view that considers happiness irrelevant in respect of the value of an act).

Hence the two parts of the *Problems of Ethics* can be accommodated. The moral imperative "Be ready for happiness" is based on an empirical-scientific view of human nature. And the ethics of desire and the ethics of demand may be synthesized by the ethics of love, according to which acting morally means acting freely, without coercion, i.e. acting playfully, in accordance with the essence of human nature. For Schlick held in agreement with Schiller the essence of humanity is playing. As he writes, also in "Vom Sinn des Lebens":

If we wish to find a meaning in it (i.e. in life G.A.), we must seek for *activities* which carry their own purpose and value within them, without any extraneous goals; activities, therefore, which are not work, in the philosophical sense of the word. [...]

There really are such activities. To be consistent, we must call them *play*, since that is the name for free, purposeless action, that is, action which in fact carries its purpose within itself. We are thereby not lending it any new or surprising meaning, but merely repeating what was perfectly clear to at least one great mind [...]. For in his *Letters on the Aesthetic Education of Men*, Friedrich Schiller utters the following words:

For, to declare it once, and for all, Man plays only when he is in the full sense of the word a man, and *he is only wholly Man, when he is playing*." (Schlick 1927/1979), 114–115)[18]

Thus Schlick attempted to console the ethics of desire with the ethics of demand in a traditional naturalistic way, by claiming that the demands must be grounded in human nature. So, by replacing the ethics of duty with the ethics of kindness Schlick's attitude is of a moralist, but of a scientific moralist.

Acknowledgements I thank very much the editors and the referees for their valuable comments and suggestions for corrections. This paper was written with the support of the 112542 research project of the National Research, Development and Innovation Office of Hungary (NKFIH).

References

Ayer, Alfred J. 1936/1971. *Language, Truth and Logic*. Harmondsworth: Penguin Books.

———. ed. 1959. *Logical Positivism*. New York: The Free Press.

Brentano, Franz. 1889/1969. *The Origin of Our Knowledge of Right and Wrong*. London: Routledge.

Hegselmann, Rainer. 1984. Logischer Empirismus und Ethik. In *Moritz Schlick, Fragen der Ethik*. Herausgegeben und eingeleitet von Rainer Hegselmann, 7–46. Frankfurt am Main: Suhrkamp.

Iven, Mathias. 2006a. Lebensweisheit. Editorischer Bericht. In *Moritz Schlick, Lebensweisheit Versuch einer Glückseligkeitslehre. Fragen der Ethik*. Herausgegeben und eingeleitet von Mathias Iven, 19–40. Wien, New York: Springer.

———. 2006b. Fragen der Ethik. Editorischer Bericht. In *Moritz Schlick, Lebensweisheit. Versuch einer Glückseligkeitslehre. Fragen der Ethik*. Herausgegeben und eingeleitet von Mathias Iven, 335–344. Wien, New York: Springer.

Moore, George E. 1903/1993. *Principia Ethica*. London: Routledge.

———. 1912. *Ethics*. New York: H. Holt and Co.

Mormann, Thomas. 2010. Zwischen Weisheit und Wissenschaft – Schlicks weites philosophisches Spektrum. *Grazer Philosophische Studien* 80 (1): 263–285.

Prichard, Harold A. 1912. Does Moral Philosophy Rest on a Mistake? *Mind* 21: 21–37.

Ross, William D. 1927. The Basis of Objective Judgements in Ethics. *International Journal of Ethics* 37: 113–127.

———. 1930/2002. *The Right and the Good*. Oxford: Clarendon Press.

Rutte, Heiner. 1991. Ethics and the Problem of Value in the Vienna Circle. In *Rediscovering the Vienna Circle: Austrian Studies on Otto Neurath and the Vienna Circle*, ed. Th. Uebel, 143–167. Dordrecht: Kluwer.

[18] Though this quote is not from the *Problems of Ethics*, I do not think it is very strained to suggest that Schlick still held to this view published only 3 years earlier.

Schiller, Friedrich J.C. 1795/1965. Letters upon the Aesthetic Education of Man. In *Literary and philosophical essays*, The Harvard Classics. Vol. 32, ed. Ch.W. Eliot, 209–303. New York: P.F. Collier and Son Company.

Schlick, Moritz. 1908. *Lebensweisheit*. München: C. H. Beck'schen Verlagsbuchhandlung.

———. 1913/1979. Is There Intuitive Knowledge? In *Moritz Schlick: Philosophical Papers: Volume Two (1925–1936)*, ed. H.L. Mulder and B.F.B. van de Velde-Schlick, 141–152. Dordrecht: D. Reidel.

———. 1926/1979. Experience, Cognition, and Metaphysics. In *Moritz Schlick: Philosophical Papers: Volume Two (1925–1936)*, ed. H.L. Mulder and B.F.B. van de Velde-Schlick, 99–111. Dordrecht: D. Reidel.

———. 1927/1979. On the Meaning of Life. In *Moritz Schlick: Philosophical Papers: Volume Two (1925–1936)*, ed. H.L. Mulder and B.F.B. van de Velde-Schlick, 112–129. Dordrecht: D. Reidel.

———. 1930/1939. *Problems of Ethics*. New York: Prentice Hall.

Siegetsleitner, Anne. 2010. Schlicks *Fragen der Ethik* und die vorherrschende Sicht logisch-empiristischer Ethik. In *Logischer Empirismus, Werte und Moral. Eine Neubewertung*, ed. A. Siegetsleitner, 131–156. Wien, New York: Springer.

———. 2014. *Ethik und Moral in Wiener Kreis*. Wien, Köln, Weimer: Böhlau Verlag.

Soames, Scott. 2018. *The Analytic Tradition in Philosophy. Volume 2: A New Vision*. Princeton and Oxford: Princeton University Press.

Stevenson, Charles L. 1937. The Emotive Meaning of Ethical Terms. *Mind* 46: 14–31.

Chapter 6
Making Logical Positivism Less Logical: The Case of Schlick and von Mises

Adam Tamas Tuboly

Abstract In this paper, I argue for the following three theses: (1) Schlick never overcame the problems of "traditional philosophy" and thus tried to find a way to embrace them from his own point of view. He did not dismiss them thus, but actually held them in high esteem and regarded them as serious intellectual riddles. (2) Though Schlick's broad interests never came close to the breadth of Neurath's general scholarly concerns, both of them shared a comprehensive attitude towards questions of life in general; nonetheless, Schlick's interest had a quite peculiar theoretical and intellectualist twist, contrary to Neurath's practice-oriented, explicitly value-laden and socially engaged conception. From Neurath's perspective, Schlick thus emptied out the whole field. (3) Finally, I will argue that Schlick was not the only one who tried to streamline the movement's thinking and hammer out the deviations (which manifested themselves in allegedly simplistic, formalistic tendencies): Richard von Mises's major work on positivism was written (and read by many) in a similar vein, that is, as a reinvention of classical philosophical questions in a peculiarly positivist/empiricist framework. The main point, or the narrative that I will try to reconstruct, is therefore that there are *different ways* of making logical positivism *less logical*, in order to demonstrate its power of embracing classical questions and problems in new settings.

Keywords Moritz Schlick · Richard von Mises · Logical positivism · History of philosophy · Values

A. T. Tuboly (✉)
Institute of Philosophy, Research Centre for the Humanities, Budapest, Hungary

ITD, Medical School, University of Pécs, Pécs, Hungary

C. Damböck, A. T. Tuboly (eds.), *The Socio-Ethical Dimension of Knowledge*, Vienna Circle Institute Yearbook 26, https://doi.org/10.1007/978-3-030-80363-6_6

6.1 Introduction

In the 1920s, logic was a much-debated subject, and the members of the Vienna Circle, which was then just forming, welcomed Rudolf Carnap as their Messiah. Herbert Feigl (1969/1981, 61–62) described Carnap's first appearance at the Circle in 1925 as follows: "Everyone was impressed with [Carnap's] logical proficiency and scientific competence. Several of us felt that here was a very atypical (at the time!) philosopher who explained his logical reconstructions in the manner of an engineer describing the workings of a machine." After all, the Circle's empiricism was called 'logical' for a reason.

Nonetheless, as is well-known nowadays, certain members of the Circle also practiced classical or traditional fields of philosophy. As testified by a number of publications, like Karl Menger's *Morality, Decision, and Social Organization* and Viktor Kraft's *Foundations for a Scientific Analysis of Values*, some members of the movement had quite comprehensive views on how to *embrace* traditional philosophy (including ethics, values, life, art, and politics) and its problems. In this paper, I will contrast two versions of this de-logicalization approach against the general background of logical empiricism: the path pursued by the leader of the Vienna Circle, Moritz Schlick, and that taken by Richard von Mises, who acted as a bridge between the Vienna and Berlin groups.

I will argue for the following three theses: (1) Schlick never overcame the problems of "traditional philosophy" and thus tried to find a way to embrace them from his own point of view. He did not dismiss them thus, but actually held them in high esteem and regarded them as serious intellectual riddles. (2) Though Schlick's broad interests (from logic to life and culture, from physics to the arts and politics) never came close to the breadth of Neurath's general scholarly concerns (from pictorial education to economics, from sociology to education), both of them shared a comprehensive attitude towards questions of life in general; nonetheless, Schlick's interest had a quite peculiar theoretical and intellectualist twist, contrary to Neurath's practice-oriented, explicitly value-laden and socially engaged conception. From Neurath's perspective, Schlick thus emptied out the whole field. (3) Finally, I will argue that Schlick was not the only one who tried to streamline the movement's thinking and hammer out the deviations (which manifested themselves in allegedly simplistic, formalistic tendencies): Richard von Mises's major work on positivism was written (and read by many) in a similar vein, that is, as a reinvention of classical philosophical questions in a peculiarly positivist/empiricist framework.

The main point, or the narrative that I will try to reconstruct, is therefore that there are *different ways* of making logical positivism *less logical*, in order to demonstrate its power of embracing classical questions and problems in new settings.

First, some words are in order about the choice of thinkers. While it is true that other members of the Vienna Circle – such as Edgar Zilsel, Philipp Frank, Felix Kaufmann and Otto Neurath – tried to make logical positivism less logical, none published works that were as systematic and easily approachable as that of Schlick and von Mises. Focusing the discussion on these two philosophers thus makes it possible to keep the story short and straightforward (on similar issues in Philipp Frank, see Tuboly 2021).

6.2 Traditional Philosophy and Schlick's Long-Standing Interest in Ethics, Values, and Life

6.2.1 The Status and Nature of Traditional Philosophy

Hans Reichenbach (1931/1978, 359) claimed that the "new scientific philosophy [...] has offered unexpected solutions to a number of old problems and has raised several novel questions whose existence had been overlooked at earlier times." Logical empiricism, from Reichenbach's perspective, thus did not just solve old puzzles – usually by dissolving them with logical devices – but raised *new questions* as well. At the time, much of traditional philosophy was overcome by the new scientific endeavors. In a letter to Felix Meiner, the publisher of *Erkenntnis*, Carnap wrote that "some parts of traditional philosophy will be rejected by us as unscientific."[1]

Nonetheless, Schlick claimed that since philosophy is not a science, it does *not* work with determinate and cognitive statements. Philosophy can therefore not change the field of problems – instead, "there will be no further need to talk of 'philosophical problems', since *all* questions will be dealt with philosophically; that is, in a clear and meaningful way" (Schlick 1930/1979, 160). This caused the boundaries between the traditional and the new philosophy to become blurry: there are only problems and these problems should be met by means of the methods of philosophy. Given his different conception of philosophy, Schlick not only declined the invitation to be an editor of *Erkenntnis*, but practiced philosophy according to his own ideals, which he communicated in a clear and unmistakably programmatic paper, titled "The Future of Philosophy".[2]

In his recollections on the Vienna Circle, Karl Menger (1994, 70) claimed, for example, that "contrary to what has been said about them, [the members of the Circle] were not altogether ignorant of the history of philosophy." While he mentioned Viktor Kraft, Rudolf Carnap, Hans Hahn, Felix Kaufmann, Otto Neurath and Herbert Feigl, he had Schlick in mind especially. According to Menger (1994, 70), "Schlick was widely read in the philosophy of the ancient Greeks, of the Renaissance, and of the modern age. In particular, he had carefully studied Kant, the neo-Kantians, and Husserl."

Schlick wasn't just well-read in the history of philosophy (as would have been required of any university professor of philosophy in the 1920s), but also had a

[1] Rudolf Carnap to Felix Meiner, April 26, 1930 (HR 013-41-69; Hans Reichenbach Nachlass).

[2] See Reichenbach (1930, 3), Schlick (1930/1979) and Schlick to Reichenbach, June 8, 1930 (Moritz Schlick Nachlass).

particular *interest* in traditional philosophy. During his Viennese years, between 1922 and 1936, he gave (at least) six historically-oriented philosophy courses, seven on ethics and moral philosophy, two on the problems of *Weltanschauung* and three on questions of culture in relation to ethics and morality (Stadler 2001/2015, 320–322). He also held lectures on Schopenhauer and Nietzsche in Rostock and Kiel during many semesters. Though only a minor part of these lectures has been published, their contents testify to Schlick's deep understanding of the historical approach to philosophical problems.[3]

Even though Schlick was interested in traditional philosophy (its proponents and topics), he never wanted to build up *traditional* systems.[4] He never aimed to turn the scholarly gathering around him into a *school of philosophy*, and for this reason (among others, of course) he disliked the famous manifesto of Carnap, Hahn and Neurath (1929/1973), fearing that the paper's style and line of argumentation would act as a dangerous, one-dimensional step towards school-building (Mulder 1968, 390). It should nevertheless be mentioned that when discussing the new linguistic method of philosophy in his posthumously published aphorisms, Schlick (1962, 31) claimed that "[t]his is our new school". Presumably, he didn't find being a member of a school or constructing one that annoying after all, at least behind the scenes.

Becoming a school was one problem, but what also caused Schlick significant headaches was the fact that some members of the Circle – i.e. Carnap and Neurath – denied that they were engaged in philosophy at all. While Schlick always aimed at *reforming* philosophy within its proper boundaries, getting rid of philosophy altogether was a path with which he never associated himself. For example, in a posthumously published paper, written for the 1937 Descartes Congress in Paris, he distanced himself from everyone in the so-called left wing of the Circle, and in the process made quite the statement.[5] First of all, he argued that criticizing metaphysics was not a new tendency and definitely not a particularly important achievement: "Looking backward and observing how Western thought has evolved for thousands of years, it is easy for us not to be metaphysicians, and we have no right to be proud of it" (Schlick 1937/1979, 492). He thus distinguished two options: on the one hand, a philosopher's work could belong to the sphere of non-metaphysics, given that

[3] These lectures were published in Schlick's collected works under the editorship of Mathias Iven. See Schlick (2013); MSGA II/5.1. The lectures are analyzed (with respect to a general conception of ethics) in Ferrari (2018, especially 377–382). Schlick's first comprehensive history of philosophy course was held in Kiel, and he returned to this course scheme from time to time. When he taught the same comprehensive course (focused mostly on ancient philosophies) in Vienna in 1931, Schlick sought to authorize the class notes for publication. Though they never appeared, Schlick noted that these lectures should replace his previous works (like *Form and Content*), which is indicative of his move, step by step, towards becoming a traditional philosopher. I am indebted to Martin Lemke for this information.

[4] Nontheless, in some briefs to Reichenbach and Carnap form the late 1920s, he seems to convey the idea that he tried to accomplish a certain form of system-building. I thank Massimo Ferrari for pointing this out.

[5] This paper is not an authorized publication; see the editorial introduction to "Metaphysiker und Dogmatiker" in Schlick (2013, 467–472); MSGA II/1.2.

Schlick recognizes that metaphysics is meaningless; in the fashion of Wittgenstein, philosophers should not talk about metaphysics, but instead philosophize in a non-metaphysical way.[6] On the other hand, there are the anti-metaphysicians who relentlessly and loudly fight metaphysics wherever they find or smell it; and the anti-metaphysicians see metaphysics everywhere. For Schlick (1937/1979, 492), shouting 'metaphysics' continuously meant that "such a man has not yet conquered the metaphysics in himself, nor risen above it: his arrogant gestures and sounding words betray only fear of metaphysics."

It is quite easy and appealing to think of Neurath when reading this passage. A few sentences before, Schlick stated that "[c]ertain members even have a marked antipathy to the word 'philosophy', which they wish to proscribe in favour of other terms, such as 'unitary science'" (ibid.). On a previous page in the same article, he even complained that some members of the Circle make fun of traditional philosophy and "prohibit the term 'philosophy' [...] replacing it by colourless and unlovely expressions, such as 'unitary science'" (Schlick 1937/1979, 495). There is no doubt that Schlick had Neurath's attitude in mind. Neurath knew this, and these notes conditioned the way in which he remembered Schlick. When Carnap accused Neurath of not being able to handle criticism in the 1940s, the latter recalled Schlick's paper and his dislike of the term 'unified science': "I think adding esthetical judgement is usually not just criticism – it tells about the person's biography, not about the work in question. [...] DO YOU THINK THAT THIS WAY OF SCHLICK TO TELL INTERNATIONAL READERS ABOUT MY POSITION WITHIN OUR MOVEMENT SHOULD BE CALLED PROPER?"[7]

Schlick (1937/1979, 494, 492) indeed made some harsh remarks in his paper; he wrote that the attitude of the people in question (a.k.a. Neurath) was not objective, but emotional, and he thus "decline[d] to make common cause with them", and claimed that "[w]hat a man does through fear [i.e. rejecting metaphysics] nearly always produces a ridiculous effect." For example, when Carnap started to use Neurath's left-wing/right-wing vocabulary (to describe the inner structure of the Circle) in the mid-1930s, Schlick drew his attention to the danger that his works would become *ridiculous* by using the Neurath jargon.[8] Being associated with Neurath and his moves thus inevitably meant *ridicule* in Schlick's eyes. Furthermore, for Schlick, Neurath's activist fight against metaphysics and his *Index verborum prohibitorum* were merely jokes that caused more harm than good for the movement.[9] Neurath would have said the same about Schlick, of course: the questions are who was louder, who had a bigger audience, or whether they had any *common* audience at all.

---

[6] As Wittgenstein wrote to Waismann (July 20, 1929), "'Renunciation of metaphysics!' As if that were something new! What the Vienna school has achieved, it ought to show not say." Quoted in Iven (2015, 110).

[7] Neurath to Carnap, September 28, 1945 (RC 102-55-12); see letter 37 in Cat and Tuboly (2019, 683).

[8] Schlick to Carnap, January 20, 1935 (RC 029-27-07).

[9] For a recent reconstruction and reevaluation of Neurath's famous *index*, see Reisch (1997).

Regarding the latter, Neurath and Schlick presumably did not often share an audience, given that Neurath was at his best when he delivered big podium speeches in front of hundreds of people. On the other hand, even though Schlick was not known as the most entertaining speaker before university audiences (Menger 1994, p. 56), he always managed to gather some groups of students around him and published regularly in leading journals. Schlick was thus known internationally for his scholarly achievements, while Neurath was the good doctor who helped people with housing issues and improved their living conditions on various fronts.[10] Not that Neurath was unsuccessful at publishing his ideas. "I cannot complain that I did not succeed in publishing what I wanted to publish," he wrote to Bernhard Reichenbach (the brother of Hans Reichenbach, who lived in England as a communist refugee), "on the contrary, I am rather a spoiled child, usually I get f[i]rst my contracts and then [I] write [...]." Neurath was also known for his theoretical achievements (especially in economics), but his reputation in intellectual circles was not based on these – in fact, "the discussion on [his] work [was] meag[er], very meag[er], in no proportion to the sale of [his] books."[11]

Nevertheless, Schlick also reserved some critical words for Carnap. In the paper he wrote for the very first issue of *Erkenntnis*, which dealt with the significance of 'the new logic', Carnap (1930/1959, 133) announced a "*new, scientific method of philosophizing*", namely logical analysis (emphasis in the original). "Logic is no longer merely one philosophical discipline among others," he stated, clearly and unmistakably, "but we are able to say outright: Logic is the method of philosophizing." Though the details often changed, Carnap never abandoned his ideal that logic was the *method* of his activities. Schlick was not happy with this view: he thought that logic is merely a language and *cannot* be the method of philosophy. The fact that they were merely using a certain *language* would also protect the members of the Circle against much of the usual criticism, since "[t]o blame us for [using logic] would be just as unreasonable and unjust as to accuse Plato of having written his dialogues in Greek" (Schlick 1937/1979, 497). This was a fundamental difference between Schlick and Carnap: logical empiricism was logical for Carnap because its *method* was formal logical analysis, which eventually became the "logic of science", while Schlick considered logical empiricism to be logical because it was based on logical insights – and formulated in a logical and exact language – that resulted in a consistent and rational activity: there is nothing particularly or inevitably formal here.

Though Schlick did not say much about his reservations regarding logic, there are some hints that may guide us through his remarks. The term and practice of "logic of science" implied that if logic were to be viewed as the method of

[10] When the American journalist Devere Allen (1933, 236) visited Vienna to experience the socialist experiment in social life, he noted in his report that "the savants who wish to find the soundest and at the same time the most vivid work in this new realm of thought [i.e. pictorial education], will have to sit at the feet of Dr. Otto Neurath." For a long time, Neurath's fame was based on this type of activity, both in England and in the United States. For more on this, see Burke, and Tuboly (ms.).

[11] Neurath to Bernhard Reichenbach, May 4, 1945 (ONN).

philosophy, then philosophy would be restricted purely to (natural and mathematical) science, "as if questions of daily life were not accessible to or worthy of philosophical treatment" (Schlick 1937/1979, 498–498). But for Schlick, this was simply not the case. While he (1962, 22) also thought that "[p]hilosophy is sick, [and] its only cure is logic," philosophy was much more than just the logic of science or even applied logic. "[E]thics for us," he wrote, "is a philosophical task, and we know that the clarification of moral concepts is infinitely more important for man than all the problems of theory" (1937/1979, 497). The "us" referred to the "Vienna group", though presumably not to Carnap, Frank, Hahn, and Neurath, but rather to anyone who would be inclined to stand with Schlick against those who 'ridiculed' philosophy.

So far so good; as we have seen, Schlick tried to distinguish himself from all the other major figures of the Circle by pointing out that their distorted conception of philosophy made a discussion of the relevant traditional questions impossible. The issue is more complex, however. Both Neurath and Carnap considered classical and practical questions to be important but placed them *outside of the philosophical-theoretical domain*. In order to contrast Schlick's conception with theirs and to delimit his version of logical empiricism, we therefore need to devote a few brief paragraphs to the ideas of Neurath and Carnap.

6.2.2 Carnap, Neurath and Traditional Problems

One of the usual objections against logical empiricism was that its philosophy or logic was too restrictive, formal, abstract and impractical; however, everyone who now works on logical empiricism knows that "the icy slopes of logic" (to use the Circle's own phrase, which was recently reintroduced by George Reisch) were not that frosty after all. Nevertheless, the phrase stuck and Carnap adopted it in a letter to Neurath, describing his own role in the movement while also justifying his moves.[12] Although Carnap was well aware of the problematic nexus of values, life, morality and the social implications of thinking, his philosophical projects *per se* consisted in another type of work. He expressed this idea in his correspondence and private writings: he often claimed that he had significant knowledge of classical and traditional questions, but that he was interested in these only as a layman in the context of human affairs.[13]

Even though Carnap, in his *philosophical* writings after the 1930s, did not put much effort into scrutinizing classical or traditional philosophical questions about the nature and philosophy of life, culture, society, values and even epistemology, he

---

[12] Carnap to Neurath, February 10, 1932 (RC 029–12-63).

[13] The closest we get to this idea is in the Preface to *Aufbau* (Carnap 1928), and in his short article on theoretical questions and practical decisions (Carnap 1934). For context and additional details, see, for example, Dewulf (2021).

never abandoned his *interest* in these problems.[14] In *To the Icy Slopes of Logic*, George Reisch (2005) discusses in detail how Carnap intervened in social matters after he arrived in the United States. Nevertheless, none of these interventions, decisions, commitments and activities displayed any clear sign of being *based* on his logical empiricist thinking. Even if there were a strong and internal relation between his theoretical thinking and practical actions, Carnap never made it explicit and never tried to propagate his philosophy by showing its practical force (André Carus (2007) and Christian Damböck (2017) recently took on this task).

Although Carnap apparently endorsed some weak form of the notion that taking the logical empiricist standpoint would also be useful and expedient for the practical matters of everyday life, others went even further (and in the opposite direction, as it were), arguing that practical issues might, in fact, revise theoretical concerns. Neurath (1932/1983, 93) claimed, for example, that Carnap's talk of "primitive protocol language" and "phenomenalist language" which require no further and external justification, his "remarks might induce younger people to search for this protocol language, and this easily leads to metaphysical digressions". Although metaphysics essentially cannot be defeated by arguments, its detractors insisted on physicalism in its most radical version. Furthermore, Neurath also often pressed the point that logical empiricism provided a particularly *democratic* and *pluralist* standpoint in social debates and matters. I do not mean to meticulously reconstruct the quite ambiguous and rambling web of Neurath's beliefs; instead, I will mention only one aspect of interest for the present discussion.[15]

Though Neurath had a quite dialectical picture of theory and practice (Uebel 2000; Cartwright, Cat, Fleck, and Uebel 1996), he never missed a chance to emphasize that actions and decisions outweigh arguments and theories. "However seriously we analyse our case, a 'decision' must be taken, which cannot be substituted by any 'account'" (1941/1983, 216). We always have to make decisions, even – or especially – in argumentative situations. "Debaters on comprehensive scientific problems," Neurath continued, "are in the end like lawyers who have to take a side." We all aim to strengthen our position, sometimes simply by weakening others' position, "but no judge is in the chair."

Yet we nevertheless still use theories and arguments; the question is what type of arguments we choose to utilize. "All doctrines which start with definite and absolute propositions [...] can basic[al]ly support an intolerant attitude."[16] Fallible premises, which encode possibilities, alternatives etc., might at least suggest tolerant actions, even if this is not a logical or practical necessity. These relevant alternatives are documented by what Neurath (1919/1973, 150–151) calls "utopia", by which he meant "all orders of life which exist only in thought and image but not in reality." In

[14] The last stage in Carnap's transformation from classical philosopher to logician of science was his departure, in the mid-1930s, from epistemology. On this phase and its significance, see Uebel (2018).

[15] Sandner (2014) and Howard (2019) give the most detailed account of how political matters intertwined with Neurath's philosophy.

[16] Neurath, "Argumentation and Action," p. 4, ONN.

other words, they describe alternative ways of things and are the "constructions of social engineers." Neurath's social engineer constructs alternative and possible ways of life, and we need to choose our way from among the variously sized packages. Which brings us back to decisions and actions.

As Friedrich Stadler (2001/2015, 282) has noted, Neurath was a 'practical cosmopolitan,' and he was known as such internationally. Many of his works (even the widely read ones) were not philosophical in nature, and his reputation was based on these: his museum directorships in Vienna, and his assistance in establishing further museums in The Hague, Chicago, Mexico and Moscow; his pictorial education work in elementary and secondary schools; his work on tuberculosis in the United States; his organizational efforts in the Viennese settlements movement and later in Bristol; and finally, his role in the socialization of the Bavarian Republic. As he claimed in a letter to Carnap, "I think, if universities should be interested in me, it would be in connection with particular problems of social sciences, unified science or visual education,"[17] and not because of any particular theoretical and *philosophical* achievements (such as holism, fallibilism, or a new theory of protocols).

As will be shown, for Carnap, traditional and practical matters were to be pursued in a *non-academic sphere*, in a non-philosophical manner, and for Neurath they *replaced* academic settings: theory becomes practice, based on decisions, actions and proposals, often about such classical topics as happiness, modes of living, and the nature and scope of democracy. For both Neurath and Carnap, traditional problems and areas of philosophy fell outside of the newly defined territory of philosophy.

6.2.3 Ethics, Values, and Culture: A Comprehensive View of Philosophy

How did Schlick conceive of this notion? I will not go into the content and argumentative structure of Schlick's ethical writings; my goal is philosophically more modest, and what matters here are the contexts and underlying motivations.[18]

Schlick distinguished two ways of doing philosophy: (1) pursuing philosophy as an academic discipline while "contend[ing] on the doctrinal battlefield", and (2) simple philosophizing. Their difference can be characterized, for example, by their results: the first activity yields bits of knowledge, which leads professionals to pursue it in departments and institutes. The second way of philosophizing, however, as

[17] Neurath to Carnap, September 21, 1941 (RC 102-55-18).

[18] Schlick's ethical ideas are discussed in detail by Siegetsleitner (2014) and by Gergely Ambrus in his chapter in this volume. Ferrari (2016) also contrasts Carnap and Schlick, while drawing "a different picture" of Schlick, emphasizing that the cultural and ethical wing of the Circle was as prominent as the theoretical-scientific one. Compare with Mormann (2010), who also argues for an ethical reading of Schlick. Recently, Andreas Vrahimis (2020) has investigated Schlick's more broadly socio-cultural concerns, especially in relation to the challenge of *Lebensphilosophie*.

Schlick (1937/1979, 491; 1994, 3) wrote, yields *wisdom*, and he had a very definite opinion of what this looks like:

> Not everyone is disturbed by the question, whether life has a meaning. Some – and they are not the unhappiest – have the child's mind, which has *not yet* asked about such things; others *no longer* ask, having unlearnt the question. In between are ourselves, the seekers. We cannot project ourselves back to the level of the innocent, whom life has not yet looked at with its dark mysterious eyes […]. (Schlick 1927/1979, 112, original emphases)

Having a fresh, innocent, and young mind is thus a prerequisite in the process of approaching the traditional philosophical questions. On the other hand, being critical, and dealing with criticism in a proper way, is also a characteristic of philosophizing: Schlick even talked about "true philosophy" as the opposite of academic philosophy. It is at least questionable who among the members of the Circle would talk about "true philosophy" at all, but Schlick had no problem with doing so: "I believe," he stated in his planned 1937 Paris lecture, "that we are all entitled to be proud of the fact that our thoughts are the outcome of a long historical development of the human spirit" (Schlick 1937/1979, 495). And as was already noted, Schlick had a deep knowledge of history and philosophy and was able to connect his way of thinking to classical figures and topics: for him, "the true father" of logical positivism was not Frege, Russell, Poincaré or Duhem (as it was for Carnap and Neurath), but Socrates, "the first true philosopher" (1937/1979, 496). Socrates asked questions of meaning and tried to clarify the definitions of notions instead of asking factual questions, engaging in mysticism, or pursuing metaphysics. Yet as is well-known, Socrates was not the most often cited or referenced philosopher among members of the Circle.

Schlick's distinction should not be seen entirely as being about the *content* of philosophy, since he made quite negative and satirizing remarks about institutional philosophy – instead, it can also be viewed from a formal point of view, that is, how, when and for what reason someone pursues philosophy. While *Naturphilosophie* and logical investigations might be important for a smaller audience, ethical and cultural treatises about traditional questions are written for a wider audience. In his 1911 inaugural lecture, Schlick (1911/1979, 114) even claimed that if philosophy was about to give a unified, general and comprehensive scientific view of the world, it would also have to deal with culture, and not just with scientific theories. Though it could be argued that the "unity of science" program aimed to deliver just such a comprehensive structure, questions of life, art and literature did not find their way into the *International Encyclopedia of Unified Science*.[19]

For Schlick, cultural questions, the conditions of human life, etc., "touch upon problems of value which everyone can think about: *tua res agitur* ['it concerns you'], quite unlike specifically scientific questions, which as such really stir only a minority profoundly." While Carnap and Neurath also thought that ethical and value

[19] In the case of the arts, this was a highly contentious issue, though not for philosophical reasons; see Dahms (2019).

questions are important for everyone, they expressed this only as non-philosophers.

How can philosophy help in this regard? In the very same lecture at Rostock, Schlick argued that "the true task of philosophy is at all times the same: its goal consists in the harmonious fulfillment of spiritual life, so far as this is attainable by intellectual means." Besides stating this deep intellectual and spiritual goals, however, Schlick noted that the exact means to achieve this satisfaction of the soul vary, since "the spiritual life of the age [...] determine[s] the means that this mission must employ to approach its goal" (Schlick 1911/1979, 107).

In the early 1930s, the most relevant period of the Vienna Circle for the present purposes, Schlick claimed that philosophy is not a science and does not give us any true statements about the world. On the other hand, philosophy is not even the glue that keeps together the achievements of science, for example by

> adding the most fundamental truths to form the base of the whole structure, and the most universal ones to crown its top. The laying of the foundation is believed to be the work of a certain 'part' of philosophy called Epistemology, and the construction of the towering summit the supreme object of another 'part' of it called Metaphysics. (Schlick, "Philosophy as pursuit of meaning," Schlick Nachlass, A58)

This passage could even be read as very harsh self-criticism. In the above-mentioned 1911 lecture, philosophy Schlick posited two goals with regard to the natural sciences: completion downwards, by "securing the foundations on which the special sciences rest", and completion upwards, by providing worldviews, that is, through the pursuit of metaphysics (Schlick 1911/1979, 109–110). In the 1930s, however, philosophy was no longer a super-science, a distinguished form of knowledge, with a specific and secured epistemic status. Nevertheless, it was still "the alpha and omega of all scientific knowledge", since this particular *activity* aimed to unearth the meaning of all (scientific) statements (Schlick 1930/1979, 157).[20]

If philosophy is a special activity of pursuing meanings, elucidations and clarifications, it would not seem very problematic to consider traditional philosophy to be a legitimate field for the logical empiricists: they merely need to clarify the meanings of its relevant statements and problems. As Schlick (1949, 3) noted in his last courses, "the task of philosophy of nature is thus to interpret the meaning of the propositions of natural science; and therefore the philosophy of nature is not itself a science, but an activity which is directed to the consideration of the meaning of the laws of nature." This could be easily applied to chemistry, biology, psychology, etc., but what happens when we consider other questions, such as ethics, values, life or culture – in other words, traditional philosophy?

Let us first take a look at ethics. While he did not see philosophy as a science, Schlick (1930/1939, 1) did consider ethics to be one: "ethics is a system of *knowledge*, and nothing else; its only goal is the truth. [...] [T]he questions of ethics, too, are purely theoretical problems." Philosophers who seek truth and only the truth can

[20] The most detailed discussion of Schlick's epistemological concerns during the 1930s is to be found in Friedl (2013).

clarify ethical propositions as scientific propositions; pursuing ethics does not require any specific method or cognitive ability. In his late aphorisms, Schlick (1962, 5) claimed that there is no such thing as "geisteswissenschaftliches und naturwissenschaftliches Denken", that is, different modes of thinking for the humanities and the natural sciences, "there is only true and false thinking". Since ethics is a science like physics, the philosophy of ethics can be pursued along the very same lines as the philosophy of physics. When approaching ethics as a philosopher, the only danger is that one might become a moralist or a preacher and not a pure seeker of truth.[21]

Regarding culture, a very similar attitude can be detected, for example in Schlick's 1911 lecture. The aim of philosophy regarding culture is descriptive: "the philosopher can only point to new values and lay bare their connections with life, he cannot newly create them. Philosophy can only bring out valuations already embedded in human nature, and define the necessary appropriate paths of development for them" (Schlick 1911/1979, 116). Philosophers are thus not in a position to create new values or new paths of life out of the blue: they can only point out already existing, but implicit, ambiguous or hardly explorable ways. Philosophy has to "understand" the world and culture; it is an intellectual venture, contrary to art and religion, which are emotionally oriented. In order to engage in this intellectual adventure, scholars have to pursue philosophy in its most distinguished sense, which

> knows no other authority; it must grant nothing that cannot be squared with this, whether it be the opinion of the majority, or even that of the greatest and best. [...] [F]or values can be purposefully and innovatively confronted only if we are not dominated by them, if they constitute no absolute authority, if we are *free* in regard to them, so that we can as it were detach ourselves from them, in order to weigh them. (Schlick 1911/1979, 115, original emphasis)[22]

This value-free picture of philosophy was heavily debated in the Circle. We know that Neurath and Philipp Frank tried to defeat this idea, and Carnap (1963, 23) later recalled that together with Schlick, many members of the Circle tried to neutralize philosophy as a theoretical discipline. But it could be argued as well that Schlick was not that neutral, after all: in his conception of the meaning of life, he quite explicitly stated what he believed, and which type of life deserved to be lived: namely a youthful one.

"Youth" was an important notion for Schlick: one of his last manuscripts, entitled *Philosophy of Youth*, consists of three parts: "Play, the Soul of Youth", "Beauty, the Countenance of Youth", and "Nobility, the Heart of Youth" (see the Schlick Nachlass, A75A). Schlick claimed that the nexus of work and of goal-oriented

[21] Note that this is exactly the opposite of what Neurath (who often spoke about the "professor neutrality" of others) thought about truth-seeking and policy-making. But at one point even Susan Stebbing (1941/1948, x) claimed that "to think in abstractions, when one's concern is moral philosophy, is to fail as a philosopher."

[22] Schlick agreed with Dilthey that there were two reasons behind the broad interest in philosophy around the turn of the century, namely Nietzsche and socialism. By the phrase "opinion of the majority", Schlick meant socialism, and by "the greatest and best" the individualism of Nietzsche.

actions petrifies modern life. Working life is valueless since "the content of existence consists in the work that is needed in order to exist" (Schlick 1927/1979, 114). If we live only to do some work that keeps us alive for further work, then "mere living, pure existence as such, is certainly valueless," and since the only function of work is to realize some purpose and purposes are finite things, work "delimit[s] us" (1962, 7). This does not sound like a neutral philosophical stance.

If purposeful actions delimit us, then we must look for something infinite instead: "Infinite is the play," Schlick (1962, 7) claimed, and he also made clear that play is to be understood as a theoretical, spiritual activity. After listing some technological achievements of the modern era, Schlick commented, in 1911, that "it is often forgotten that all these things are means only, obviously not final ends, for the latter are always spiritual in nature, they lie in the soul of man" (1911/1979, 117).[23]

The pursuit of values and cultural questions is inextricably interwoven with the actual sciences. On the one hand, practical and traditional philosophical questions "are ultimately the most important" ones even for scientists, "since they are everywhere the primary ones" (Schlick 1911/1979, 114). This sounds like a highly dogmatic and cryptic note about why scientists should listen carefully to philosophers, but we can give it some sense by pointing out the following. In the same lecture, Schlick (1911/1979, 105) argued that even if scientists reject philosophy, they are already engaged in philosophy: "In their verdicts of rejection the doubters are thus doing precisely what they have sought to forbid, namely to philosophize". It is thus impossible to get rid of philosophy, at least in the above-mentioned sense of "philosophizing"; institutional philosophy, on the other hand, can be rejected and overcome quite easily.[24]

That being said, according to Schlick (1962, 14), "a thinker, who is *only* a philosopher, cannot be a *big* philosopher". What else does a philosopher need to be? In another aphorism, he (1962, 15) tells us that "who does not understand any of the special sciences and do philosophy like that, is only a dilettante." Besides the quite internal field of philosophy, a wise thinker should also be familiar with the questions and methods of the sciences: since the aim of philosophy is to develop a unified and comprehensive view of the world, starting from only one science would result in a one-sided and thus distorted picture. Hence the entanglement of philosophy and science.

Consequently, Schlick was a very traditional thinker: logical empiricism was never meant to be a restriction of his inquiries, and he engaged himself in all the classical fields, such as philosophy of life, culture and ethics, as well as political

[23] It should be noted, however, that in a 1936 lecture, Schlick (1936/1979a, 513) seemingly accepted a more practical consideration. He said that "[t]he true grounds of man's humility and pride, his joy and sorrow, his value and disvalue, are not to be sought in the theoretical, epistemic constitution of the world, but simply and solely in his mode and manner of utilizing the mastery of his world that has been granted to him."

[24] Philipp Frank (1946) claimed the same idea of rejecting philosophy and worldviews is a form of philosophy and worldview in his *Foundations of Physics* monograph.

philosophy.[25] I do not mean to get into the question of whether he indeed contributed anything original in these fields. In his famous review of Schlick's *Philosophical Papers*, which also functioned as a catalyst for the re-evaluation of logical empiricism in the 1980s, Michael Friedman (1983/1999, 17) argued, for example, that if we consider Schlick's "On the Meaning of Life," it is perhaps no cause for regret that his earlier and later studies on life and culture are not included in the two volumes of his *Philosophical Papers*.

While it can be debated whether Schlick's ideas were naive or uninteresting in the face of the strict logical requirements of analytic philosophy, we should not forget that Schlick was one the thinkers who started to chart the path for analytic and scientific philosophy. Friedman (1983/1999, 17) has expressed this nicely: what we see in Schlick's papers is "[c]haotic, analytic philosophy in the making." As such, what matters here is that Schlick presented one possible line for making logical empiricism less logical, and scientific or analytic philosophy more comprehensive.

A note should be added here about a possible comparison between Schlick and Neurath. Though Schlick was the son of an upper-class factory manager and Neurath that of a university teacher, their cultural and social development took quite different paths. According to Heinrich Neider's recollections,

> Schlick, the soft-spoken tall man, supposed to be of aristocratic parentage, lived in the style of the grand bourgeoisie – he used to ride on horseback before his lectures – loved to travel, gave musical parties and dinners in the style of cultured rich people of the bourgeois era; Neurath, the consciously non-bourgeois revolutionary, disregarded the petit-bourgeois forms in home and dress (mostly bare-headed, he sometimes wore a cap to show his solidarity with the working class). (Neider 1973, 47)

As a result of this rather sharp difference, Schlick never invited Neurath socially, because he "cannot bear [Neurath's] loud voice" (ibid., 48). Not that Neurath would have wanted to visit Schlick.

This cultural difference also manifested itself in how both thinkers envisioned the aim and development of the philosophy of culture and life. Whenever he could, Neurath resorted to concrete items: museums, charts, buildings, reconstructions, practical aids and so on. He even went to cold, dark and underdeveloped regions, filled with dust and coal deposits to see how happiness might actually be achieved for as many people as possible.[26] Theories were important for him, of course, but life started with actions. As he wrote to Bernhard Reichenbach, "[i]nstead of reading concrete facts (FRAZER, MALINOWSKI etc) even the scientifically minded Germans, prefer to read the Generalities of JUNG, FREUD etc. Theories are fine, but you should know the primitive material a little."[27] Philosophy (whatever that might be exactly) should thus start from empirical considerations, from the works

[25] Schlick's most detailed presentation of his views on culture and politics is his posthumous booklet on nature and culture ("Natur und Kultur"), see Schlick (1952). For the only available analysis of this topic, see Schleichert (2003).

[26] On Neurath's town-planning activities, see Hochhäusl (2011) and Nikolow (2004).

[27] Neurath to Bernhard Reichenbach, September 12, 1943 (ONN/289).

of those who engaged in fieldwork and encountered actual people with concrete wishes. Theoretical considerations will naturally affect our actions, but philosophy should not be done *ex-cathedra*.

Schlick, on the other hand, opted for the theories and ideal circumstances that make the intellectual journey into the field of life and practical matters possible. He pursued his ethical research and his own philosophy of life [*Lebensphilosophie*] *after* and *beyond* his 'straggling official work.'[28] In his correspondence, he often indicated that his works were progressing under "the auspicious conditions" available in California, and later in Dubrovnik: "In this paradise, I work only on the philosophy of culture which I find pretty important at the moment. I write these lines [sitting] on a rock above the seas in the delectable fresh morning – it's beauty is beyond expression."[29]

6.3 Von Mises and the Less Logical Positivism

The other example, which I will only sketch briefly, is that of Richard von Mises, especially his 1939 book, *Kleines Lehrbuch des Positivismus: Einführung in die empiristische Wissenschaftsauffassung*, and its revised translation, *Positivism: A Study in Human Understanding* (published in 1951).[30]

Von Mises' book, as he emphasized elsewhere, (1940/1964, 524), is not a simple treatise on positivism, evaluating it from a neutral standpoint; instead, "it is the positivist who speaks, who argues, who describes the world, and, above all, within the world the intellectual efforts of men." This is an important sentence: it expresses the view that scholars make intellectual efforts and have needs, and that these can be achieved from a strict positivist viewpoint.

Von Mises was connected to the Vienna Circle through various personal connections, most importantly with Frank and Neurath. Philosophically, the most important link between Frank, Neurath and von Mises was Ernst Mach's philosophy. For Neurath's *Einheitswissenschaft* series, von Mises (1938/1987) not only wrote a supportive and programmatic pamphlet on the occasion of the centenary of Mach's birth, but he also carried out Mach's program in the most detailed and consistent manner. But what exactly was that program?

In his *Analysis of Sensations*, Mach claimed, as quoted by von Mises (1938/1987, 186), that "I only wish to find a standpoint in physics which one does not have to give up as soon as one wants to look beyond it into the domain of another science."

---

[28] Schlick to Reichenbach, March 18, 1926.

[29] Schlick to Carnap, September 19, 1931 (RC 029-29-16) and April 16, 1935 (MSN).

[30] Unfortunately, this reevaluation of logical empiricism has not yet extended to the philosophy of Richard von Mises. The only exception is the work of Reinhard Siegmund-Schultze (2004, 2006, 2007, 2018) who has published numerous interesting and enlightening papers not just on von Mises' scientific results, but also on his philosophical ideas and context. I will therefore only consider one point in this paper.

Though this comprehensive, unificatory picture was epistemologically driven by Mach, it had a linguistic twist for Neurath (1946/1983, 230), who stated that because of their "Machian tradition", the members of the First Circle "tried to pass from chemistry to biology, from mechanics to sociology without altering the language applied to them." While von Mises also applied this linguistic reading, he emphasized the need and "search for a consistent *unified picture* of the world transcending the limits of the special sciences" (ibid.).

One of the most important elements of this unified picture is the continuity of life and science: von Mises's expressed his more permissive but still unified positivist view of the world through the introduction of the connectability idea. Contrary to Carnap (who claimed that metaphysics is often based on the violation of linguistic rules), von Mises thought (similarly to Neurath) that language is always in the making: though many of its rules are explicit and obvious, many important restrictions and customs are either implicit, or continuously change in the face of new experiences. As von Mises (1951, 72) noted, "[w]e do not think it possible to anticipate the rules of language in any exhaustive manner before knowing the sentences that will have to be tested by them in order to decide their admissibility or inadmissibility."

Therefore, von Mises proposed another criterion of meaningfulness and thus a different criticism of metaphysics: instead of speaking about syntax, verification or any of the usual suspects, he talked about "connectability".[31] Sentences are connectible if they are in agreement or are compatible with the apparent stock of linguistic rules. The sequence of words, "On the 32nd of January" is not connectible since the word "January" is regulated as having only 31 days. This is seemingly a weaker criterion than Carnap's, and it also has slightly different consequences: according to von Mises, many of the classical metaphysical sentences that Carnap declared to be meaningless (e.g. Heidegger's and Hegel's statements), "cannot be dismissed by simply pointing out that, according to the rules of ordinary language, one cannot assign a predicate to the subject 'nothingness'" (von Mises 1951, 74). If this usage on the part of metaphysicians complies with the assumed rules of language, then their sentences are connectible: they only have to make them explicit.[32]

And von Mises' draws his conclusions from the above considerations: "If one wants to reject theories because they are not connectible with a certain class of linguistic rules, then this is an arbitrary decision or a judgement of value which cannot be justified logically." Consequently, the Vienna Circle's "absolutist distinction" between meaningful and meaningless statements 'is replaced by a more relative one' (von Mises 1940/1964). Von Mises thinks that metaphysics aims to give a picture of the world and is hence a form of scientific activity, though more dogmatic

[31] The idea of "connectibility" is quite similar to Neurath's notion of "isolated statements", which are statements that are not related at a certain point to the body of our accepted scientific statements. They hence seem to be metaphysical, while the possibility is always kept open that, at a later point, they could be connected.

[32] It is a topic for further research whether (contrary to the "Überwindung") Carnap's later principle of tolerance and his framework relativism in "Empiricism, Semantics and Ontology" might be related to von Mises's idea.

than the positivist one, and that the rules for its language usage are more restrictive and narrow, often confined only to the particular given work. Though he regards it as problematic, his theory of language actually permits metaphysics.

Nevertheless, von Mises rejects a wider domain of activities, which he calls "negativism" in contrast to "positivism." By positivism, he means someone "*who, when confronted by any problem in life, reacts in the manner in which a typical contemporary scientist deals with his problems of research*" (von Mises 1953/1964, 539, original emphasis). This vague characterization amounts to the idea that all the problems we face both in science and life should be approached from the same experiential base, using the tools of reason. Negativists are those who try to restrict the role and usage of reason, aiming for a special synthesis by means of extra-intellectual means.[33] Negativism does not group together all the usual suspects: neither skepticism nor believers are included in the negativist camp; von Mises seems to be permissive in this regard as well.

This "permissivism" also extended to the domains of art and poetry. Von Mises (1940/1964, 528) claimed that "every artist provides in his work a sort of theory of the subjects he deals with," – that is, a certain proto-theory about the world and people's place within it – "the work of art is a hint, an allusion to an experience, just as a scientific theory." Though all reviewers of von Mises's book criticized this idea, it clearly indicates his intention: he wished to put forward a general, and, especially, a unified picture of the world that we would not have to give up as soon as we move from physics and applied mathematics to the social sciences and the arts.

There are many interesting things to be discussed regarding von Mises, but what we have before us here is *another possible route that could be taken in order to make positivism a more appealing movement*. Von Mises' approach focused less on traditional questions; he mainly touched upon the meta-level of traditional problems and approached them from a birds-eye view. Contrary to Neurath, both von Mises and Schlick agreed on the intellectualization of these fields, but they found their ancestors in different thinkers: Schlick in Socrates and his method of relating morals to theoretical knowledge, and von Mises in the positivists, mainly in Mach. As a result, while Schlick called his position 'Consistent Empiricism' (1936/1979b, 459), von Mises's stance might be described as 'Consistent Positivism'.

One of the reviewers, Morton White, described the approach of von Mises as follows, while also summarizing the intentions of the paper:

> Now von Mises is a representative of the less formalistic wing of logical positivism [...]. Like Comte, von Mises defends a positivism which is large and loose-fitting, covering as it does almost all corners of intellectual activity. [...] His brand of philosophy should be attractive to many who deplore (wrongly, I think) what they call the 'thinness' of Carnap's logistic and the linguistic subtleties of the later Wittgenstein. Von Mises' positivism *does* have a certain solidity, richness, and three-dimensional quality which the others lack because he addresses himself to 'Big Questions' surrounding the social sciences, religion, poetry, art, ethics and the law. (White 1952, 155, original emphasis)

[33] What von Mises calls the "positivist" and the "negativist" conception are quite similar to what Hans Hahn (1930/1980) calls "world-*affirming*" and "world-*denying*" philosophies.

One more note should be added: in 1946, Willem Frederik Zuurdeeg, a theology and philosophy student, published his dissertation, titled "A Research for the Consequences of the Vienna Circle Philosophy for Ethics." In this, unfortunately much neglected, book, he analyzed three members of the Circle, namely Neurath, Schlick and von Mises; as Zuurdeeg (1946, 22–23) admitted, he chose these figures mainly for contingent historical reasons, since only their general works had appeared before the Second World War and could thus not be accessed in the Netherlands. As this paper has shown, Zuurdeg was, however, also right for theoretical reasons: Neurath, Schlick and von Mises presented a general, comprehensive and unified picture of 'less logical positivistic thinking.'

Acknowledgments This paper was supported by the MTA BTK Lendület *Morals and Science* Research Group, the MTA Lendület Values and Science Research Group, and by the MTA Premium Postdoctoral research scholarship. I am indebted to the Carnap Archive at Pittsburgh (Rudolf Carnap Papers, 1905-1970, ASP.1974.01, Special Collections Department, University of Pittsburgh), and to the Wiener Kreis Archiv (Rijksarchief in Noord-Holland, Haarlem, The Netherlands) for the permission to quote the archive materials as "MSN" for the Moritz Schlick Nachlass and "ONN" for the Otto Neurath Nachlass (all rights reserved). The translations from the archive files are mine. I am indebted to Christian Damböck, Massimo Ferrari, Johannes Friedl, Martin Lemke and Reinhard Siegmund-Schultze for their comments on the first version of this paper.

References

Allen, Devere. 1933. The World's Stake in Austria. *The North American Review* 236 (3): 213–237.

Burke, Christopher; Adam Tamas Tuboly (ms.) *Otto Neurath in Britain: Speaking Broken English Fluently*. Unpublished manuscript.

Carnap, Rudolf. 1928. *Der logische Aufbau der Welt*. Berlin: Weltkreis.

———. 1934. Theoretische Fragen und praktische Entscheidungen. *Natur und Geist* 2 (9): 257–260.

Carnap, Rudolf, Hans Hahn, and Otto Neurath. 1929/1973. Scientific Conception of the World: The Vienna Circle. In *Empiricism and Sociology*, ed. M. Neurath and R.S. Cohen, 299–318. Dordrecht: Reidel.

Cartwright, Nancy, Jordi Cat, Lola Fleck, and Thomas Uebel. 1996. *Otto Neurath: Philosophy Between Science and Politics*. Cambridge: Cambridge University Press.

Carus, André. 2007. *Carnap and Twentieth-Century Thought*. Cambridge: Cambridge University Press.

Cat, Jordi, and Adam Tamas Tuboly, eds. 2019. *Neurath Reconsidered: New Sources and Perspectives*. Cham: Springer.

Dahms, Hans-Joachim. 2019. Logical Empiricism and Art: The Correspondence Otto Neurath/Meyer Shapiro. In: Cat and Tuboly (2019): 471–88.

Damböck, Christian. 2017. *»Deutscher Empirismus« Studien zur Philosophie im deutschsprachigen Raum 1830–1930*. Dordrecht: Springer.

Dewulf, Fons. 2021. Carnap's Opposition to Logic of the Giesteswissenschaften. In *Young Carnap in an Historical Context: 1918–1935*, ed. Ch. Damböck and G. Wolters, 55–73. Cham: Springer.

Feigl, Herbert. 1969/1981. The *Wiener Kreis* in America. In *Herbert Feigl: Inquiries and Provocations: Selected Writings 1929–1974*, ed. R.S. Cohen, 57–94. Dordrecht: D. Reidel.

Ferrari, Massimo. 2016. Wiener Gefühle. Fragen der Ethik zwischen Schlick und Carnap. In *Wissenschaft und Praxis. Zur Wissenschaftsphilosophie in Frankreich und Österreich in der ersten Hälfte des 20. Jahrhunderts*, ed. Ch. Bonnet and E. Nemeth, 5–28. Dordrecht: Springer.

———. 2018. Nature und Freiheit: Zum Kant-Verständnis des Wiener Kreises. In *Akten des XII. Internationalen Kant-Kongresses*, ed. V.L. Waibel, M. Ruffing, and D. Wagner, 377–396. Berlin/Boston: de Gruyter.

Frank, Philipp. 1946. *Foundations of Physics*. Chicago: University of Chicago Press.

Friedl, Johannes. 2013. *Konsequenter Empirismus. Die Entwicklung von Moritz Schlicks Erkenntnistheorie im Wiener Kreis*. Wien/New York: Springer.

Friedman, Michael. 1983/1999. Moritz Schlick's *Philosophical Papers*. In *Positivism*, ed. Reconsidering Logical, 17–34. Cambridge: Cambridge University Press.

Hahn, Hans. 1930/1980. Superfluous Entities, or Occam's Razor. In *Hans Hahn: Empiricism, Logic, and Mathematics*, ed. B. McGuinness, 1–19. Dordrecht: D. Reidel.

Hochhäusl, Sophie. 2011. *Otto Neurath – City Planning: Proposing a Socio-political Map for Modern Urbanism*. Innsbruck: Innsbruck University Press.

Howard, Don. 2019. Otto Neurath: The Philosopher in the Cave. In: Cat and Tuboly (2019): 45–65.

Iven, Mathias. 2015. Er 'ist eine Künstlernatur von hinreissender Genialität': Die Korrespondenz zwischen Ludwig Wittgenstein und Moritz Schlick sowie ausgewählte Briefe von und an Friedrich Waismann, Rudolf Carnap, Frank P. Ramsey, Ludwig Hänsel und Margaret Stonborough. *Wittgenstein-Studien* 2015 (6): 83–174.

Menger, Karl. 1994. *Reminiscences of the Vienna Circle and the Mathematical Colloquium*. Dordrecht: Kluwer.

Mormann, Thomas. 2010. Zwischen Weisheit und Wissenschaft. Schlicks weites philosophisches Spektrum. *Grazer Philosophische Studien* 80: 263–285.

Mulder, Henk. 1968. Wissenschaftliche Weltauffassung. Der Wiener Kreis. *Journal of the History of Philosophy* 6: 386–390.

Neider, Heinrich. 1973. On Otto Neurath. In *Empiricism and Sociology*, ed. M. Neurath and R.S. Cohen, 45–49. Dordrecht: D. Reidel.

Neurath, Otto. 1919/1973. Utopia as a Social Engineer's Construction. In *Empiricism and Sociology*, ed. M. Neurath and R.S. Cohen, 150–155. Dordrecht: D. Reidel.

———. 1932/1983. Protocol Statements. In *Otto Neurath: Philosophical Papers 1913–1946, ed. R.S. Cohen and M. Neurath, 91–99. Dordrecht: D. Reidel.*

———. 1941/1983. Universal Jargon and Terminology. In *Otto Neurath: Philosophical Papers 1913–1946*, ed. R.S. Cohen and M. Neurath, 213–229. Dordrecht: D. Reidel.

———. 1946/1983. The Orchestration of the Sciences by the Encyclopedism of Logical Empiricism. In *Otto Neurath: Philosophical Papers 1913-1946*, ed. R.S. Cohen and M. Neurath, 230–242. Dordrecht: D. Reidel.

Nikolow, Sybilla. 2004. Planning, Democratization and Popularization with ISOTYPE, ca. 1945: A Study of Otto Neurath's Pictorial Statistics with the Example of Bilston, England. In *Induction and Deduction in the Sciences*, ed. F. Stadler, 299–329. Dordrecht: Kluwer.

Reichenbach, Hans. 1930. Zur Einführung. *Erkenntnis* 1: 1–3.

———. 1931/1978. Aims and Methods of Modern Philosophy of Nature. In *Hans Reichenbach: Selected Writings: 1909–1953*, ed. M. Reichenbach and R.S. Cohen, vol. 1, 359–388. Dordrecht: D. Reidel.

Reisch, George. 1997. Epistemologist, Economist ... and Censor? On Otto Neurath's Infamous Index Verborum Prohibitorum. In *Perspectives on Science*, vol. 5, 452–480.

———. 2005. *How Cold War Transformed Philosophy of Science: To the Icy Slopes of Logic*. New York: Cambridge University Press.

Sandner, Günther. 2014. *Otto Neurath. Eine politische Biographie*. Wien: Zsolnay.

Schleichert, Hubert. 2003. Moritz Schlick's Idea of Non-Territorial States. In *The Vienna Circle and Logical Empiricism: Re-evaluation and Future Perspectives*, ed. F. Stadler, 49–61. Kluwer, Dordrecht.

Schlick, Moritz. 1911/1979. The Present Task of Philosophy. In *Moritz Schlick: Philosophical Papers: Volume One (1909–1922)*, ed. H.L. Mulder and B.F.B. van de Velde-Schlik, 104–118. Dordrecht: D. Reidel.

———. 1927/1979. On the Meaning of Life. In *Moritz Schlick: Philosophical Papers: Volume Two (1925–1936)*, ed. H.L. Mulder and B.F.B. van de Velde-Schlik, 112–129. Dordrecht: D. Reidel.

———. 1930/1939. *Problems of Ethics*. New York: Prentice-Hall, Inc.

———. 1930/1979. The Turning-Point in Philosophy. In *Moritz Schlick: Philosophical Papers: Volume Two (1925–1936)*, ed. H.L. Mulder and B.F.B. van de Velde-Schlick, 154–160. Dordrecht, D. Reidel.

———. 1936/1979a. Meaning and Verification. In *Moritz Schlick: Philosophical Papers: Volume Two (1925–1936)*, ed. H.L. Mulder and B.F.B. van de Velde-Schlick, 456–481. Dordrecht: D. Reidel.

———. 1936/1979b. The Universe and the Human Mind. In *Moritz Schlick: Philosophical Papers: Volume Two (1925–1936)*, ed. H.L. Mulder and B.F.B. van de Velde-Schlick, 499–513. Dordrecht: D. Reidel.

———. 1937/1979. The Vienna School and Traditional Philosophy. In *Moritz Schlick: Philosophical Papers: Volume Two (1925–1936)*, ed. H.L. Mulder and B.F.B. van de Velde-Schlick, 491–498. Dordrecht: D. Reidel.

———. 1949. *Philosophy of Nature*. New York: Philosophical Library.

———. 1952. *Natur und Kultur. Aus den Nachlaß des Autors, herausgegeben von Josef Rauscher*. Wien/Stuttgart: Humboldt-Verlag.

———. 1962. *Aphorismen*. Wien: Im Selbstverlag des Herausgebers (Blanche Hardy Schlick).

———. 1994. The Main Ideas of the Theory of Values. In *Ethics and the Will*, ed. B. McGuinness and J. Schulte, 1–6. Dordrecht: Springer.

———. (2013). *Nietzsche und Schopenhauer (Vorlesungen)*. Band 5.1 der Moritz Schlick Gesamtausgabe. Edited by Mathias Iven. Wien/New York: Springer.

Siegetsleitner, Anne. 2014. Ethik und Moral im Wiener Kreis. *Zur Geschichte eines engagierten Humanismus*. Wien/Köln/Weimar: Böhlau Verlag.

Siegmund-Schultze, Reinhard. 2004. A non-conformist longing for unity in the fractures of modernity: towards a scientific biography of Richard von Mises (1883–1953). *Science in Context* 17 (3): 333–370.

———. 2006. Probability in 1919/20: The von Mises-Polya-controversy. *Archive for History of Exact Sciences* 60 (5): 431–515.

———. 2007. Philipp Frank, Richard von Mises, and the Frank-Mises. *Physics in Perspective* 9 (1): 26–57.

———. 2018. Applied Mathematics versus Fluid Dynamics: The Catalytic Role of Richard von Mises (1883–1953). *Historical Studies in the Natural Sciences* 48 (4): 475–525.

Stadler, Friedrich. 2001/2015. *The Vienna Circle*. 2nd ed. Dordrecht: Springer.

Stebbing, Susan. 1941/1948. *Ideals and Illusions*. London: Wattson & Co.

Tuboly, Adam Tamas. 2021. Understanding Metaphysics and Understanding Through Metaphysics: Philipp Frank on Scientific Theories and Their Domestication. In *Logical Empiricism and the Natural Sciences: From Philosophy of Nature to Physics*, ed. S. Lutz and A.T. Tuboly, 401–421. London/New York: Routledge.

Uebel, Thomas. 2000. *Vernunfkritik und Wissenschaft: Otto Neurath und der erste Wiener Kreis*. Vienna/New York: Springer.

———. 2018. Carnap's Transformation of Epistemology and the Development of His Metaphilosophy. *The Monist* 101 (4): 367–387.

von Mises, Richard. 1938/1987. Ernst Mach and the Scientific Conception of the World. In *Unified Science: The Vienna Circle Monograph Series originally edited by Otto Neurath, now in an English edition*, ed. B. McGuinness, 166–190. Dordrecht: D. Reidel.

———. 1940/1964. Scientific Conception of the World: On a new Textbook of Positivism. In *Selected Papers of Richard von Mises: Volume Two: Probability and Statistics, General*, ed. Ph. Frank et al., 524–529. Providence: American Mathematical Society.

———. 1951. *Positivism: A Study in Human Understanding*. Cambridge, MA: Harvard University Press.

———. 1953/1964. The Role of Positivism in the XX Century. In *Selected Papers of Richard von Mises: Volume Two: Probability and Statistics, General*, ed. Ph. Frank et al., 537–547. Providence: American Mathematical Society.

Vrahimis, Andreas. 2020. "The Vienna Circle's Responses to *Lebensphilosophie*" In: *Logique et Analyse* 253: 43–66.

White, Morton. 1952. Positivism Less Logical [Review of von Mises 1951]. *The Kenyon Review* 14 (1): 152–157.

Zuurdeeg, Willem F. 1946. *A Research for the Consequences of the Vienna Circle Philosophy for Ethics*. Utrecht: Kemink en Zoon N.V.

Chapter 7
Leo Apostel and Rudolf Carnap: The Development of Logical Empiricist Ethics in Post-War Europe

Fons Dewulf

Abstract In this article I will discuss two twentieth century philosophers, Rudolf Carnap and Leo Apostel, who aimed to remove ethics from philosophy. Carnap's ideas about value judgments amount to a well-known, classical account of non-cognitivism in twentieth century meta-ethics. Although Apostel's ideas are unknown, I will show that they are a direct result of his interaction with Rudolf Carnap and develop Carnap's position in an interesting way. Uncovering Apostel's attempt to develop Carnap's views on ethics is not only interesting because it sheds new light on Carnap's position, but also because it sheds light on the politically activist reception of logical empiricist philosophy in the post-war period. Contrary to what most commentators have argued, logical empiricism did not entirely lose all of its political potential after the second World War. In Belgium, through the activity of Leo Apostel, it re-acquired some of its interbellum political agenda.

Keywords Carnap · Apostel · Scientific Ethics · Non-cognitivism · Values

7.1 Introduction

How can philosophy as a discourse help to distinguish what is right from what is wrong to pursue in life? In a time of the crisis of climate, migration and the increasing popularity of nationalist politics the urge to produce philosophical aid to the seemingly insurmountable social conflicts that beset society seems pressing. It requires audacity to resist this urge, to resign from philosophy – one of the most powerful traditional tools of rationality – the responsibility of ethical guidance. In

F. Dewulf (✉)
University of Ghent, Ghent, Belgium
e-mail: Fons.Dewulf@ugent.be

C. Damböck, A. T. Tuboly (eds.), *The Socio-Ethical Dimension of Knowledge*, Vienna Circle Institute Yearbook 26, https://doi.org/10.1007/978-3-030-80363-6_7

this article I will discuss two twentieth century philosophers who resisted this urge, Rudolf Carnap and Leo Apostel. The former's ideas about value judgments amount to a well-known, classical account of non-cognitivism in twentieth century meta-ethics. The latter's ideas on the relation between philosophy and ethics are virtually unknown, although, as I will show below, they are a direct result of his interaction with Rudolf Carnap and arguably develop Carnap's position in an interesting way. Apostel's attempt to develop Carnap's views on ethics has until now remained hidden from sight. Uncovering the attempt is not only interesting because it sheds new light on Carnap's ethical position, but also because it sheds light on the politically activist reception of logical empiricist philosophy in the post-war period. Contrary to what most commentators have argued (Reisch 2005; Howard 2003), logical empiricism did not entirely lose all of its political potential after the second World War. It did not entirely die out in Europe to become a stale, politically neutral logic of science in the United States. In Belgium, through the activity of Leo Apostel, it re-acquired some of its interbellum political agenda.

7.2 Leo Apostel Visits Rudolf Carnap

Leo Apostel (1925–1995) is an unknown, but intriguing figure in twentieth century philosophy. He grew up in Antwerp and entered the philosophy program of the Free University of Brussels in 1945, right after the liberation of Belgium. During his philosophy education Apostel had classes with the Comtian positivist Eugène Dupréel, but he was influenced the most by a young professor at Brussels, Chaïm Perelman, who supervised Apostel's dissertation on introspection in 1949 (Apostel 1989, 13). Before the Second World War, Perelman had been the most important local contact of Carl Hempel, Paul Oppenheim and Kurt Grelling, when they were collaborating privately on logical issues in Brussels in the period between 1934–1939. Perelman invited Grelling and Oppenheim to meetings of the philosophy department in Brussels, and reviewed Oppenheim and Hempel's book, *Der Typusbegriff*, which was written during their Brussels collaboration.[1] Unsurprisingly, Perelman introduced Apostel to logical empiricist philosophy, and especially the work of Rudolf Carnap. Apostel read *Der logische Aufbau der Welt* in 1946, which he immediately thought was a masterpiece and which, according to his own account afterwards, "enflamed his passion for Carnap" (Apostel 1986/2011, 480). In his autobiography Apostel writes that in those years he conceived himself as a logical empiricist (Apostel 1989, 15). It is clear that he understood logical empiricism at the time from a continental point of view: Apostel read the *Aufbau* within the context of German neokantian philosophy, relating the work to Husserl, Dilthey and Cassirer.

[1] Grelling mentions Perelman's work on their book to Oppenheim in a letter from 2 March 1940, Carl Hempel Papers (CH), Box 44, folder 2, Archives of Scientific Philosophy (henceforth ASP), Special Collections, Hillman Library, University of Pittsburgh, Pittsburgh. Quoted by permission of the University of Pittsburgh. All rights reserved.

He also interpreted logical empiricism as a total attempt to destroy traditional philosophy by erecting a new one (Apostel 1989, 15). Apostel thus understood the activist potential behind the logical empiricist program early on: one can overcome ideological conflict by abandoning traditional metaphysics and adhering to the neutral logical investigation of science. In 1949 Apostel started working as the research assistant of Perelman, while writing a PhD dissertation on the concept of causality within logical empiricist philosophy. In 1951 Apostel was granted funds from the B.A.E.F. (Belgian American Education Foundation) to follow the courses of Carnap at Chicago University during the academic year 1951–1952. Apostel visited Carnap in Chicago during the fall term of 1951, and he would later go to Yale to pursue courses with Carl Hempel (Apostel 1989, 24–27). Apostel's contact with Carnap during this period had a great impact on him. In his autobiography he writes:

> When I met Carnap, I felt he was everything I had imagined he should be: he was deep, clear, considerate, modest, giving all his attention to his students and his work, explaining with care the difficult things he was doing. (Apostel 1989, 24)

In a lecture on Carnap's philosophy which Apostel gave in 1986, Apostel even notes that "he loves Rudolf Carnap."[2]

> It is difficult to make you love Carnap as much as I do, and to produce in you as much adoration for him as I have. Still, this is the hidden purpose of my lecture. (Apostel 1986/2011, 466)

According to Apostel's notes taken during his Chicago visit and also according to his testimony in the autobiography, Apostel and Carnap discussed the notions of causality and physical necessity at length, entertaining the possibility to express necessity through modal operators (Apostel 1989, 25).[3] However, in his 1986 lecture on Carnap, 35 years after the interaction, Apostel does not focus on these discussions, but talks about his interaction with Carnap concerning ethics.

> When I became more acquainted with Carnap, I was allowed to visit him for an entire afternoon on my own every two weeks. Sometimes he also gathered people in the evening – some sort of parties – where we could talk more freely. I once asked Carnap "Why are your views on ethics apparently so underdeveloped? After all, in everything you do, I sense a very clear ethical and political motivation." This was impossible to miss: Carnap was a socialist and a radical pacifist. […] However, his ethical and political stance was never mentioned in his courses. He upheld a strict distinction between his coursework and his ethical motivation. That is the reason why I asked this question. He replied: "You should consider me as a member of a movement. All my friends that formed this movement with me, each have a part to play. The person who informed me of ethical and political problems

[2]The lecture was the first in a series of lectures for the alumni of the Ghent philosophy program, in which retired professors would talk about their greatest philosophical influences. Apostel worked diligently to prepare the lecture. In his archives there are multiple preparatory versions in manuscript form (Archief Leo Apostel (henceforth ALA), Universiteit Gent, folder 293). In the remainder of the paper I refer to a version of the lecture that was recently edited and published by Ronald Commers in (Commers 2011). This version is based on a transcript of an audio recording of the event. All the translations from the original Dutch are my own.

[3]Apostel typed down some of his conversations with Carnap on physical necessity, see ALA, folder 281.

> was Otto Neurath. Ever since I felt how deeply I agreed with him in all ethical and political contexts I stopped dealing with those issues. I trusted he would do so. However, I know that I have not done enough on that subject. I have defended that ethical judgements have no cognitive meaning, but this position is underdeveloped, one could say much more. You seem to be interested in ethical issues. Would you not like to continue your PhD under my supervision focussing on ethics? Stevenson's *Ethics and Language* contains too few insights in the structure of languages, in semantics and syntax. Don't you feel like working on those issues?" (Apostel 1986/2011, 468)

Both in his autobiography and in the lecture, Apostel says that he declined Carnap's offer to continue his PhD in Chicago on ethics. However, he mentions two different reasons. In the autobiography, it is Apostel's own weak psychological disposition which is invoked to explain the refusal to continue work with Carnap.

> I knew my weakness and was not sure whether I would be able to endure this stressful atmosphere. Moreover Carnap overwhelmed me. I had to admire him from afar. My relation to him in a certain sense could have become self-destructive. His presence, his greatness crushed me. (Apostel 1989, 26)

In the lecture, Apostel invokes disappointment about the fact that Carnap did not find Apostel's work on causality interesting enough and instead wanted him to work on ethics.

> And now, Carnap told me that I could best collaborate in the movement by working on ethics. [...] I did not do this, because I could not withdraw myself from my fascination with causality. [...]
>
> But, I should have done what Carnap advised me to do. He was right. (Apostel 1986/2011, 468)

It is difficult to assess from the available sources whether Carnap really offered Apostel the opportunity to continue his PhD in Chicago, and also whether Carnap actually wanted Apostel to work on ethics. There are no sources contemporary to the events which can confirm this. In a letter, dated 21 February 1952, Ina Carnap writes to Carl Hempel about Apostel, as someone "to whom Carnap regularly gave extra time" and whom "Carnap thinks is very intelligent, maybe a little on the slow side – but that may be due to the language".[4] Carnap clearly considered Apostel an interesting student, potentially also a unique student since Apostel understood Carnap's work within its European context. Whatever offer Carnap made or did not make to Apostel, the most important fact is that Apostel understood Carnap's philosophy as part of a more general movement which aimed to change society through philosophy, and Apostel deeply desired to be part of it.[5]

In his 1986 lecture he talks at length about the logical empiricist movement as an active engagement "to bring science into society, so that the people can take destiny in their own hands" (Apostel 1986/2011, 470). To exemplify this, he discusses Neurath's ISOTYPE project, situates logical empiricism in the context of Red

[4] Ina Carnap to Carl Hempel, 21 February 1952, CH, Box 11, folder 1, ASP.

[5] Ronald Commers has already discussed Apostel's ideas on ethics within the context of interbellum philosophy, including John Dewey's pragmatism and the Vienna Circle's scientific World Conception (Commers 2010, vol. 2, Ch. 2).

Vienna, goes over the parallel developments in the Bauhaus school and defends the claim that the distinction between analytic and synthetic judgments is the "socio-political expression of the option to develop a philosophy which can have a politically transformative function" (Apostel 1986/2011, 472).[6] The discussion of the logical empiricist movement in the lecture ends in existential terms:

> This [political project] was the movement. It was one enormous fight, outside and inside. [...] You probably noticed, and I cannot deny it: my deepest sympathies and friendship go out to this movement. I am it and I want to be it. That is philosophy. I do not say this in a non-committal way. (Apostel 1986/2011, 475)

Apostel deeply sympathized with Rudolf Carnap's ethos in particular, claiming that Carnap was not in the least focused on dominating others through his philosophical discourse. Unlike most other philosophers who desired the power to claim what is right or wrong, Carnap, in Apostel's evaluation, was set on liberating others, making them choose between a variety of equally possible frameworks. Apostel characterized this ethos with a line from Nietzsche's poem *Der Einsame* in *The Gay Science*: "Verhasst ist mir das Folgen und das Führen – nothing better exemplifies Carnap" (Apostel 1986/2011, 483).

Whatever the actual relationship was between Carnap and Apostel, Apostel understood himself as a member of the logical empiricist movement which aimed to transform society. To Apostel this movement was not an extinct project; it was alive and he would become an exponent in its European rebirth.[7] After his return to Belgium in the summer of 1952, Apostel finished and subsequently defended his doctoral dissertation *Law and Causes* (*La Loi et les Causes*). Between 1953 and 1955 he first became a post-doctoral assistant to Perelman, and then he joined the renowned psychologist Jean Piaget in Geneva between 1955 and 1956 as a philosophical collaborator. During this post-doctoral period, Carnap encouraged Apostel to write to him:

> You need not feel that you can only write to me when you are able to produce a masterpiece of scientific accomplishment. We all try to work on something, and sometimes we succeed quickly, sometimes it takes longer. [...] We shall always be glad to hear from you, how you are, how you live, what your position at the University is – in short, the human aspects of your life as well as your theoretical work! (Carnap to Apostel, 17 July 1954, Rudolf Carnap Papers (RC), box 88, folder 1, ASP)

---

[6] Apostel's lecture was never translated to English, even though its interpretation of Carnap's philosophy resonates closely with the current state-of-the-art research on Carnap. E.g. Apostel discusses the relation between the *Aufbau* and Kant, Dilthey, Husserl and even Hans Driesch (Apostel 1986/2011, 481–483).

[7] In the 1960s Apostel wrote an extensive course on Logical Empiricism for his students at Ghent University, including a history of the movement, discussing the migration to the United States. He noted that after the second World War logical empiricism had almost grown extinct in Europe. He also remarked that in its American context the movement lacked "a scientific ethics", although certain strands of such an ethics could be found in American pragmatist thinkers like Dewey. Apostel's course was written to revive logical empiricism and its ethical agenda for his Belgian students. See, "Course on Logical Empiricsm", p. 38, ALA, folder 176.

Their correspondence, which is kept in the Carnap papers at University of Pittsburgh, shows that Apostel kept Carnap updated about his various intellectual developments and research projects. He wrote to Carnap about an empirical survey on scientists' ideas about science which he had developed with Piaget. Apostel equally updated Carnap about his advances in formalizing necessity and lawfulness.[8] In 1956 Carnap supported a permanent appointment of Apostel as professor in Logic at Ghent University through a letter of recommendation.[9] Because Apostel was an atheist, his appointment at Ghent was boycotted by the Catholic professors of the Faculty of Arts & Philosophy, which resulted in several months of delay. In April 1957, however, Apostel became professor in logic at Ghent University. By then he had already established a wide-ranging network of contacts, bridging disciplines (philosophy and psychology) and regions (Europa and the United States). He had also co-founded the journal *Logique & Analyse* and published a wide-ranging set of papers.[10] In European philosophy Apostel had acquired a good reputation. This can be seen from his participation in the infamous 1957 conference on analytic philosophy in Royaumont which brought together the most important philosophers from the English (including Quine, Strawson and Austin) and the French speaking world. During the conference Apostel attempted to bridge the gap between ordinary language philosophy and European structuralist, comparative semantics – taking up a typical logical empiricist point of view, that any philosophical analysis of empirical language should start from the latest scientific insights in the structure of empirical languages (Apostel 1962).

In the academic year 1958–1959, only one year after his appointment at Ghent, Apostel accepted an invitation by Henry Johnstone to lecture at Penn State University as a visiting professor (Apostel 1989, 55). At first, Apostel saw the invitation as an opportunity to continue his philosophical career in the United States: he also alerted Carnap to the possibility of meeting each other again.[11] However, during this year, something changed in Apostel's perception of himself. He did not work out a way to continue his career in the US, explicitly avoiding the APA meetings, and he did not find a way to visit Carnap in Los Angeles. Instead, he decided to forego an international career altogether (Apostel 1989, 61).

> I realised that in the U.S.A. an academic was only an isolated specialist (in the best cases). I wanted more: I wanted to live and propagate "the philosophical life" and this required that I go back to a country where I could be more than only a writer or a pedagogue but also – as I hoped in my naive enthusiasm – an agent in the transformation of a culture. (Apostel 1989, 61-62)

In line with Apostel's own understanding of the logical empiricist ethos, he did not want to become a disciplinary "school philosopher", engaged in closed-off

[8] Apostel to Carnap, undated, probably 1956, RC, box 27, folder 1, document 1, ASP.

[9] Apostel to Carnap, 25 December 1956, RC box 27, folder 1, document 3, ASP, and Faculteitsraadverslag, 25 April 1956, Archief Faculteit LW, 6A1_09, Universiteitsarchief Gent.

[10] A full bibliographical list can be found at http://www.leoapostel.ugent.be/bibliografie/index.html

[11] Apostel to Carnap, 20 September 1958, RC box 27, folder 1, document 2, ASP.

scholarly debates. His goal was to transform the culture in which he lived. Upon his return to Belgium in August 1959, he was immediately granted with an opportunity to act on his intentions.

On 29 May 1959, the Belgian Parliament voted positively on a major reform of the entire education system. Among many other changes, the reform now installed the official possibility for non-religious students in public secondary schools to pursue a course in "non-confessional morals". According to the Belgian constitution every student is obliged to pursue two hours of moral training every week. Until 1959, however, students who did not want to pursue religious (mainly catholic) education had no proper alternative to religious courses. The law of 1959 was an important progress for liberal pressure groups in Belgian society to introduce non-Catholic moral training into public education. The reform, however, did not foresee any university training program for the teachers of the non-confessional course. So, in reality, this implied that the new course would be taught by non-specialized teachers – a sub-optimal situation according to many liberal pressure groups. This situation resulted in increasing political pressure on the Ministry of Education to create a new University program that aimed to train teachers in non-confessional morals. Apostel, who was an important member of the Belgian Freemasons and was acutely aware of the political situation, seized this opportunity to serve his own agenda: he aimed to secure funds to think through Carnap's views on ethics and execute a full-fledged scientific research program that would transform traditional philosophical ethics into a scientific ethics.

7.3 Carnap's Views on Ethics

In his autobiography Carnap reiterated a point of view on value judgments which he had held ever since the 1930s: unconditional value judgments (statements that say an action or a thing is good or valuable in itself) should not be interpreted as having any truth value. One cannot refute or confirm such statements based on empirical information about the world. However, such statements do play a positive role through upbringing, education and persuasion in motivating people to act in a certain way (Carnap 1963, 80). Carnap subscribed to a conception of life according to which valuation was the domain of one's *Lebensgefühl*, the stance of an individual towards the driving goals in her own life, typically expressed in poetry or music and strictly separated from science as the system of theoretical knowledge (Mormann 2007, 135). In Carnap's view, scientific knowledge could still guide one's actions, since scientific knowledge yields the most reliable information of how to materially realize certain goals. However, scientific knowledge could not help a person to determine which goals or values to strive for (Carnap 1934/2013, 178). As a consequence, questions like "Should I aid the government in killing my Jewish fellow citizens?" cannot benefit from scientific information in any way. Ultimately, the answer to such question depends on the practical decision that an individual has to make for herself (Carnap 1934/2013, 176). Therefore, scientific philosophy, as the logical investigation of scientific languages, can offer no guide in ethical decision making.

Carnap's position on ethical problems often provoked dismay or alarm. As Carnap reports in his autobiography, his former colleague at the University of Prague, Oskar Kraus, believed that Carnap's position on value statements was a danger to the morality of youth (Carnap 1963, p. 81). In a letter to Carl Hempel, written on 13 August 1942, Carnap laments that many people in the United States had similar problems with his position:

> Often people, although they no longer have any theoretical argument against our position in this point [the strict distinction between knowledge and decision] are dissatisfied, because they do not see on what basis we can reject say the value actions [sic] of the Nazis. If I were to discuss with a Nazi, I should first try to find out whether we could find a common aim even if rather remote and general. If so, there could be a theoretical discussion concerning suitable means to that end. If not, then there is no theoretical discussion possible. What remains is simply appeal, persuasion, education, or whatever we wish to call it. This may, of course, be combined with theoretical considerations; I should try to show how the life of the whole of humanity would look in the future if the Nazis had their way (which is a scientific consideration); and then I should make an appeal to reject that future state of affairs and therefore the way proposed by the Nazis. (Carnap to Hempel, 13 August 1943, CH Box 11, folder 1, ASP)

In this letter to Hempel, Carnap shows that a non-cognitivist, like himself, can still discuss ethical issues with a Nazi, although not with the intent to prove that the Nazi's actions are wrong – this is precisely beyond the realm of proof. The discussion would consist of three parts. First, Carnap and the Nazi have to discuss which aims they share. In itself, such a discussion would be difficult: Carnap and the Nazi would have to figure out which aims they actually have – something that is not immediately apparent to any person. Carnap leaves unmentioned how one could in fact pursue this part of the discussion. But once they have this figured out, the second part of the discussion could ensue: an investigation into the best means to achieve possible common aims. Here, scientific knowledge would be of paramount importance. The third part covers the aims that the interlocutors do not share: now, they attempt to persuade each other through rhetoric. Again, scientific knowledge can be helpful in this persuasive exercise. Carnap thinks his persuasive attempt to change the Nazi's valuation could be improved by using science to show the Nazi what life would be like if he had his way – in the hope that the Nazi would come understand that he does not actually desire what he says that he desires.

In a recent paper A.W. Carus similarly defended that Carnap had an overarching conception of rationality covering both knowledge and values, where "values ultimately determine what we define as reason, but the reason thus provisionally defined informs and enlightens our values" (Carus 2017, 174). This dialectic relation between values and knowledge was never articulated by Carnap himself. As we will see below, Leo Apostel's articulation emphasizes the social processes that have to be put in place for such a dialectic to take place and become a model of reasoning on values.[12]

[12] Similarly, Carus (2017, 178) emphasized that Carnap's ideas on practical reason crucially reflect their embeddedness in political interaction.

In his autobiography, written some 10 years after the letter to Hempel, Carnap similarly emphasized that his non-cognitivist position was compatible with a 'scientific humanism'. This entailed the idea that human beings are themselves responsible for the improvement of the conditions of their life, and that this deliberate self-improvement might best come about by acquiring scientific knowledge of the world (Carnap 1963, 82–83). Using science to improve "the internal and external situation of one's own life" is not entailed by the non-cognitivist position: Carnap cannot prove that one should value the use of science for the improvement of life, or for the resolution of ethical disagreements. This is a valuation, and thus not a theoretical position which can be true or false, neither is it a philosophical position. It is a stance that a person can take, and that Carnap aimed to advance in all levels of society. In the Vienna Circle Manifesto, this was expressed in the following way: "We are witnessing how the spirit of the scientific world-conception penetrates in growing measure the forms of personal and public life, of education, of childrearing, of architecture, and how it helps shape economic and social life according to rational principles" (Carnap et al. 1929/1973, 317). The stance of scientific humanism implied that one would no longer follow what ideology or religion says that one should do, or be engaged in a debate on ideology or religion. Instead, a person or organization investigates what they aim for as transparently as possible, and then they use scientific knowledge to execute these aims (Uebel 2012, 141–142). In 1934 Carnap also expressed this stance using a Marxist metaphor:

> We reject this narcotic (of philosophical and religious metaphysics to discuss values). However, if someone else prefers to enjoy their pleasures, then we cannot refute this theoretically. But this does not mean that we are indifferent to how people decide on this matter. (Carnap 1934/2013, 179)

Carnap's scientific world conception was not indifferent to politics in the broadest sense. Carnap's stance implied the promotion of a revision of how people in society came to decide on value-laden issues and how they act upon their decisions. Unfortunately, Carnap rarely investigated how this promotion could be pursued most rigorously. Moreover, Carnap had no real clue how scientific knowledge exactly could be used within this new process of decision-making. In an interview of 1963, he states the following:

> begin with clearing up the differences in views about facts and only when one has reached agreement in the most important points of this sort start with the discussion of the real value questions. I believe that in this way philosophy will not help us decide the value questions themselves in one way or another, but it will help us to find a clearer basis for the discussion of these questions. (Quoted from Uebel 2012, 143)

This gives a rough sketch of the process. First, find out the differences in conditional and unconditional value judgments. Second, use factual, scientific knowledge to clear up the differences in conditional value judgment's. Lastly, after clarification, enter a phase of persuasion. Anyone who wants to pursue Carnap's stance, has two challenges. On the one hand, one has to flesh out a model of how decision making in society comes about without the use of ideology or religion and clarify in which way scientific knowledge and logical clarification can be used in this model. On the other hand, one has to find out how to promote the use of science in societal and personal decision-making, and demote the use of ideology or religion. As long as these two challenges are not met, the stance of scientific humanism remains mere hand waving.

7.4 Apostel's Science of Morals

Apostel took up both challenges in 1959 after his definitive return to Ghent University. He aspired to create a scientific program that could bridge scientific work at the university, political decision-making at the ministries and ethical education in secondary school. In Apostel's vision, the new program would become a motor to revise how people and organizations in society made their decisions, removing the use of ideology and religious discourse from that process. He called his proposed program *science of morals* [moraalwetenschap].[13] At the faculty meeting of 7 March 1960 the problem concerning the training of teachers for the non-confessional morals course in secondary education came up. Apostel seized this opportunity to introduce his own program *science of morals* to resolve the teachers-issue.[14] Although Apostel during the initial meeting never mentioned his true intentions behind the program, all of his colleagues welcomed the initiative. Apostel's plan was subsequently handled by an internal commission, where the debate focused on the autonomy of the program. The report of the committee states that some colleagues initially questioned Apostel's ambition to start an entirely new section with its own scientific research unit.[15] Instead, they suggested that the new program could be a sub-specialization in the already available philosophy program. This would imply that the program had no research autonomy – a suggestion that was dangerous to Apostel's wish to create a new program with its own research unit. Apostel did not want the *science of morals* to become part of a philosophy program. On the one hand, he believed, just like Carnap, that ethical judgments are not the subject matter of philosophy proper. On the other hand, he wanted researchers and students in the program to perform empirical research (cf. Infra) – the philosophy department would not be the right place to execute that plan. In its final report the committee argues in favor of Apostel's autonomous program:

> The curriculum in Philosophy is both too broad and too narrow. Of course, philosophy students are offered ethics, but also a substantial number of more general courses, which are, no doubt, of interest, but at the same time too specialised to serve the preparation to this

[13] Although "moraalwetenschap" is traditionally translated as "moral science", I find the use of the adjective in that translation deceptive, seemingly distinguishing it from immoral or amoral science. Since the program is meant to train students to scientifically study moral behavior, and use this study to overcome ethical disagreements, I find "science of morals" more suitable as a translation.

[14] Faculteitsraadverslag, 7 March 1960, Archief Faculteit LW, 6A1_013, Universiteitsarchief Gent.

[15] Verslag van de speciale Commissie Moraalwetenschap, bijlage, Faculteitsraadverslag, 8 November 1961, Archief Faculteit LW, 6A1_015, Universiteitsarchief Gent.

> special kind of education [in non-confessional morals]. Moreover, the philosophy curriculum is too narrow. A number of courses in psychology, biology and sociology, which are essential to the moral scientist are lacking. (my translation; ibid.)

The committee noted that the teachers in non-confessional morals "should encourage confidence in the young adolescents through the teacher's human understanding" and that the teacher should help the adolescent develop a value-scale for life. These skills might imply a training in psychology, but do not warrant the inclusion of biology and sociology per se, and certainly not the development of research skills in these fields through seminars, as the committee strongly advised in its report. The inclusion of a doctoral degree into the plans is equally strange within the limited motivation to set-up a program for teachers in secondary schools. Apostel, who was the head of the commission, wanted to create something else than a university program for teachers, but strategically hid most of his ideas about the program until after the rector and minister had agreed to proceed with his sketches. By October 1963, Apostel's strategy was successful: the new, autonomous program *Science of Morals* had been established and during the celebratory inauguration Apostel finally presented his vision on the scientific and social function of the program, which most likely surprised the rector, the dean and other fellow colleagues who had helped Apostel to set it up. His inaugural speech which explicitly addressed the new students, was published as *Pluralist Foundations of the Science of Morals* [Pluralistische Grondslagen van de Moraalwetenschap].[16]

In the introduction Apostel articulated how most of his colleagues thought about the program: on the one hand it served as a new "practical organ" which had arisen from the need for trained teachers of non-confessional morals, and on the other hand the program was made up of a mere "encyclopaedic collection of facts" concerning the psychology of adolescents which teachers would use in their educational practice (Apostel 1965b, 69–70). Apostel intended to present to "the general public and the future students how the science of morals had the aforementioned practical and encyclopaedic nature, but how this nature should necessarily give rise to a new kind of scientific work" (Apostel 1965b, 69). Apostel's envisioned new science faced a crucial paradox: scientists can only describe the world as it is through their methods, not as it should be, whereas the philosophers mistakenly believe that the study of values is precisely what distinguishes their field from the sciences (Apostel 1965b, 71). Consequently, there can never be a science of values. In this traditional view, scientific knowledge and valuation stand in opposition. For Apostel, this opposition has to be dissolved.[17]

---

[16] All subsequent translations of this text are my own.

[17] One can recognize a dialectical materialist ideal in Apostel's phrasing of the problem, but also in his proposed solution to turn the reigning hierarchy between science and values on its head. Already in his student years Apostel had read Marx. Between 1959 and 1963, the period in which he fully articulated his ideas on ethics, Apostel was in close contact with Jaap Kruithof, a Marxist inspired philosopher who started as a professor in philosophy at Ghent University in 1959. For a detailed analysis of the Marxist influences at Ghent in the 1960s and 1970s, see (Christiaens 2016). During the negotiations over the new program, Kruithof always supported Apostel, but he did not

> Until now, the ethics of a group has been formed by poets, politicians and prophets; subsequently the vision of these individuals has been rationalised by the individual systems of separate philosophers. Next to the apparatus that sets the goals of the group actions, there used to be the apparatus that sets the means to the group actions: essentially, science. [...] We could call this the linear and hierarchical view on culture: the goals control the means and not the other way around (hierarchical) and the apparatus of control cannot be controlled itself (linear). [...] We should abandon this structure for a cyclic and multilateral one. (Apostel 1965b, 71)

According to Apostel, a revision of decision-making in society is at stake in the new science of morals. Along the lines of Carnap, Apostel takes for granted that philosophers have no rational resource to find out what are the right values to have or the right norms to install. At best, philosophers had previously rationalized the ideas of poets, prophets and politicians. At worst, philosophers had projected their inner, psychological conflicts as eternal structures of humanity or the universe. "Much of the classic ethical literature belongs to a dangerous, emotionally appealing, but rationally obscuring genre" (Apostel 1965b, 84). Consequently, philosophers should abandon their program to set out the rules for the actions of the group. Moreover, politicians, poets and prophets – Carnap's narcotic of ideology and religion – should equally cease to guide group action. Apostel calls this hierarchy where ideology or religion guide group action an "unconscious method": a collective cannot properly control itself, but relies on specific individuals to guide its course.[18] The collective, as it were, is under a narcotic. Instead, science as the apparatus that traditionally only yielded the means to realize the goals, should be used to help the collective set its own goals and thus abandon the narcotic.

> In the emancipation of the science of morals one can see the mutation of human culture which still announces itself only silently, but which in fact implies that humanity in group takes destiny in its own hands by choosing its own goals autonomously and change these goals after mutual impersonal deliberation, using the only form of collective thinking, which is scientific thinking. Such deliberation is the goal of the science of morals. (Apostel 1965b, 72)

Apostel never created an ethical theory as a philosopher. This is exactly what one cannot do, if you take up Carnap's stance. What remains to do, is to take up the stance of scientific humanism and transform the culture of decision-making, and this cannot be executed on the level of ideas, but has to be implemented on the level of institutions. Changing the model of decision-making in society is a social

initiate any of the discussions. During all the meetings with the faculty members and the rector, Apostel always presented the program as his own conception. During the inauguration of the program Kruithof agreed with Apostel that "the aim of the program is to train young students in competent scientific research" (Kruithof 1965, 15). Kruithof himself was mainly interested in the social and historical opportunities the program offered for Belgian culture.

[18] In the essay "Theoretical Questions and Practical Decisions", Carnap similarly discusses the unconscious use of ideological and religious theories as the wrong tools to use in social conflicts, since their theories distract and confuse people (Carnap 1934/2013, 179). Carnap believed that the logical investigation of scientific languages offered conceptual tools that no longer confused people in their deliberation over social issues (Richardson 2007, 304–305).

process, which has to be initiated through certain institutional changes. In Apostel's mind, the creation of the program had a direct two-fold institutional effect. On the one hand, it would introduce teachers in secondary schools who would train students to make decisions differently than before. Students would no longer be taught to adhere to an available value scheme, like humanism or Catholicism. Instead, they would be taught to investigate within themselves what they exactly valued, how their current values had been formed and how to discuss their differences in values among each other. This changes how education in a society disciplines citizens to discuss ethical issues (Apostel 1965b, 72). On the other hand, Apostel believed the role of the university in society could also be changed: students in the *science of morals* could bridge the gap between all kinds of social science research at the University and social organizations (e.g. unions, political parties, ministries, etc.). These new students would be the beginning of an entirely novel kind of scientific researcher, who is constantly both theoretical and practical.

In Apostel's view the changes in education practices and university research would engender a complete overhaul of decision-making in society. "The science of morals can exist, must exist, but it does not exist yet. Whomever understands the enormous cultural-historical mutation that is hidden underneath the institutionalization of the science of morals, should not be discouraged by temporary stagnation or defeat" (Apostel 1965b, 72). This institutionalization would on the one hand introduce teachers in secondary education, and on the other hand doctors at the university trained specifically for building interdisciplinary bridges.[19] Once both started introducing the new way of decision making on other levels of society, the process would only accelerate and eventually install itself permanently. Carnap's first problem, how to transform the decision-making culture, had thus been solved by Apostel.

The integration of scientific knowledge into the deliberation over values was the second challenge that Carnap bequeathed Apostel, and the bulk of his inaugural lecture discusses it in great detail. As Carnap describes in his letter to Hempel, the first part of any deliberation should consist in the search for disagreement over values within a collective. The first task for the moral scientist, in Apostel's text, is to descriptively chart existing moral beliefs in society (Apostel 1965b, 73–79). Several questions pop up within the task: how to distinguish moral beliefs, and how to distinguish the various systems (of groups/classes) in which these beliefs operate? How can one empirically ascertain moral beliefs: is a questionnaire, an interview, or a historical method more suited? Lastly, how should the moral scientist distance her own moral beliefs from her subject of investigation? How can one train oneself to remain as neutral as possible? These four challenges make up the *factual basis* of the science of morals – and it shows why the students had to become acquainted with biological, sociological, psychological and historiographical methods to empirically map moral beliefs. As Carnap advised in his letter to Hempel, a focus on the conflicts and tensions in moral beliefs should be central. Apostel writes

[19] Apostel focused heavily on the interdisciplinary nature of his proposal. Although one might think that Otto Neurath's work might have been an inspiration on Apostel in this regard, there is no indication that he was familiar with Neurath's work, at least in the 1960s.

similarly: "conflicts [over values] give rise to the possibility to influence, and to the necessity of helping the conflicted, moreover, they point out to places where reflection can become useful to practice" (Apostel 1965b, 78).

Once conflicts in moral beliefs have been identified and empirically studied, the moral scientist can use the points of conflicts as levers for action. Apostel's paradigm for the transformation of conflicts into levers for change, is the advertisement industry (Apostel 1965b, 83). Advertisement companies align production activities of a company with the study of the group targeted by the production, the consumers. These companies first investigate what certain consumers value in products, then show to the production companies what they need to produce to entice a maximal number of consumers, and, after production, the advertisement company uses its knowledge about the consumer valuation to advertise the product in the best possible way. After the launch of the new product, the consumer population is then re-assessed and possible changes in consumer valuation are communicated to the production-level once more, and so the cycle repeats itself.

The science of morals does exactly the same thing, but it has "the total task of humanity and society as its enterprise" (Apostel 1965b, 83). If a group intends to apply an action to society, the moral scientist must be able to supply the following information: which persons in the leadership would cooperate, which persons in the group are mostly keen on action, which subgroups would be willing to collaborate, which interests might delay the action, which channels of communication are the most efficient in the group, which needs in the group could be used to pursue the action most efficiently, and which parts of the action are the easiest to implement and could help other parts (Apostel 1965b, 83). Most of this information follows from the *factual basis*. The task of the moral scientist is to implement the action in an efficient way within existing social organisations. How to proceed will not be transparent from the start. Only the execution of these ideas on initially, small scales (small communities) will yield the essential experimental information of how such implementations might work in a variety of contexts (Apostel 1965b, 83). This is exactly why the start of the program is also precarious: as long as no one starts to execute the research program, no one can ever attain the necessary experimental knowledge to achieve successful transformations in society. Apostel repeats this paradox several times throughout the text: only someone who was trained in the program, will be competent to teach the program, because this person will have executed the kind of *research and action in society* that the program intends to pursue (Apostel 1965b, 84, 124). The solution to the paradox also entailed that foreign universities had to be encouraged to implement similar programs: although the program was created in Ghent due to the practical teachers-problem, it could only succeed through international expansion and collaboration. Similar empirical research had to be conducted in other countries and concerning other social conflicts, if the empirical part of the program was to achieve useful generalizations (Apostel 1965b, 123).

The advertisement-paradigm is the crux of Apostel's integration of science and the deliberation about values: the moral scientist maps the moral beliefs in society, locates the conflicts, clarifies possible misunderstandings of conflicts, shows where collaboration is possible on common ground and offers the most efficient social tools to bring collaboration about – potentially creating entirely new social institutions to bring about the desired effects (Apostel 1965b, 83). Afterwards, the whole research cycle repeats itself. There is, however, one crucial distinction between the advertisement agent and the moral scientist. Whereas the former has the aim to maximize consumption, the moral scientist's aim is a harmonization of values in the targeted community. Although moral scientists, according to Apostel, would be trained to remain neutral towards the various value-conflicts in the targeted community, the aim of their activities was not. At the core, the enterprise's goal was a harmonization of conflicting groups with the aid of scientific knowledge about the conflicts between those groups. Previously, religion and ideology had attempted to structure the values of groups in society. Now, Apostel envisioned, a new kind of scientific research would take over this role. This presupposed that the researchers in the program would not intervene in the social and political conflicts with their own value scheme, but that they would only intend to harmonize the already present conflicts. This is, of course, not a politically neutral scientific enterprise, just as an advertisement-agency is not neutral towards consumption or profit-maximization.

After his presentation of the advertisement-paradigm, Apostel offered five viewpoints which would help future moral scientists to execute their work. The viewpoints were to serve as focus points for the empirical research that had to be developed in the program. The first viewpoint concerns the logical analysis of belief systems. Mapping how moral beliefs are logically connected to each other within the various belief systems in a society, can aid the moral scientist in understanding where people have actual or illusory conflicts over values. These conflicts can be present between any kind of social group in a given society, e.g. different generations, different classes, etc. Only the empirical research on groups in society would show what the actual conflicts in a society were. Apostel believed that the moral scientists had to develop modal logics which could be applied to the moral judgments that they empirically ascertained in groups of society. Four logical motives could then be used to transform the moral belief systems in society: vagueness, incompleteness, incoherence and contradiction. The moral scientist can confront the actual, moral belief systems with the formalizations – and use the divergences as levers to change the value judgments involved, e.g. by showing in an effective way to target groups that two or more different choices are contained within one vague notion which these target groups use in their judgments (Apostel 1965b, 86).[20]

A second viewpoint covers "existential perspectives", a term which Apostel borrowed from the research of Abraham Edel.[21] An existential perspective on a norm investigates the kinds of entities or processes that have to exist, if the norm is to have

[20] Throughout the text Apostel adheres to a strict non-cognitivist position, constantly emphasizing the necessity to make choices. Logical clarification, which is the only operation philosophical discourse is left with, cannot help to decide, but it can help to articulate which are the available value judgments to choose from. In this sense, Apostel shares the voluntarist idea behind logical clarification of Hans Reichenbach and Rudolf Carnap (Richardson 2006).

[21] Apostel's most important reference point is (Edel 1955). For a discussion of the relation between Apostel's and Edel's views on ethics, see Commers (2010, vol. 2, 109–172).

the capacity to guide actions. E.g. take the norm "proliferate the victory of the working class over the bourgeoisie". If there is no working class or bourgeoisie as two identifiable groups in society, or if they are not necessarily entwined in a mutual conflict, the norms become meaningless as a guide for action. In this way, one can investigate almost any moral belief on the entities or process that it stipulates, and this in turn can be checked by scientific investigations (Apostel 1965b, 90–93). Such information can then be communicated back to target groups to help them reassess value judgments.

As a third viewpoint Apostel discusses "the explanatory method". This encapsulates all empirical research about the causes and consequences of accepting moral systems in a society. The acceptance of certain systems engenders instability or destruction of a society. E.g. The goal to erect enormous statues on the Easter Island to honor the Gods may have caused deforestation and possibly civil war. The moral scientist has to be aware that changing the goals of certain groups can have an enormous impact on society at large, potentially destabilizing it entirely or marginalizing parts of it. Therefore, the moral scientist has to have psychological, sociological and economic knowledge on the effects that a change to the goals of groups can have. Furthermore, the moral scientist also has to be aware of how people in society come to accept their goals within a moral system – most new moral systems are only accepted because they are mostly congruent with older ones, and this can create a situation where people have certain values in an inauthentic and thus unstable way (Apostel 1965b, 98–99). If moral scientists do not investigate all of this, their action research will be severely hampered.

The fourth viewpoint investigates how ethical systems operate on human behavior (Apostel 1965b, 100–114). Here, Apostel attempts to recuperate traditional normative ethics: one can study utilitarianism, hedonism or deontology as rules that guide human beings. Depending on the kinds of rules that people use in their daily life, they will start to act differently. There will be various feedback loops: the outcome of certain actions can change the kind of system which motivates people to act in a certain way. Investigating these feedback cycles will benefit the moral scientist to understand how moral rules and practices shift over time.

As the last viewpoint Apostel discusses the functions of ethical language. Here, Apostel investigates the potential use of language analysis for the moral scientist. Since an ethical judgment is an actual instrument that people use in deliberation over moral issues, the moral scientist has to know the variety of possible functions that an ethical judgment can have in specified, discursive contexts. To that end, the moral scientist can use the various philosophical language analyses of ethical judgments in an empirical investigation of how ethical statements function in real human discourse.[22] In this section, Apostel aligns emotivism with the research program of

[22] One of the first students of the new program, Ronald Commers, executed this idea, investigating how the "language of ethics is a never ending process of human discourse about the good" (Commers 2009, vol. 1, 9). Commers investigated how three distinct ways to talk about the good arose within different sociohistorical contexts. For a full discussion of the historical methodology behind this materialist approach to ethical discourse, see Commers (2010).

the moral scientist: "Emotivism, a position which seemingly makes any scientific ethics impossible, can be transformed into the source of an important section in scientific ethics, if only one is prepared to study concrete emotions, judgments and their linguistic counterparts within their full context" (Apostel 1965b, 120). Exactly once one understands that ethical judgments are not true or false, but express emotions and attitudes which drive human action, one can use an investigation of ethical judgments to help influence which actions people make within a society.

These five viewpoints all concern logical and empirical information that the moral scientist can use to analyze and dissolve ethical disagreements in groups, following the paradigm of the advertisement agency. Before his students would become the "Mad Men" for a better society, Apostel believed that they had to be transformed into a new kind of human being (Apostel 1965b, 125). This personal transformation was also part of the program's political aim, i.e. a harmonization of social conflicts. The students should be able to take up the role of a *judge* concerning their own norms. They should also have to know the technologies to implement changes into social reality, being as cunning as a *politician*. Furthermore, they have to assume a *scientific ethos*, removing themselves from their own moral consciousness, and they should also have a constant *therapeutic* urge to use the moral conflicts in others as levers for personal and social transformation (Apostel 1965b, 110–111). Combining the role of politician, psychotherapist, scientist and judge, Apostel's ideal moral scientist had much to learn. Yet, Apostel believed that two fundamental processes would realize his ideal: the students were to undergo personal and collective psychotherapy, so that they were aware of their own moral conflicts, and the students were to be positioned as interns in organizations like the industry, the military, prison, political administrations and banks. In these various contexts, the student would personally investigate the reality of social conflicts and get intimately acquainted with them (Apostel 1965b, 125). The ambitions for the program from Apostel's point of view were incredibly high: the students had to realize a scientific institution at the university capable of transforming the core of societal decision-making. The creation of a scientific, non-ideological revolution was the goal. Such ambition was not proportionate to Apostel's actual, social means.

7.5 Conclusion

The scope and wealth of suggestions for research in Apostel's text is daunting. With it, we are far removed from the preliminary sketches that Rudolf Carnap produced in his lifetime. The text fully develops the challenges that Carnap never took up. First, Apostel proposes to institutionalize new scientific programs at universities which will train teachers and researchers who can transform the way decision-making is done in society. In this way, Apostel answered a crucial problem of Carnap, how to actually initiate the transformation of society – it could not be done by simply maintaining a non-cognitivist position *as an academic philosopher*. One had to transform the non-cognitivist position into an institution that is *effective in society* – exactly what a purely theoretical stance can never be. Second, Apostel worked out the many ways in which scientific information could be used by researchers to guide deliberation in society, mainly through the advertisement agency as paradigm and the five viewpoints on future empirical and logical research.

However, this ambitious vision also brought Apostel in conflict with his colleagues: the science of morals was to be a scientific research program, not a preparation for the moral training of secondary school students. In his lecture, Apostel heavily protested against the idea that the new program would become "propaganda of the non-confessional movement" (Apostel 1965b, 122). The program and its students should be neutral towards religious values: these are values that some persons in a society have and that sometimes engender conflict with the values of non-confessional members of society. Dislodging these conflicts cannot be done by claiming that the non-confessional values are the correct ones to have – such claim would remain within the traditional model of ethical deliberation where one side was right and the others wrong. Instead, these conflicts had to be dislodged through science-aided deliberation.

Before Apostel began his lecture during the inauguration his colleagues had presented the program in terms that clearly conflicted with Apostel's view. The Dean of the Faculty, René Derolez, thought the program primarily aimed to train teachers, and these teachers would replace religious education with non-confessional guidance: "the first task of the teacher is to offer psychological assistance to his pupils; he should assist them in building a value-scale to account for their actions" (Derolez 1965, 8). In a similar vein, Emiel Leemans, classical philologist and also the head of the philosophy department, declared to the first generation of students that: "No one can hope that our students will reach any scientific mastery in the offered domains. They will never be able to perform any independent, scientific research. Our program is aimed at moral training, focused on fostering wisdom and responsibility" (Leemans 1965, 13). The head of the University, Jean-Jacques Bouckaert, similarly did not conceive of the program as a scientific enterprise: "The aim of the program is to give students a chance to look at moral problems from all perspectives which are compatible with academic education" (Bouckaert 1965, 5). Apostel's colleagues all believed the program was meant to train teachers who would give moral guidance to their students from a non-religious point of view, which was exactly what the liberal pressure groups had been asking the Minister of education for. None of the University leaders were aware of Apostel's broader goals to transform Belgian society.

Except for his closest collaborator, Jaap Kruithof, no one during the inauguration mentions Apostel's ambition to introduce a new scientific discipline which aims to start a transformation in societal decision-making. From the various addresses held during the opening celebration, it is clear that Apostel's strategy to hide the true vision behind the program was almost doomed to fail. If the students would not get the proper scientific training by the professors and if there would be no institutional support for the student's internships, there would never be any actual empirical research. Between January and October 1963 Apostel had been confronted with the enormous difficulty to enthuse sociologists, biologists, and economists to collaborate and support the necessary empirical element of his program. Most attempts were in vain. The archival folder "Foundation of a Bachelor, Master and Aggregate in the science of Morals 1961–1969" contains many letters from the dean Derolez to rector Bouckaert about the problem of assigning professors to the various courses in the new program.[23] One of the new integration courses, "General Anthropology: philosophy, psychology, history and biology", which aimed to integrate perspectives from all these disciplines in discussing ethical conflicts in society, still had no official teacher by 1969, 6 years after the launch of the program. The students who had signed up on Apostel's promise that they would spearhead a new kind of science, were cheated: they never received proper training in empirical research.

Since the courses of Apostel's new program had to be supported by existing funds, the students would *de facto* pursue introductory courses offered in other faculties – nothing like the research seminars that the committee behind the program had strongly advised in its reports. By September 1965, Apostel had a nervous breakdown. After his institutional struggle to create the program between 1960–1963, he had attempted to create an interdisciplinary faculty where his new research program would be properly supported. The struggle to create this new institution is a story in itself, since Apostel, between April 1963 and July 1965 mobilized all available political forces outside and inside the university to create his imagined interdisciplinary faculty. The full story behind this attempt cannot be told here; it involves a whole set of political negotiations, including ministers, policy makers, university administration and captains of industry. In the end, Apostel failed, and after 1965 he gave up all of his ambitions to influence the course of his surrounding society. In a letter to Lucien De Coninck, a biologist who had collaborated actively on the creation of the science of morals program, Apostel writes: "I am deeply disappointed about the developments in the science of morals department; our plan utterly failed."[24] His disappointment cut deep. After September 1965, ethical topics in general were removed from Apostel's interests. Whereas he had still taught a course on scientific ethics[25] between 1958 and 1963 and published on ethical issues (Apostel 1961, 1963, 1965a), like euthanasia, he limited himself after 1965 to strictly logical and epistemological subjects.

In his 1986 lecture on Carnap Apostel lamented that "Carnap was right" - he should have pursued a PhD on ethics at Chicago University. More than 20 years after the foundation of the science of morals program Apostel was still convinced that the transformation of ethics from a philosophical discourse into a scientific discourse was necessary, even though his own attempt at such a transformation had stalled. Despite Apostel's enormous disappointment, the program of the science of morals persisted. It did not function as a lever to change the way Belgian society

[23] Folder 4A2_61020, Universiteitsarchief Gent.

[24] Apostel to De Coninck, undated, probably 1969–1970, ALA, folder 530.

[25] In this course Apostel argued that ethics should be transformed from a philosophy to a science. Possibly the course served as a preparatory work to his text, *Pluralist Foundations of the Science of Morals*. "Cursus Ethica 1961–1962", ALA, Folder 20.

solved ethical and political conflicts, and it did not engender systematic empirical research on value conflicts in Belgian society. For more than 50 years, it largely functioned as a training program for teachers in non-confessional morals, the purpose for which Apostel had gotten institutional support to create the program, but resolutely not why he had designed it. In this regard, Apostel's politically laden logical empiricist philosophy ultimately suffered the same fate as philosophy of science in the United States: it became an academic enterprise that had little to no effects on extra-academic developments. In his 1986 lecture Apostel promised in his introduction to discuss the future of logical empiricist philosophy: "I would like to clarify why I think Rudolf Carnap still lives on. I have read that logical empiricism and the Wiener Kreis are dead. To me their philosophy has yet to begin, to me their philosophy continues. I want to end with a plea to take up their tradition once more, although many believe it to be dead" (Apostel 1986/2011, p. 466). By the end of the lecture there was no time remaining to pursue this line of thought. As throughout his career, Apostel's promise to articulate the political potential of logical empiricism remained just that, an unfulfilled promise. It died with Apostel in 1995.

Acknowledgements This research was supported by the Research Foundation – Flanders (FWO).

References

Apostel, Leo. 1961. De Concrete Grondslagen van de Moraal in Het Onderwijs. *Dialoog* 1: 3–37.

———. 1962. Le Champs Semantique de l'Incertitude. In *La Philosophie Analytique*, 188–248. Paris: Edition de Minuit.

———. 1963. L'euthanasie Comme Problème Moral et Philosophique. *Morale et Enseignement* 12: 13–35.

———. 1965a. Organisation et Éthique. *Revue Universitaire de Science Morale* 3: 3–23.

———. 1965b. Pluralistische Grondslagen van de Moraalwetenschap. In *Problemen Rond de Moraalwetenschap*, 68–125. Gent: Rectoraat Rijksuniversiteit Gent.

———. 1986/2011. Leo Apostel over Zijn Leermeester Rudolf Carnap. In *Jeugd Die Niet Voorbijgaat*, ed. R. Commers, 465–493. Brussels: Academic Scientific Publisher.

———. 1989. *A Life History: Leo Apostel*. Gent: Communication and Cognition.

Bouckaert, Jean-Jacques. 1965. Toespraak Door Prof. Dr. J.J. Bouckaert. In *Problemen Rond de Moraalwetenschap*, 5–6. Gent: Rectoraat Rijksuniversiteit Gent.

Carnap, Rudolf. 1934/2013. Theoretische Fragen Und Praktische Entscheidungen. In *Der Wiener Kreis, Ausgewählte Texte*, ed. Ch. Damböck, 175–179. Stuttgart: Reklam.

———. 1963. Intellectual Autobiography. In *The Philosophy of Rudolf Carnap*, ed. P.A. Schilpp, 3–84. La Salle: Open Court.

Carnap, Rudolf, Hans Hahn, and Otto Neurath. 1929/1973. The Vienna Circle of the Scientific Conception of the World. In *Otto Neurath: Empiricism and Sociology*, ed. M. Neurath and R. Cohen, 299–318. Dordrecht: D. Reidel.

Carus, André W. 2017. Carnapian Rationality. *Synthese* 194 (1): 163–184.

Christiaens, Wim. 2016. *Het Gents Marxisme*. https://www.academia.edu/24802681/Het_Gentse_marxisme.

Commers, Ronald. 2009. *Kritiek van Het Ethisch Bewustzijn: Van Liefde Met Recht En Rede*. Vol. 1. Den Haag: Acco.

———. 2010. *Kritiek van Het Ethisch Bewustzijn: Van Liefde Met Recht En Rede*. Vol. 2. Den Haag: Acco.

———. 2011. *Jeugd Die Niet Voorbijgaat/Denkweg in Terugblik*. Brussels: ASP.

Derolez, René. 1965. Toespraak Door Prof. Dr. R. Derolez. In *Problemen Rond de Moraalwetenschap*, 7–10. Gent: Rectoraat Rijksuniversiteit Gent.

Edel, Abraham. 1955. *Ethical Judgment: The Use of Science in Ethics*. New York: Free Press.

Howard, Don. 2003. Two Left Turns Make a Right: On the Curious Political Career of North American Philosophy of Science at Midcentury. In *Logical Empiricism in North America*, ed. A. Richardson and G. Hardcastle, 25–93. Minneapolis: University of Minnesota Press.

Kruithof, Jaap. 1965. De Plaats van de Moraalwetenschap in de Universiteit van Morgen. In *Problemen Rond de Moraalwetenschap*, 24–67. Gent: Rectoraat Rijksuniversiteit Gent.

Leemans, Emiel. 1965. Toespraak Door Prof. Dr. E. Leemans. In *Problemen Rond de Moraalwetenschap*, 11–14. Gent: Rectoraat Rijksuniversiteit Gent.

Mormann, Thomas. 2007. Carnap's Logical Empiricism, Values, and American Pragmatism. *Journal for General Philosophy of Science* 38 (1): 127–146.

Reisch, George. 2005. *How the Cold War Transformed Philosophy of Science: To the Icy Slopes of Logic*. New York: Cambridge University Press.

Richardson, Alan. 2006. Freedom in a Scientific Society: Reading the Context of Reichenbach's Contexts. In *Revisiting Discovery and Justification*, ed. J. Schickore and F. Steinle, 41–54. Dordrecht: Springer.

———. 2007. Carnapian Pragmatism. In *The Cambridge Companion to Carnap*, ed. M. Friedman and R. Creath, 295–315. Cambridge: Cambridge University Press.

Uebel, Thomas. 2012. Carnap, Philosophy and 'Politics in Its Broadest Sense. In *Rudolf Carnap and the Legacy of Logical Empiricism*, ed. R. Creath, 133–148. Dordrecht: Springer.

Chapter 8
The Cult of Genius and Its Critics: Edgar Zilsel and Otto Neurath

Günther Sandner

Abstract Edgar Zilsel and Otto Neurath were among the few intellectuals who subjected the concept of genius to critical examination during the first half of the twentieth century. Although they lived during the same time and their intellectual and political biographies cross at numerous points, Zilsel and Neurath worked on their examinations at different times. Zilsel first dealt with the topic in 1918 in his remarkable book, *Die Geniereligion (The Religion of Genius)*, and then again in 1926, in more detail, with a work on the origin of the genius concept: *Die Entstehung des Geniebegriffes* (*The development of the idea of Genius*). Neurath, on the other hand, addressed the topic only after he emigrated (in 1934 resp. in 1940) and in a different historical context: he dealt with the figure of the genius in connection with his reflections on the causes of National Socialism and its positive ideological acceptance within wide parts of German (and Austrian) society. Despite different starting points and, at least partly, different cognitive interests, their texts and thoughts on the concept of genius have a lot in common, that is, despite the different thematic, cultural, and political contexts of their reflections, both Edgar Zilsel and Otto Neurath understood that the idea of genius was not only a cultural question but also a political one.

Keywords Vienna circle · German *Sonderweg* · Religion of genius · Cult of personality · National Socialism · German climate

G. Sandner (✉)
Institut Wiener Kreis, Universität Wien, Vienna, Austria
e-mail: guenther.sandner@univie.ac.at

C. Damböck, A. T. Tuboly (eds.), *The Socio-Ethical Dimension of Knowledge*,
Vienna Circle Institute Yearbook 26, https://doi.org/10.1007/978-3-030-80363-6_8

8.1 The Cult of Genius

The idea of genius was a very influential concept at the turn of the twentieth century. History writing at this time focused frequently on the depiction of great, genius-like men. The biographies and life stories of inventors and discoverers, researchers and statesmen, emperors and generals, as well as religious founders and artists, were popular and successful. Both fictional and factual biographies of geniuses were met with great interest by the reading public (Köhne 2014, 58–113). This interest in genius was not a new phenomenon, however, having already experienced a heyday during the Romantic era. In the nineteenth century there were, for instance, biographies of Johann Wolfgang von Goethe that praised their object as an unreachable great man; and Goethe himself had celebrated Shakespeare as a singular genius many decades before. Many biographies of geniuses written by poets, scholars, and scientists were published at the turn of the twentieth century. In *Genius and Character* (*Genie und Charakter*, 1900) historian of literature Robert Saitschick portrayed Shakespeare, Lessing, Goethe, Schiller, Schopenhauer, and Wagner. A publisher's list of genius biographies from 1900 – *Geisteshelden* – included (in addition to those mentioned above), Carlyle, Darwin, Jahn, Luther, Mozart, and many others (Köhne 2016, 116).

The celebration of genius-like individuals was not only a cultural but also a social and political phenomenon. Books on genius focused not only on poets, composers, scientists etc. The idea of genius was also an important element in the field of social organization, its hierarchy and order. Only one of the many examples is the three-volume history of the *Wandervogel* Movement by the controversial writer and philosopher Hans Blüher (1912a, b, c), in which he describes a male association characterized by outstanding, charismatic and ingenious leaders. Politically, the affirmative discourse on genius was *anti-democratic* in nature. It often went hand in hand with an elitist criticism of the masses, which was an important part of intellectual culture. Right-wing intellectuals such as the Austrian economist and sociologist Othmar Spann strongly polemicized against democratization and egalitarianism (Sontheimer 1962, 215).

Only a few intellectuals, such as the almost forgotten Julian Hirsch (1914), recognized the problem and started a critical discussion. Among those who subjected the concept of genius to critical examination during the first half of the twentieth century were Edgar Zilsel and Otto Neurath. Although they lived during the same time and their intellectual and political biographies cross at numerous points, Zilsel and Neurath worked on their examinations at different times. Zilsel first dealt with the topic in 1918 in his remarkable book, *Die Geniereligion (The Religion of Genius)*, and then again in 1926, in more detail, with a work on the origin of the genius concept: *Die Entstehung des Geniebegriffes* (*The development of the idea of Genius*). Neurath, on the other hand, addressed the topic only after he emigrated (in 1934 resp. in 1940) and in a different historical context: he dealt with the figure of the genius in connection with his reflections on the causes of National Socialism and, if you will, its positive ideological acceptance within wide parts of German (and Austrian) society.

Despite different starting points and, at least partly, different cognitive interests, their texts and thoughts on the concept of genius have a lot in common. Moreover, they knew each other and had an intellectual exchange over a long period of time. In the interwar years, Neurath occasionally referred to Zilsel's sociological approach to the phenomenon of genius (Neurath 1930/1981, 364; 1931/1981, 499–500), and he even did so in a wartime essay (1941/1981, 912). Additionally, there are also references in his correspondence. In a letter to Bernhard Reichenbach (a brother of Hans Reichenbach), he wrote "that one could not find a hypothesis, which can use the term 'GENIE' consistently" and added: "There is a good paper by Zilsel on that subject. He has shown, that 'GENIES' appear, where certain occupations are acknowledged as proper by the ruling group" (Otto Neurath to Bernhard Reichenbach, April/May 1944, 12).[1] It is surprising, therefore, that Otto Neurath never explicitly referred to Zilsel's books and articles in his later writings that addressed this topic. Neither are there references to genius in the Neurath-Zilsel-correspondence between 1934 and 1940,[2] and there is no collection of earlier letters between the two.

8.2 Edgar Zilsel and Otto Neurath

There are only a few biographical texts on Edgar Zilsel such as those of Johann Dvořák in the introduction of his edition of *Die Geniereligion* (Dvořák 1990, 7–40) and the biographical chapter in his book on Zilsel (Dvořák 1981, 19–31). Additionally, there are an introductory essay by Wolfgang Krohn and Diederick Raven in their collection of Zilsel's essays, *The Social Origins of Modern Science* (Krohn & Raven 2000), and the memories of his son (Zilsel 2004), which were published in Friedrich Stadler's collection of essays, *Vertriebene Vernunft* (*Exiled Reason*). The latter focused on the years in exile in the USA, where Zilsel eventually committed suicide in 1944. Only a few years ago, Christian Fleck (2015, 251–294) examined Edgar Zilsel's years in the USA.

So who was Edgar Zilsel? Born in Vienna in 1891, the son of a lawyer, he worked as an actuary after finishing his studies at the University of Vienna (philosophy, mathematics, and physics, graduating with a doctorate in philosophy in 1915). Later, he passed his teaching certification exam, started working as a secondary school teacher, and took a leave of absence in 1922/23 to teach courses at the *Volkshochschule* (adult education center) and lead specialist study groups (Filla

[1] Correspondence Otto Neurath and Bernhard Reichenbach, Papers Otto and Marie Neurath, *Handschriftensammlung*, Austrian National Library (Ser.n. 1223–39 and 1223–40).

[2] Correspondence Otto Neurath and Edgar Zilsel (Microfiche), Papers of Otto Neurath, Institute Vienna Circle, University of Vienna (Source documents: Noord-Hollands Archief, Haarlem, Wiener Kreis Archief, Inv.-Nr. 321).

2001, 369). In 1923, he submitted his habilitation thesis at the University of Vienna. The basis of his thesis was his two-part manuscript, "Beiträge zur Geschichte des Geniebegriffes" (Contributions to the History of the Concept of Genius). The habilitation committee was divided in its opinion of his thesis (Stadler 1997, 569–571). In particular, the philosopher Robert Reininger and the educationalist Richard Meister were against Zilsel. In their view, Zilsel's text focused too strongly on economy, was much too rationalistic and had, in the end, almost nothing to do with what they defined as philosophy. Most probably, however, there were not only purely scientific reasons. Meister, for instance, was a firm opponent of the progressive school reform advocated by Zilsel. Also, both Reininger and Meister were part of a secret society-like, anti-Semitic interest group (called "Bärenhöhle" or "bear cave") whose purpose was to prevent Jewish, liberal and Marxist scholars from being hired at the University of Vienna, as the work of Klaus Taschwer, in particular, has demonstrated (Taschwer 2015). The left Jewish intellectual Edgar Zilsel was thus apparently a victim of this effort to exclude scholars from academic life on the basis of their worldview and family background. Zilsel eventually withdrew his *Habilitationsschrift* and this prevented him from pursuing a career at the University (Dvořák 1990, 8–13; Taschwer 2015, 125–127).

During the interwar period, Zilsel worked as an instructor at the adult education centres in Ottakring, Simmering and Leopoldstadt, as well as holding teacher-training courses at the Pedagogical Institute of the City of Vienna. He also participated in the discussions of the Vienna Circle and is often associated with the left wing of this group, together with Philipp Frank, Hans Hahn, Rudolf Carnap and Otto Neurath. According to the Manifesto "The Scientific World-Conception", he was not a member but among the "authors close to the Vienna Circle" (Stadler & Uebel 2012, 107).

Between 1934 and 1938, he worked at various secondary schools (because his work in the adult education sector was no longer funded under the Austrofascist regime). In the year of the *Anschluss*, in 1938, he fled with his wife and son to England, and from there, travelled to the USA in 1939 (first to New York and later to the West Coast), where he committed suicide on March 11, 1944 (Dvořák 1990, 13).

The German philosopher of science, Wolfgang Krohn (1985), has divided Edgar Zilsel's work into three groups: first, there are the publications on the origins of the modern "ideal of personality" (*Persönlichkeitsideal*) and the cult of genius; second, there are the sociological analyses of ideologies; and third, there are his contributions to the origins of modern science. It can hardly be overlooked that, while these fields can be considered separately from one another, when taken together they reveal a clear program of research.

Otto Neurath was born in Vienna in 1882. He studied philosophy and economics among other subjects in Vienna and primarily in Berlin. Neurath first became known as a sociologist and, above all, as an economist. For instance, he is considered to be the founder of "war economics" and was a leading representative of the socialization theory after the end of the First World War. In Munich, he was appointed by the

Bavarian government as president of the Central Economic Office and started – but eventually failed – to socialize the economy (Sandner 2014, 122–143).

In the 1920s, he became a leading member of the Vienna Circle – to describe him as a "philosopher" is problematic – and a visual educator: he was the initiator and propagator of a picture language that originally bore the somewhat cumbersome name of "Vienna Method of Pictorial Statistics" and would later in the mid-1930s become known worldwide as *Isotype* (International System of Typographic Picture Education) (Burke et al. 2013).

Neurath was extremely active politically. He gave courses at the social democratic workers' university (*Arbeiterhochschule*), and wrote regularly for socialist periodicals (Sandner 2019). Consequently, he emigrated in February 1934 and lived first in The Hague from 1934 to 1940, and then in Oxford (where he founded the Isotype Institute) from 1941 until his death on December 22, 1945.

Edgar Zilsel and Otto Neurath had many things in common: Both were associated with the Vienna Circle, both were active in popular and workers' education, both were convinced socialists and compared notes with leading socialist politicians and political intellectuals such as Otto Bauer, the chief-theoretician of the Social Democratic Workers' Party (SDAP) of Austria. The point of relating these short biographies is to illustrate the different stages of life they were in when they worked on their concept of genius.

8.3 Zilsel: Religion of Genius and Cult of Personality

Edgar Zilsel published his book *Die Geniereligion* (*The Religion of Genius*) in 1918. It was a time of radical change. In the same year, the Republic of German-Austria was founded. For the first time in Austrian history, there was universal suffrage and a number of progressive social policy measures (such as the eight-hour working day) were realized. The labor movement based its program on Marxist theories, and as a mass movement together with the Social Democratic Workers' Party (SDAP), it was, without doubt, a powerful political player. In February 1919, the SDAP won the first post-war elections. At the same time, conservatism dominated in the Austrian provinces and, even in Vienna, right-wing intellectuals heavily criticized parliamentary democracy both before and after 1918. For right-wing conservatives such as Joseph Eberle (and many others) a political system that was dependent on the will of the masses was not capable of leading a community to order and achievement. Only a dictator could do that. "Eberle questioned the intelligence of the masses and lamented the weakness of a democratically elected government" (Wasserman 2014, 40). The anti-democratic right strictly believed in the need for a *Führer*, as Kurt Sontheimer pointed out in his classic book on antidemocratic ideas in the Weimar Republic (Sontheimer 1962, 268–269). In this respect, the political debates in the young Austrian Republic were comparable. Obviously, there was not only a *Red Vienna* (Gruber 1991) but, especially in the intellectual field and in the academy, there was also a *Black Vienna* (Wasserman 2014).

Although not explicitly in his work on the concept of genius, Zilsel would continue to refer to this intellectual climate throughout his writings, especially in his articles for the theoretical journal of the social democratic movement, *Der Kampf (The Struggle)*. His very critical double-review of Karl Jaspers (*Die geistige Situation der Zeit*) and Hans Freyer (*Revolution von rechts*) is only one example (Zilsel 1932).

Edgar Zilsel planned his monograph *Die Geniereligion* as part of a larger intellectual project. There was an announcement in the original edition of 1918 which indicated the publication of a second, "historical-critical" part in the near future, including a table of contents. It took until 1926, however, until a different book, *Die Entstehung des Geniebegriffes*, appeared. In its preface, Zilsel announced further publications on the topic (1926, v) but it remained his final book.

The book *Die Geniereligion* is divided into three parts: "The Dogmatics of the Religion of Genius" (Die Dogmatik der Geniereligion); "Genius Worship and the Genius Ideal" (Die Genieverehrung und das Genieideal); and "On Metaphysical Values. The Ideal of Objectivity" (Über metaphysische Werte. Das Ideal der Sache). The first two parts are divided into several subsections, in which the author presents the key terms of what he characterizes as the dogmatics of genius. These terms include posterity and the contemporary world, the brotherhood of genius, the notion of profundity, objectivity versus personality (der Gegensatz Sache-Persönlichkeit), the relation of genius worship and tolerance, the ideal of personality (Persönlichkeitsideal), hero worship and finally the significance of the genius ideal for human culture. Written in an entirely ironic or sometimes polemical tone, the book primarily criticizes the widespread admiration of genius of the time based on Romantic ideology that, according to the thesis, necessarily entails contempt for the masses.

In the subsection "Views of Posterity and the Contemporary World" (Nach- und Mitweltvorstellungen), Zilsel examines how the genius's work is essentially directed toward posterity, and not his contemporaries. The genius's significance can only be recognized by later generations. Closely entwined with this emphasis on posthumous fame, the image of the misunderstood genius, unrecognized by his contemporaries, proves upon closer examination, as Zilsel points out, to be the exception: most of the artists and scholars who were highly respected in 1918 had been also admired in their own day. This posthumous fame offers the ingenious artist or author some solace for the apparent lack of recognition. The public, on the other hand, reacts with religion-like admiration to the suffering of the genius in the face of his contemporaries' scorn. The misunderstood genius can thereby become a martyr. Such an analysis supports Zilsel's identification of the worship of genius as a religion.

"The Brotherhood of Genius and the Notion of Profundity" (Die Brüderschaft der Genies und der Tiefenbegriff) form a further part of the dogmatics of genius. Was there a brotherhood of genius? Many geniuses failed to respect each other, as Zilsel illustrates with the help of several examples. How does one recognize a genius? Since the masses were "genius blind," in the opinion of the dogmatists of genius, and the geniuses themselves were not always able to recognize one another as such, a third class of people were needed – the experts, the connoisseurs, the literati:

> They are superior to mediocre people. They themselves are unproductive, but they mediate between their men and the masses, so some of the glorious radiance of the genius shines on them: one could call them the priests of the religion of genius. They are described, if not as ingenious, then at least as congenial. (Zilsel 1918/1990, 87)

This group often dismisses the discord among geniuses themselves as merely superficial, for geniuses are bound together by something entirely different: profundity (*Tiefe*). The concept of profundity, according to Zilsel, provides both the kernel and the protective shell of the brotherhood dogma. Thus, geniuses are united in profundity. For Zilsel, however, this profundity (of thought, of feeling, etc.) is nothing other than a term to conceal the lack of (academic) standards and surrender in the face of the question of truth.

> Whoever values profundity more than truth perhaps makes a virtue out of necessity and elevates the lack of any objective standard to a principle; in this way, it is quite easy to rhapsodize about profundity. (Zilsel 1918/1990, 94)

But when people speak of profundity, Zilsel suggest, they are referring to the thinker, not the thoughts. It turns out

> that the reasoning underlying the dogma of the brotherhood of genius is scarcely better than that of the dogmas of posterity and the contemporary world. We are now justified in doubting whether humanity can be so conveniently divided into the two classes of genius and the masses like the somewhat crude concept of the great man would like us to assume. (Zilsel 1918/1990, 100)

Because genius enthusiasm is always enthusiasm for an exceptional individual, a personality (*Persönlichkeit*), and never a matter or a cause (*Sache*), the genius worshipper, according to Zilsel, is *per se* the biased person (*unsachlicher Mensch*). But how are genius worship and tolerance related? The genius worshipper's concept of tolerance is ambivalent, for while tolerance is shown towards the genius, opponents are greeted with nothing but utter intolerance. At least as long as it does not concern another genius, a hostility between geniuses, like the one that existed between Friedrich Nietzsche and Richard Wagner. Genius worship, according to Zilsel, is based on the ideal of personality (*Persönlichkeitsideal*):

> It is no surprise that the superior detective, the misunderstood artist, and the great inventor and naturalist are so popular in our films because they are also exceptional individuals (*Persönlichkeiten*) who, to the great satisfaction of the audience, can look down on their allies and opponents one moment with contempt and the next with compassionate love. (Zilsel 1918/1990, 133)

The concept of genius is thus based on a neglect of objectivity (*Sachlichkeit*) in favour of the experience of personality (*Persönlichkeitserlebnis*). What does this cult of personality mean for society and culture?

> At first, the enthusiast believes the world has been turned upside down when operetta composers fare better than philosophers; then he feels it is morally superior to have tradesmen who want to spend their hard-earned money on the 'Merry Widow' attend a performance of the Ninth Symphony instead; and, finally, he is filled with indignation that a people go to

> war out of petty interests rather than for Goethe, Beethoven, and Kant. To an unbiased mind, it seems that the tradesmen should spend their money on what makes them happy, that petty interests are the only ones for which petty-minded people will risk their lives, and that there is no delusion more dangerous and inhumane than taking something physical away from one's fellow man in order to pay him back with something metaphysical. If questions of war and peace were considered more in terms of "superficial" reasons of utility, if questions of the political and social order were regarded with less contempt for the masses and less metaphysics of personality (*Persönlichkeitsmetaphysik*), then this would be to the greater good of every human society. (Zilsel 1918/1990, 185)

Genius, however, can only serve as an ideal and role model if it is not unattainable:

> Therefore, it appears that rather than focusing on the contempt for the masses, the far more important task today is to lend all those countless fellow humans ridiculed as mediocre by an unclear enthusiasm the *courage* to live life once again according to our harmless and carefree amusements – even when they are bourgeois – and to let dispassionate people (*sachliche Menschen*) dare to show their faces without fear of being accused of being superficial. (Zilsel 1918/1990, 190)

Zilsel's preference for objectivity (*Sachlichkeit*) over personality (*Persönlichkeit*), which is diametrically opposed to Karl Kraus's aphorism "Good opinions are worthless. It depends on who has them", raises a crucial question.[3] How should the value of a matter or a cause be judged? What are the criteria for determining "value"? Contrary to a powerful trend of his time, Zilsel's answer is clearly negative: nature is incapable of answering this question for us. For there are no values in nature. Those who stumble through the world of things and classify them according to their measure of merit miss their mark and inflict violence on living nature. Man alone assigns values, and thus all his value judgments are subjective and relative. Yet values can only be obtained by critical, rational, and objective (*sachliche*) examination. Zilsel warns against absolute values, however. Those who do not express absolute enthusiasm (for the genius), do not harbor absolute contempt for the masses.

In his later book of 1926, *Die Entstehung des Geniebegriffes*, Zilsel analyzed the development of the idea systematically from ancient world to early capitalism. The book is written less polemically, and the focus on the factual is expressed also in its written style. Although he published no more books on genius afterwards, Zilsel addressed the subject in a number of political essays. He discussed the relationship between the *Führer* (which he paralleled in some respects at least with the genius) and the masses, between individualistic and collective perspectives on history (Zilsel 1930a); and he analyzed the metaphysics of the genius as a sociological phenomenon (Zilsel 1930b) as well as an expression of Romantic ideology (Zilsel 1933). Interestingly, there are no references to the concept of genius in his articles on National Socialism, which he published pseudonymously (Richter 1933a, b).

[3] "Gute Ansichten sind wertlos. Es kommt darauf an, wer sie hat" (Kraus, 1919/1986, 224).

8.4 Neurath: Genius and German Climate

It seems that Neurath, on the other hand, first took up the topic during his exile in England. Strictly speaking, however, this is only half the truth. Already in 1910, Neurath edited and translated together with his wife, feminist and socialist Anna Schapire-Neurath, Francis Galton's book *Hereditary Genius* into German (Galton 1910). The German title was *Genie und Vererbung (Genius and Heredity)*. It is often credited as a co-translation but Anna Schapire-Neurath had long since established herself as a professional translator, and Neurath's English at the time was still rather poor, so it was, most probably, her translation. In this book, Galton – in contrast to the Austrian philosopher Otto Weininger, for example – linked genius and ingenuity with heredity. In their introduction, both Otto Neurath and Anna Schapire-Neurath agreed with the author's thesis on genius and heredity but praised also his view on eugenics (Neurath & Schapire-Neurath 1910). The interesting feature about this is that these ideas do not show up in the later work of Anna Schapire-Neurath nor Otto Neurath despite their favorable foreword to the translation of Galton's book.

Otto Neurath's discourse on genius in his exile years was very different from this. Although he has an extremely long list of publications, there are hardly any texts among the hundreds of titles that bear any connection to this topic. The only exception are late newspaper articles. And yet this topic plays an enormously important role in his letters during his English years and in some of his unpublished texts. The term he used to talk about this topic was "German climate", which referred to an intellectual climate and tradition that, according to Neurath, had undergone a specific development in Germany (Sandner 2011, 2014, 284–290). The debate on a German *Sonderweg* (special path) comes to mind when reading his texts, although Neurath never used the term himself (and, in fact, it was introduced only after his death). Neurath wanted to show that certain attitudes and opinions had been able to develop in German literature and philosophy that, while not necessarily leading to National Socialism, had served as a base for it and could help explain why the Nazi-ideology was accepted so readily and positively by wide sections of society.

In a certain sense, Neurath was also reflecting on his own story or biography. A distinctive characteristic of his exile correspondence was that he was continually asking friends and acquaintances about their shared past in Vienna (Sandner 2014, 287–288). It could perhaps be said that he is thinking here of Ernst Mach's critique of the concept of self and its continuity, stability, and unity and applying it to his own biography.

In the exile years, Otto Neurath concerned himself with National Socialism. He had been observing for years from the outside, so to speak, first from the Netherlands and then from England. Otto Neurath's published texts offer only a small glimpse into the extent to which the reasons for the rise of National Socialism occupied his mind. While he did publish two articles pseudonymously in an emigrant newspaper, *Die Zeitung – Londoner deutsches Wochenblatt* (*The Newspaper – The London German Weekly*) (Schlosser 1944; Feltner 1945), the true importance of this topic to him first becomes apparent in the correspondence of this time. It offers a clear picture of the intensity and the amount of energy he invested into advocating his thesis and trying to convince other people of it. His correspondence with German socialists Carl Herz and Bernhard Reichenbach, with former Viennese friends such as Dora Lucka or Josef Frank, with his sister-in-law Rosa Schapire, and with many others is full of reflections on the "German Climate."[4] In addition, there exists a whole series of unpublished texts in which he addressed this issue and dealt with it in a variety of ways.[5]

In brief, his thesis was as follows: the rise and rule of National Socialism and its great support and acceptance among the people were the result of a specifically "German Climate." According to Neurath, the two main features of this "human climate" – which, incidentally, was the title of one of the two articles mentioned above – were the idea of the genius and the categorical imperative. Throughout the history of culture, the figure of the genius had always been the embodiment of the extraordinary, the metaphysical, and sometimes even the demonic and was generally exempt from the rules of society. A genius could do things that ordinary people were not able or not allowed to do. He often associated the cult of genius specifically with Goethe – both as the author of *Faust* as well as a genius-like figure in the history of literature. The other, closely related side was Kant's categorical imperative in which, as Neurath emphasized, the individual's fulfillment of duty was valued higher than his own conscience. He was convinced that this mix of genius cult, on the one hand, and duty and obedience, on the other, created the foundation for the positive acceptance of National Socialist ideology.

Published in February 1945, the article "Das Genie und der kategorische Imperativ" ("The Genius and the Categorical Imperative") is an unusual, essentially literary text (Feltner 1945). It focuses on three protagonists – Karl, Arthur, and Anna. The story is told in the first person from the point of view of Arthur (Arthur Feltner being one of the author's pseudonyms), but it is largely a conversation between the three in direct speech. Neurath, who clearly shares Karl's views, is obviously recreating a situation that he himself has experienced innumerable times as his letters, as previously mentioned, bear witness. Karl is Otto Neurath's middle name, and he wrote several articles under the name of Karl Wilhelm in 1918/19. At one point, Karl says: "For the extraordinary person, deemed so by the others, the normal rules do not apply. He is the genius person, the demonic person, the person with intuition." And the others? "They have their 'categorical imperative,' the feeling that they have to violate their own self if moral law requires it" (Feltner 1945, 7).

[4] Otto Neurath to Josef Frank (1219/5), Otto Neurath to Dora Lucka (1221/37-42), Otto Neurath to Bernhard Reichenbach (1223/40), Otto Neurath to Carl Herz (1220/11), Otto Neurath to Rosa Schapire 1224/2–6). All documents are part of the Papers Otto and Marie Neurath, *Handschriftensammlung*, Austrian National Library.

[5] Neurath, Otto: "Das menschliche Klima Deutschlands" (204/ K. 65); "Eine Aussprache über das deutsche Klima" (204/K. 66); "Deutschlands menschliches Klima I: Der geniale Mensch und der kategorische Imperativ" (204/ K. 67). All documents are part of the Otto Neurath Papers, Noord-Hollands-Archief, Haarlem, Wiener Kreis Archief); "Vom deutschen menschlichen Klima" (Papers Otto and Marie Neurath, *Handschriftensammlung*, Austrian National Library, Ser.n. 42.317).

Of course, Neurath's analysis of the "German Climate" was not limited to Goethe or Kant; he also repeatedly quoted at length from authors whose world view was either closely related to that of National Socialism or who could be instrumentalized in this way, like Otto Weininger, Friedrich Nietzsche, Paul de Lagarde, Julius Langbehn, Hans Grimm, Oswald Spengler or Houston Stewart Chamberlain. His attack, however, struck at the heart of German literature and philosophy. But what exactly did Neurath's human climate (be it German or Austrian or English) mean? He was very skeptical about the idea of equating a human climate with a national character. However, he insisted that there was a close connection between certain features and attitudes expressed in philosophy or literature and the behavior of people. The human climate was a kind of ensemble of different attitudes and patterns of behavior. The fact that the climate of a country influenced the people living in it did not necessarily entail that all the people had read the respective literature. Of course, this was not the case. However, it did allow the literati to draw on these thoughts to justify their actions while the masses were made acquainted with these thoughts through the newspapers or in school. In one of his unpublished papers, he describes this mechanism as follows:

> It seems a more likely hypothesis that kind and friendly German boys and girls may become influenced by the over emphasis on certain things in the tradition of the 'best' German literature, and so be indirectly prepared for Nazidom, rather than that we subscribe to the Vansittart theory,[6] that on an average Germans are butcherbirds by birth. (Otto Neurath: Education in Occupied Germany – Intricate Problems. Otto Neurath Papers, Nord-Hollands-Archief, Haarlem, Wiener Kreis Archief 205/K. 74)

Neurath's thesis received only very limited support. Even the editors of *Die Zeitung (The Newspaper)* were careful to distance themselves in an introductory paragraph and hardly any of the people with whom he corresponded fully shared his views – some, in fact, were violently opposed. The correspondence with Rosa Schapire is only one of many examples.[7] Neurath, however, insisted on his view and wished to convince all those who were skeptical or opposed to his thesis on German climate.

The thesis of a specific "German climate" that led along a *Sonderweg* (special path) to National Socialism has certain weaknesses. If the Austrian climate was different, why did its people support National Socialism–not less than the Germans? Although Neurath seems to have been fully aware of Austria's role in National Socialism, the concept of the genius did not play any role in what he called the "Austrian human climate" (Schlosser 1944, 9). If a "human climate" develops over generations – could there be any chance for a short-term re-education? Against the background of his experiences with National Socialism, Neurath wanted, at any

[6] Robert Lord Vansittart, former Permanent Under-Secretary of the British Foreign Office, expressed the view in his books *Black Record* (1941) or *Bones of Contention* (1945) that the majority of Germans were "unsuited to democracy" und that the outbreak of the Second World War could be attributed to their historically conditioned, exceptional aggressiveness.

[7] Correspondence Otto Neurath and Rosa Schapire, Papers Otto and Marie Neurath, *Handschriftensammlung*, Austrian National Library, 1224/2–6.

rate, to reconsider the entire mental, cultural and intellectual tradition. Neurath's approach here was radical in the truest sense of the word: he traced the inhuman, often irrational and metaphysical, back to its roots, with the aim of not only drawing straight lines to totalitarianism but also of showing which dispositions facilitate acceptance of National Socialist thinking and prevent the inhuman from being met with firm opposition. Of course, much of what fell "victim" to this attempt was considered "holy" by friends and former kindred spirits and was an essential part of their education and culture. In this way, Neurath partially maneuvered himself if not into isolation, then at least into a niche in which only a few friends and acquaintances were willing to follow him.

In the course of this analysis, Neurath also began to reconsider his own history. Especially in his correspondence with people whom he knew from his time in Vienna, he repeatedly asked his correspondents to supplement and confirm his memories. He asked about people and events in Vienna, about moods and mentalities, and over and over again, about anti-Semitism. The correspondence with his old Viennese friend Dora Lucka is probably the best example.[8] He came to see himself more and more now in his previous role as an outsider of a dominant political and cultural climate in which there was little sympathy for his views.

8.5 The Critique of Genius and Its Political Relevance

The "genius" was addressed as a godlike being, a demiurge, or Christ-like figure who, at the same time, labored in the pursuit of modern science and knowledge, wrote Julia Köhne in her book on the *Geniekult* around 1900 (Köhne 2014, 528). According to Zilsel, the discursive existence of the "genius" functioned on the basis of religious-dogmatic conditions and the postulated belief, admiration, and enthusiasm of the idolizing group. Zilsel described the "Geniereligion" as a response to de-sacralizing trends, as at once a conscious and unconscious strategy, created mostly by scholars and researchers, to justify anti-egalitarian politics and metaphysics. For him, the dominance of the "notion of the genius personality and of profundity" indicated a "severe danger" for the age (Zilsel 1918/1990, 234; Köhne 2016, 117). Otto Neurath connected the idea of the Genius directly with National Socialism but in contrast to some earlier writings, he did not refer explicitly to Zilsel's concept anymore. It was, of course, a great difference between the approaches that Zilsel published his books in 1918 and 1926, while Neurath reflected on the genius in the light of the experiences with Fascism and National Socialism. In contrast to many other émigrés, he insisted on the relation of German culture and Nazi barbarism. He rejected approaches that at first glance seemed to closely resemble his own such as

[8] Correspondence Otto Neurath and Dora Lucka, Papers Otto and Marie Neurath, *Handschriftensammlung*, Austrian National Library, 1221/22–35, 1221/37–42.

that of Robert Vansittart, although he did not completely reject the Vansittart thesis of a particular German tradition of militarism and brutality.

However, there may be a surprising closeness of (parts of) his approach to that of someone whom he repeatedly included as a typical representative of German culture: Thomas Mann. Initially, Neurath referred to the author of *Reflections of a Nonpolitical Man* (*Betrachtungen eines Unpolitischen* 1918). The later Thomas Mann, however, the emigrant, stressed the role of philosophical idealism, Romanticism and the romantic school as ideal preconditions for the Nazi ideology (Görtemaker 2005, 196–205, Mann 1945) – as Neurath did as well. Despite the different thematic, cultural, and political contexts of their reflections, both Edgar Zilsel and Otto Neurath understood that the idea of genius was not only a question of culture. It also had heavy political consequences.

Acknowledgments Austrian Science Fund (FWF): P 31500-G32. My special thanks goes to Andrea Kraus who translated parts of the text into English and to Christopher Burke for comments and discussion.

References

Blüher, Hans. 1912a. *Wandervogel: Geschichte einer Jugendbewegung. Erster Teil: Heimat und Aufgang*. Berlin: Weise.

———. 1912b. *Wandervogel: Geschichte einer Jugendbewegung. Zweiter Teil: Blüte und Niedergang*. Berlin: Weise.

———. 1912c. *Die deutsche Wandervogelbewegung als erotisches Phänomen: Ein Beitrag zur Erkenntnis der sexuellen Inversion*. Berlin: Weise.

Burke, Christopher, Eric Kindel, and Sue Walker, eds. 2013. *Isotype: Design and contexts, 1925–1971*. London: Hyphen Press.

Dvořák, Johann. 1981. *Edgar Zilsel und die Einheit der Erkenntnis*. Wien: Löcker.

———. 1990. Zu Leben und Werk Edgar Zilsels und zur Soziologie des Geniekultus. In: *Edgar Zilsel: Die Geniereligion. Ein kritischer Versuch über das moderne Persönlichkeitsideal, mit einer historischen Begründung*. Herausgegeben und eingeleitet von Johann Dvořák, 7–40. Frankfurt/M.: Suhrkamp.

Feltner, Arthur (=Otto Neurath). 1945. Das Genie und der kategorische Imperativ. In: *Die Zeitung. Londoner deutsches Wochenblatt*, No. 415, 16. February 1945: p. 7.

Filla, Wilhelm. 2001. Wissenschaft für alle – ein Widerspruch? Bevölkerungsnaher Wissenstransfer in der Wiener Moderne. In *Ein historisches Volkshochschulmodell*. Innsbruck, Wien, München: Studienverlag.

Fleck, Christian. 2015. *Etablierung in der Fremde. Vertriebene Wissenschaftler in den USA nach 1933*. Frankfurt am Main: Campus.

Galton, Francis. 1910. *Genie und Vererbung. Autorisierte Übersetzung von Otto Neurath und Anna Schapire-Neurath*. Leipzig: Klinkhardt.

Görtemaker, Manfred. 2005. *Thomas Mann und die Politik*. Frankfurt am Main: Fischer.

Gruber, Helmut. 1991. *Red Vienna: Experiment in working-class culture 1919–1934*. New York/ Oxford: Oxford University Press.

Hirsch, Julian. 1914. *Die Genesis des Ruhmes. Ein Beitrag zur Methodenlehre der Geschichte*. Leipzig: Johann Ambrosius Barth.

Köhne, Julia B. 2014. *Geniekult in Geisteswissenschaften und Literaturen um 1900 und seine filmischen Adaptionen*. Wien, Köln, Weimar: Böhlau.

———. 2016. The Cult of the Genius in Germany and Austria at the Dawn of the Twentieth Century. In *Genealogies of Genius*, ed. J.E. Chaplin and D.M. McMahon, 115–135. Houndmills, Basingstoke, Hampshire: Palgrave Macmillan.

Kraus, Karl. 1919/1986. *Aphorismen. Sprüche und Widersprüche, Pro domo et mundo, Nachts. Karl Kraus Schriften, Band 8*. Frankfurt am Main: Suhrkamp.

Krohn, Wolfgang. 1985. Edgar Zilsel zur Methodologie einer exakten Geisteswissenschaft. In *Philosophie, Wissenschaft, Aufklärung. Beiträge zur Geschichte und Wirkung des Wiener Kreises*, ed. Hans-Joachim Dahms, 257–275. Berlin, New York: De Gruyter.

Mann, Thomas. 1945. *Deutsche Hörer! Fünfundfünfzig Radiosendungen nach Deutschland. Zweite, erweiterte Ausgabe*. Stockholm: Bermann-Fischer.

Neurath, Otto. 1930/1981. Der Erfolg. Rezensionsaufsatz zu: Gustav Ichheiser, Kritik des Erfolges. Eine soziologische Untersuchung. In *Otto Neurath. Gesammelte philosophische und methodologische Schriften*, ed. R. Haller and H. Rutte, 363–367. Band 1. Wien: Hölder-Pichler-Tempsky.

———. 1931/1981. Empirische Soziologie. Der wissenschaftliche Gehalt der Geschichte und Nationalökonomie. In *Otto Neurath. Gesammelte philosophische und methodologische Schriften, Band 1*, ed. R. Haller and H. Rutte, 423–527. Wien: Hölder-Pichler-Tempsky.

———. 1941/1981. Universaljargon und Terminologie. In *Otto Neurath. Gesammelte philosophische und methodologische Schriften. Band 2, ed. R. Haller and H. Rutte, 901–918. Wien: Hölder-Pichler-Tempsky.*

Neurath, Otto., Schapire-Neurath, Anna. 1910. Vorwort und Einleitung. In *Galton* (1910): i–vii.

Raven, Diederick, and Wolfgang Krohn. 2000. Edgar Zilsel: His Life and Work (1891–1944). In *Edgar Zilsel: The Social Origins of Modern Science*, ed. D. Raven, W. Krohn, and R.S. Cohen, xix–lix. Dordrecht: Springer.

Richter, Rudolf. 1933a. SA. Philosophiert. *Der Kampf* 26: 393–402.

———. 1933b. Das Dritte Reich und die Wissenschaft. *Der Kampf* 26: 486–493.

Saitschick, Robert. 1900. *Genie und Charakter. Shakespeare, Lessing, Goethe, Schiller, Schopenhauer, Wagner*. Berlin: Hofmann.

Sandner, Günther. 2011. The German Climate and Its Opposite: Otto Neurath in England, 1940–45. In *Political Exile and Exile Politics in Britain after 1933*, ed. A. Grenville and A. Reiter, 67–85. Amsterdam, New York: Rodopi.

———. 2014. *Otto Neurath. Eine politische Biographie*. Wien: Zsolnay.

———. 2019. Science and Socialism. Otto Neurath as a Political Writer (1919–1932). In *Neurath Reconsidered: New Sources and Perspectives*, ed. J. Cat and A.T. Tuboly, 67–87. Cham: Springer.

Schlosser, F. (=Otto Neurath). 1944. "Österreichs menschliches Klima." In: *Die Zeitung. Londoner deutsches Wochenblatt*, No. 399, 27 October 1944: p. 9.

Sontheimer, Kurt. 1962. *Antidemokratisches Denken in der Weimarer Republik. Die politischen Ideen des deutschen Nationalismus zwischen 1918 und 1933*. München: Nymphenburger Verlagshandlung.

Stadler, Friedrich. 1997. *Studien zum Wiener Kreis. Ursprung, Entwicklung und Wirkung des Logischen Empirismus im Kontext*. Frankfurt am Main: Suhrkamp.

Stadler, Friedrich, and Thomas Uebel, eds. 2012. *Wissenschaftliche Weltauffassung. Der Wiener Kreis. Reprint der Erstausgabe*. Wien, New York: Springer.

Taschwer, Klaus. 2015. *Hochburg des Antisemitismus. Der Niedergang der Universität Wien im 20. Jahrhundert*. Wien: Czernin.

Wasserman, Janek. 2014. Black Vienna. In *The radical right in the Red City, 1918–1938*. Ithaca and London: Cornell University Press.

Zilsel, Edgar. 1918/1990. *Die Geniereligion. Ein kritischer Versuch über das moderne Persönlichkeitsideal, mit einer historischen Begründung*. Herausgegeben und eingeleitet von Johann Dvořák. Frankfurt am Main: Suhrkamp.

———. 1926. *Die Entstehung des Geniebegriffes. Ein Beitrag zur Ideengeschichte der Antike und des Frühkapitalismus*. Tübingen: Mohr.

———. 1930a. Weltgeschichte und Einzelmensch. In *Das Jahr 1930. Österreichischer Arbeiterkalender*, 59–67. Wien: Herausgegeben von Joseph Luitpold Stern.

———. 1930b. Soziologische Bemerkungen zur Philosophie der Gegenwart. *Der Kampf* 23 (10): 410–424.

———. 1932. Die geistige Situation der Zeit? *Der Kampf* 25: 168–176.

———. 1933. Die gesellschaftlichen Wurzeln der romantischen Ideologie. *Der Kampf 26 (3): 154–164.*

Zilsel, Paul. 2004. Über Edgar Zilsel. In *Vertriebene Vernunft II. Emigration und Exil österreichischer Wissenschaft 1930–1940. Teilband 2*, ed. F. Stadler, 929–932. Münster: Lit.

Chapter 9
Philipp Frank's Relativism: Presentation, Appreciation, and Critique

Philippe Stamenkovic

Abstract In his 1950 *Relativity: a Richer Truth*, Philipp Frank promotes a generalized "relativism", or "doctrine of the relativity of truth", requiring the specification of the meaning of the terms used in the statements one makes (their "relativization"), not only in physics but also in ethics, politics or everyday experience. This is achieved by providing the "frame of reference" relative to which the terms (and consequently statements) are expressed and on which their truth in fact depends, in analogy with Einstein's (special) theory of relativity. He grounds this application of the concept of relativity outside physics on his empiricist interpretation of Einstein's theory, together with a pragmatic and operationalist theory of meaning, which insists on the qualification, and thus relativity, of particular statements. Frank's interpretation of relativity (both within and outside physics) is fully understandable not only because of his logical empiricist stance, but also given his fight against philosophical misinterpretations of the theory of relativity, as well as against political totalitarianism. It also has particularly convincing features. Still, Frank's interpretation of relativity in physics is debatable, since Einstein's theory could just as well be read as a defense of absolutism. What is more, his application of relativity to ethics and politics is disputable, as it allows for a kind of moral relativism.

Keywords Philipp Frank · Relativity · Relativism · Absolute values · Operationalism

P. Stamenkovic (✉)
Ben Gurion University of the Negev, Beer Sheva, Israel

C. Damböck, A. T. Tuboly (eds.), *The Socio-Ethical Dimension of Knowledge*, Vienna Circle Institute Yearbook 26, https://doi.org/10.1007/978-3-030-80363-6_9

9.1 Introduction

9.1.1 Presentation

A prominent member of the Vienna Circle, Philipp Frank gathered in his short book *Relativity – a richer truth* (1950/1951b)[1] his interventions between 1940 and 1947 at the annual "Conference on Science, Philosophy and Religion" (CSPR), which took place every year between 1940 and 1951 in New York City.[2] According to Frank, this Conference originated in the widespread belief (apparently shared by some members of the Conference itself[3]) that the political and moral breakdown of the West in the face of totalitarian ideologies during World War Two had to do with "agnostic and sceptical attitudes – a disbelief in absolute values" or a "relativism" (more about these terms below) intrinsically associated with science (11). On the contrary, Frank's contributions to the Conference intend to show that the "'relativism' [of science] has not the slightest thing to do with agnosticism or scepticism, that *it is in no way hostile to the belief in ethical or democratic values*, that it is accompanying every advance in science and is nothing but a significant representation of the enrichment of human expression which is inseparably connected with our gradually increasing experience" (12, original emphasis).[4] To do so, Frank's book mainly provides a "doctrine of the relativity of truth" based on his interpretation of the concept of relativity in physics, but which he also applies to domains other than science, namely morality, politics and everyday life.[5] The concept of relativity is thus central to the book (as its title suggests), but the book also provides insights into other (although related) aspects of Frank's philosophy, most notably his endorsement of (logical) "positivism" and of the "operational" and "pragmatic

---

[1] Hereafter *RRT*. There is a German edition of the book entitled *Warheit – Relativ oder absolut?* (Frank 1952), whose title better illustrates the content of the book, as Bloor (2011, p. 429) remarks.

[2] For an introduction to the historical context of this conference, as well as to Frank's book and work, see Nemeth (2003, 2010), Reisch (2005, 220–224, 311, 316, 362, 379), (2017) and Siegetsleitner (2014, 251–264), (2017). All page numbers without reference will refer to *RRT*.

[3] See Frank's remark on p. 12: "[...] the Conference was about to jump 'out of the frying pan into the fire'. I saw that the fight against 'relativism' could easily degenerate into a fight against the spirit of modern science." For a critique of some anti-scientific opinions held in this Conference, see Nagel (1943).

[4] Frank (1949, 52) summarizes *RRT* by denying that the "'relativism' of modern science" would have been harmful to the establishment of objective values in human life. "[In *RRT* he] made an argument to prove that the 'relativism of science' has also penetrated every argument about human behavior. 'Relativism' is not responsible for any deterioration of human conduct. What one calls 'relativism' is rather the attempt to get rid of empty slogans and to formulate the goals of human life sincerely and unambiguously."

[5] Frank (1949, 230 sqq.) also mentions the "doctrine of the 'relativity of truth'", which "seemed therefore to some people a 'social danger', since it might contribute to a disbelief in the 'absolute values' of ethics."

theory of meaning"[6]. *RRT* is the main locus where Frank explicitly presents and promotes his peculiar version of relativism, although there are a few other places where he does so (1946, 1949, 52, 230 sqq.),[7] as well as rare passages where he does it less clearly and more implicitly (1957b, 143–44, 155), or indirectly by endorsing the "relativism" of other authors such as Poincaré or Mach (1949, 9, 17).

RRT has been studied in several books and articles (see the references mentioned in footnote 2), which provide a good historical overview of it. However, a detailed systematic account of Frank's relativism, and in particular of its weaknesses, is still lacking. Ignoring *RRT*, Uebel (2000, 149) even claims that Frank is not a relativist (in the sense of the Strong Program of the sociology of scientific knowledge (Bloor 1976)): this is understandable since, apart from the above-mentioned few instances, Frank's work does not display any relativism, but rather attempts to (rationally, or rather *scientifically* in the large sense of science) justify the choice of scientific theories by having recourse to extra-scientific (in the strong sense of the natural sciences) factors (see Sects. 9.3.2 and 9.4). Similarly, Howard (2003, 78) argues that Frank's "holistic, underdeterminationist variety of conventionalism provides a robustly empiricist, nonrelativist understanding of the role of social factors in theory choice, one in which theory choice is constrained by evidence – but not up to the point of uniqueness – leaving a space within which both the freely creative imagination of the scientist and social factors can play a role".

On the contrary, directly denying Uebel's (2000) claim, Bloor (2011, 428) calls Frank a "relativist", and fully endorses his conception. Bloor (2011, Ch. 10) gives a brief and fairly well informed account of Frank's relativism as expressed in *RRT*, and interestingly uses it to support his thesis of a relativistic understanding of aerodynamical knowledge in its historical development between the Cambridge and Göttingen schools. There are, however, inaccuracies in Bloor's account, and he does not criticize Frank's view. Indeed Bloor insists (2011, 429) on Frank's claim that there are no "absolutist", in the sense of non-revisable, claims in science, although this is not the main point of the book – which is about the contextualization of the claims. Similarly, Bloor somewhat strangely and inaccurately summarizes Frank's relativism as the view according to which "[scientific knowledge] is relative to whatever causes determine it. There are as many 'relativities' as there are causes. That is the point: knowledge is part of the causal nexus, not something that transcends it" (2011, 430).[8]

---

[6] Although it is studied throughout the book, the concept of relativity is mainly addressed in the first part. The second part mainly deals with other (although related) considerations, most notably: the shortcomings of the US educational system and its responsibility in bad interpretations of modern science, which are to blame (according to Frank) on its failure to integrate the sciences between themselves, and with philosophy.

[7] Strangely, Frank (1949, 52) endorses "'relativism'" and "'the relativism of science'" (with quotation marks), just after including the "relativistic" interpretation of science – along the "idealistic" and "materialistic" ones – in the "metaphysical interpretations of science" (51). He seems to oppose two acceptations of relativism: a metaphysical, wrong one; and his own.

[8] Bloor (2011, 428) also interprets Frank's (1954–55/1956a) paper on "The Variety of Reasons for the Acceptance of Scientific Theories" as a further illustration of Frank's relativism, fully endors-

In this article I will first present Frank's relativism or "doctrine of the relativity of truth", before appreciating and criticizing it in a second part. In the first part I will show how Frank's relativism basically is about specifying the meaning of the terms used in the statements one makes, which amounts to specifying the "frame of reference" relative to which their truth holds. In the second part I will show that Frank's attempt to defend relativism on the basis of the (special[9]) theory of relativity is unconvincing, since Einstein's theory could just as well be read as a defense of absolutism. I will also criticize the moral relativism which Frank derives from his "doctrine of the relativity of truth" and advocates, and claim that certain moral demands are true without any further relativizing qualifications. This will lead me to conclude on the limits of Frank's application of physical concepts such as relativity to other, non-scientific, domains (such as morality and politics).

9.1.2 *Preliminary Remarks*

Before going into the details of Frank's conception, let me make some preliminary remarks.

First, it is tempting to consider these rare instances of Frank's relativism as inspired by Einstein's theories of relativity (which Frank knew very well indeed), but one should not forget the properly interpretational elements of Frank's philosophy of science (especially his operationalistic conception of the theory), which favor such a relativistic reading, as we shall see Sect.9.3.3) – in contradistinction to *e.g.* Cassirer (1921) who would insist on the invariant features of the theory, in conformity with his idealistic and top-down approach, as it were (Stamenkovic 2015, 2017).

Second, one must keep in mind that *RRT* (gathering Frank's contributions to the CSPR) is dedicated to a large public. This may explain Frank's rather loose use of expressions such as "relativism", "agnosticism" or "skepticism" (see Sects. 9.2.1.1 and 9.2.2.4). However, this should not prevent us from thoroughly analyzing it: a good popularization book should not distort the opinions which its author expresses

ing the utilitarian conception of scientific theories displayed by Frank there – for whom the latter have either a "technological" or a "sociological" purpose (1954–55/1956a, 16), and the compromises between their various properties when deciding which one to accept are presented as akin to those made while choosing between the various properties of an aircraft while building it. However, what Bloor does not say is that although Frank recognizes that theories are underdetermined by evidence (making talk of "the [unique] truth" of a theory irrelevant (14), and providing grist for his instrumentalist mill) and that there is no "perfect simplicity" of a theory (3), Frank nonetheless ultimately aims at justifying the choice of scientific theories by extra-scientific (in the strict sense, *i.e.* political, moral or religious) factors, thanks to the "general science of human behaviour" he is envisioning, and which is supposed to incorporate the sociology of science (18).

[9]The only one Frank (1950/1951b) mentions. Frank (1957b, 155) also refers to the general theory of relativity.

in other, more academic, settings[10] – in particular in the case of such an important venue as the CSPR. Besides, Frank also expresses the same relativism in other, more academic places, as noted above. *RRT* is clearly the most detailed account of his relativism, and therefore justifies a thorough analysis.

Finally, and more importantly, taking into account Frank's previous work, as well as the historical context of the CSPR, will help better understand Frank's relativism as expressed in *RRT*, by showing his long standing fight against philosophical misinterpretations of physical theories (and in particular Einstein's theories of relativity), as well as his critique of totalitarian ideologies and political absolutism. However, it is not my purpose here to write a historical article, and I will only provide a short contextualization in the discussion section. I think that one should be allowed to criticize philosophical positions and test their normative power outside their historical setting, although this may imply anachronism to some extent.

9.2 Frank's Relativism

9.2.1 The "Doctrine of Relativity of Truth"

9.2.1.1 Relativism, Relativity of Truth and Science

Why has the relativism of science led to "the disbelief in absolute values", as Frank diagnoses in the introduction of *RRT*? Frank first mentions "a definite attitude of mind [which] has arisen" since the sixteenth century, "in science as well as in its application to human conduct", which "refuse[s] to take anything for granted":

> "Essentially the new attitude does not believe in statements the truth of which, once established, can be trusted for ever. This attitude has been called positivism, pragmatism, relativism, operationalism, and so on" (19–20).

This sentence shows the composite (and simplistic, according to Toulmin 1951) nature of *RRT*, which associates relativism with positivism, pragmatism and operationalism. Note that here Frank mentions the revisability of statements, not their contextualization – which is in fact the main point of *RRT*, as we shall see. According to Frank, this attitude, "based upon the methods of science", has been "instrumental in human progress", in "thought" as well as in "life" (*ibid.*).

Note that Frank uses the expressions of "relativism" and "the (doctrine of the) relativity of truth" synonymously. Indeed, relativism about truth can be considered as a generalized relativism from which other forms of relativism (such as epistemic or moral relativism) derive (see Sect. 9.3.2). Frank explicitly states that the relativity of truth plays the same role in ethics and in science: namely it enables to qualify

[10] Indeed Toulmin's (1951) somewhat condescending characterization of Frank's book as "logical empiricism told to children" should not be taken at face value, as Reisch (2017, 236) remarks (see Sect. 9.3.1).

general (scientific or ethical) principles by operational definitions (see *e.g.* 48). Note also that relativism seems to ground, as well as to exceed, science, in so far as it lies at the basis of the "scientific attitude", but also of "liberal Christianity and reformed Judaism", which are "offsprings of the 'relativity of truth'" (*ibid.*).

Finally, note that Frank does not seem to fully endorse these expressions of "relativism" and "relativity of truth", as his repeated use of quotation marks and of the expression "so-called 'relativism'" (*e.g.* 12, 20) show. However, he does place relativism at the basis of the "scientific attitude" (*e.g.* 21): thus, apart from the right choice of word, relativism is indeed a fundamental concept for him. In his (1949) *Modern Science and its Philosophy*, the word relativism without quotation marks is used pejoratively (*e.g.* to qualify the attitude of the Catholic Inquisition p. 91), and positively with quotation marks (to qualify his own conception, of his forthcoming *RRT*, 52). According to Bloor (2011, 430), the quotation marks of the expression "relativity of truth", symmetrical of those of "absolute truth", may indicate that Frank was conscious that both were "meaningless pseudopropositions": "absolute truth" would be meaningless for a positivist like Frank, and relativism, if taken as the negation that there are any absolute truths, would also be meaningless (as the negation of a meaningless proposition).

9.2.1.2 Relativization of Statements

In general, the "relativity of truth" applies to any statements, and is implemented by their "relativization" or "qualification", *i.e.* the specification of the meaning of the terms used in them. Equivalently, it requires the specification of the "frame of reference" (27) relative to which they are expressed and on which their truth in fact depends. To illustrate this, Frank (23) explains how the meaning of the words "above" and "below" has changed since the discovery of the antipodes (now formulated with respect to gravity in a specific framework). In order to have a "definite meaning", the statement "my head is above my feet" must be completed by a "qualification" which specifies the gravity considered at a specific place: my head is above my feet relative to the gravity at my place, or my feet are above my head relative to the gravity at the place of my antipodes.

This process of relativization applies to moral statements as well. Thus a statement like "John is wicked" has no precise meaning until it has been qualified, *i.e.* until it has been supplemented by adequate specifications (see Part I, Ch. III). For a young child it may just mean that he does not obey his parents, but as soon as he goes to school it may also mean that he does not obey his teacher. What is more, the child's parents may be criminals, in which case it is obeying them which is wicked. One might be tempted to turn to a "higher authority", like the chief of the nation, but he himself may be a criminal. Ultimately, one may want to equate "being wicked" with "being disobedient to the supreme authority", but the latter might be God's will (which needs to be interpreted, according to a method which needs to be specified), a supreme ethical law (like Kant's categorical imperative, which has "hundreds of interpretations") or the voice of our own conscience (in which case the statement must indicate the way to test it)... In the end, it is impossible to assign any definite meaning to a simple, unqualified statement like "John is wicked" without further specification.[11] Thus no "absolute truth" in the sense of an "unqualified truth" can be obtained (28).

9.2.1.3 Relativization in Physics

Concerning the relativization (or qualification) of the notions of rest and motion, the history of science (starting with Galileo and Newton) shows that it became increasingly clear that they had no meaning if we did not specify a frame of reference relative to which our statements were to be understood. Furthermore, in Einstein's (special) theory of relativity, "we learned of the contraction of yardsticks by motion, the slowing of the rate of clocks by motion, and so forth" (29). This, of course, refers to length contraction and time dilation according to the state of movement of the observer. More precisely, because of the Lorentz transformations one can write, with $\gamma = \frac{1}{\sqrt{1-\frac{v^2}{c^2}}} > 1$, v being the relative velocity between an observer (measuring an object) and the object[12]; $L = L_0/\gamma$, where L is the length of the object observed by the observer in motion relative to the object and L_0 is the proper length of the object (the length of the object in its rest frame). Similarly, if the object under consideration is a clock, $T = \gamma T_0$ expresses the fact that the observer's period of the clock T is longer than the period T_0 of the clock in its own resting frame (meaning that time passes more slowly for the observer than for the clock).

> Therefore the result of any measurement of length or time is ambiguous unless we know the motion of the yardstick and the clock which have been used to carry out this measurement. A plain statement such as 'this table is three feet long' is now incomplete and, therefore, meaningless. For this statement does not contain the description of the motion of the yardstick used. We have to replace this meager statement by a richer one. Its form has to be: 'This table is three feet long measured with a yardstick which is at rest with respect to this room.' The same fact can be expressed more briefly by stating: 'This table is three feet long relative to this room.' In this sense we say that 'length' has been 'relativized' by twentieth-century science. (p. 30)

[11] "For if we want to know whether or not in a particular case a particular John is 'wicked' we have to find out whether or not he was actually obedient to God's will or to Kant's imperative" (28).

[12] Note that both relations (length contraction and time dilation) are symmetric, *i.e.* they keep the same form whatever the point of view taken (of the object at rest and the observer moving, or the converse), precisely because of the principle of relativity (according to which the laws of nature have the same form in all inertial reference frames). Thus there is no need to specify that the observer is at rest and the object moving, or the converse: what matters is the relative speed v between them. But the length contraction takes place only in the direction of the movement, *i.e.* if both L and L_0 are measured parallel to the direction of the vector $\mathbf{v}$.

Note also that Frank makes a confusion in his example on p. 46: the table should be the longest (5 feet), *i.e.* should have its proper length, in its rest frame which is the Earth, and should be shorter (3 feet) in the reference frame of the stars which is moving relative to the earth.

Thus, this process of "relativization" or "qualification" does not imply a weakening of objectivity, on the contrary: things are more precisely described, and once the appropriate frame of reference is specified, the statement becomes "a definite assertion" (p. 30). To use a Kantian parlance (as Nemeth 2003, 121 does), we might thus say that relativization *constitutes* objectivity.

What is more, relativization also represents an enrichment of the description of scientific facts. As the theory of relativity illustrates, the history of science shows a characteristic trend of "more and more additions and qualifications in the description of [the] facts" continuously discovered (29). This relativization trend leads to an enrichment of the language we use to describe facts: "[r]elativism means the introduction of a richer language which allows us to meet adequately the requirements of an enriched experience" (30). In this way, relativism enables to take into account the complexity of natural as well as human situations with a sufficiently rich language: "[...] this so-called relativity of truth is nothing more and nothing less than the admission that a complex state of affairs [whether natural or human] cannot be described in an over-simplified language [whether scientific or ordinary]" (52).

9.2.2 Relativism in Ethics

Since the "relativity of truth" is not opposed to, but on the contrary constitutive of objectivity, Frank extends this theoretical reasoning to the moral realm: "Since the doctrine of the relativity of truth does not imperil the 'objectivity of truth', it is hard to understand how the 'relativism' of twentieth-century science could shock the belief in the 'objectivity of human values'." (47). Let us see how he proceeds.

9.2.2.1 Two Kinds of Values

Frank proposes to distinguish between two kinds of values.[13] Since his text is somewhat unclear,[14] it is worth quoting it at length:

> On the one hand we speak of the value of specific institutions, of a specific way of life, *e.g.* the value of parliamentary representation, of income taxes, of art museums and so forth. On the other hand, we speak of certain general principles as 'values', for example, the value of freedom, of religion, of democracy and so on.
>
> Speaking in the first sense we may say, for instance, that we like a general atmosphere of tolerance, of equal opportunity for everybody independent of race, creed and wealth, etc. These 'values' are recognized by a large group of people and it is hard to understand how the doctrine of the 'relativity of truth' could make these people change their preference. For the belief in these values is not based upon some abstract principle, it is not based upon whether it is possible to deduce them from a small number of hypotheses in a logical way.

[13] Apart from truth, all other values considered by Frank are moral values.

[14] Siegetsleitner (2017, 221–222) does not criticize Frank's typology, and apparently subscribes to it.

> This belief emanates rather from a general atmosphere of happiness, which for that large group of people is connected with those 'values'.
>
> [...] Both kinds of 'values' of which we spoke are preferences. In the first case, we prefer one concrete institution over some other, say parliament over dictatorship. In the second case, we prefer any one situation or action which has a particular property, over any alternative action that lacks it. If 'truth' is a 'value' for us, we will prefer a statement that is true over any alternative one which is merely comfortable. (47–48)[15]

The first kind of value is made up of preferences concerning "concrete [political] institution[s]" or modes of social, or cultural, organization. According to Frank, such values are (apparently spontaneously) "recognized by a large group of people", and "emanate from a general atmosphere of happiness", not from abstract principles. Since these values do not derive from "abstract principles" or "a small number of hypotheses in a logical way", the "relativity of truth" cannot change these values. Frank's idea seems to be that relativization can only take place when the statement to be relativized (*i.e.* qualified or specified according to a "reference frame") has a logical link with an abstract principle. It is not clear why it should necessarily be so. For example if we take, in everyday experience, the statements of "to the left (or right) of" (instead of Frank's examples of "above" and "below", which indeed refer to an abstract principle, namely gravity), such statements indeed need to be relativized ("to my left" *i.e.* "to the left relative to me", or "to your left" *i.e.* "to the left relative to you"), but such relativization does not necessary require an abstract principle (the notions of left and right can be immediately, intuitively defined).

The second kind of value seems more abstract, and has to do with general concepts such as democracy, freedom, and religion. Apparently such values are amenable to the relativity of truth, supposedly because we can identify a kind of instantiation[16] of, or conformity to, these values in a given action.

It would thus seem that values of the first kind are best understood in the common sense of concrete, intuitive and emotional (although widely shared) preferences: they are preferences for "concrete institution[s]", they are not deducible from abstract principles in a logical way, they are akin to "beliefs" emanating "from a general atmosphere of happiness", etc. On the other hand, values of the second kind should be understood in an authentic philosophical sense (with a claim to generality and objectivity), inasmuch as they prescribe theoretical or practical (in the sense of moral) norms. However there are several objections to this classification:

- First, there are values in the first group – especially the values of political institutions, or even concepts, such as "parliamentary representation" and

[15] Note that such a position runs *contra* the one expressed by P. Frank (1957b, 144) who seems ready to accept the Ptolemaic system instead of the Copernican one on grounds of social utility (as providing "a feeling of security"), or to Frank (1954–55/1956a, 15) for whom we choose scientific theories "according to our purpose" (see Sect. 9.3.2).

[16] Frank talks indeed of an action having a "particular property" (48). On the contrary, he apparently considers that values – or rather, preferences – of the first kind cannot be specified in a particular situation (but see my critique below).

"dictatorship" – which are not so concrete, and also somewhat abstract, although to a lesser extent than those of the second category (such as "democracy"). For example, parliamentary representation is a component of representative democracy.

- Second (and concomitantly), why should these somewhat abstract values of the first group be somehow less amenable to the "relativity of truth"[17] than those of the second group? For example, parliamentary representation, or equal opportunity, are not just values intuitively based on a "general atmosphere of happiness" or "tolerance" as Frank claims: they can be interpreted (and even justified) rationally by using an appropriate method of verification taking into account their factual consequences (as opposed to dictatorship, for example), *i.e.* through operational definitions (to comply with Frank's conception – see Sect. 9.2.3.2). One can ask what is meant exactly by "equal opportunity" (*i.e.* one can ask about its operational meaning) just as one can inquire about the truth (a value of the second group) or the falsity of a statement. Conversely, valuing a statement as true or untrue might also "emanate" from emotional considerations (as for values of the first group), and not necessarily by qualifying or specifying an abstract principle of truth (as Frank seems to imply).

Thus, Frank's divide between concrete and emotional values on the one side (first group), and abstract, general and objective values on the other side (second group), is not very convincing – at least for certain values (such as parliamentary representation, equal opportunity, etc.) of the first kind.

9.2.2.2 The Relativization of Moral Statements

Let us now pursue with Frank's values of the second kind (the properly philosophical values in his conception): from these "general values" we should be able to "derive specific decisions" (p. 48). In other words, our (moral)[18] values prescribe the general rules of conduct or principles which our particular actions or decisions

[17] Frank's formulation is confused here: "These 'values' are recognized by a large group of people and it is hard to understand how the doctrine of the 'relativity of truth' could make these people *change their preference* [*i.e.* their values of the first kind]" (47, my emphasis). Here, not only does Frank say that the relativity of truth does not apply to values of the first kind, but he also seems to imply that the relativity of truth changes (or can change) one's values. This contradicts the rest of his book, where he strives to show that the "relativity of truth" does *not change* the values (this would promote skepticism or agnosticism, which Frank fights), but merely *qualifies* or *specifies* them.

Besides, talking of the "relativity of truth" as applied to *another value* is a misleading formulation, because truth is already a value in itself (and the "truth of another value" does not make any sense). It would have been more correct (and simpler) to just talk of the relativity of the value in question.

[18] Here I follow Frank (48–49) who treats only moral values.

should respect. Thus, we use these values to *evaluate*[19] our decisions or actions (*i.e.* to verify their conformity with our values). The question is of course how to do this evaluation for each particular case. In other words, what do these general values mean concretely? This is Frank's answer:

> In the deduction of our concrete evaluations from some general principles, the doctrine of the 'relativity of truth' plays the same historic role as for any deductive theory in science or any other field of knowledge. When we attack the problem of how to deduce our actual valuations from general principles we have to face what we have learned from the history of human behaviour. The more we learn the more do we understand that human behaviour is more complex than we had anticipated. Therefore, the language by which we have laid down the valuations of human conduct has become more and more complex too. This means it has been 'relativized' in the same sense as the language of science. (p. 48)

Thus, relativity applies in ethics in the same way as in science: the relativization or qualification of a moral statement or of a general moral principle (rather than of a value)[20] clarifies and specifies its meaning, just as the relativization of a statement in science. In this way, moral principles (expressed in ordinary language) become "relativized" or "qualified" thanks to "operational definitions" (see below) in order to acquire a determinate factual meaning, exactly as theoretical principles or statements (49).

Indeed, by sticking to absolute general principles without qualifying them, all one can do is deduce other abstract general principles, but it is impossible to deduce any definite conclusion relevant for "an actual life situation" (50). For example, the general principle "you must not kill" is subject to many questions: is self-defense, making war or eliminating a blood-thirsty tyrant still "killing"? What is worse, without qualification, an absolute principle can, by choosing appropriate operational definitions, be used to "justify almost any line of action" (52). Thus, claiming that an ethical or political system is "absolutely true" opens the door to arbitrariness

---

[19] Frank uses the term "valuation" in the received sense of evaluation (verification of the conformity to an existing value) and not valuation (creation of the value).

[20] For the sake of rigor, it would be better to talk of:

– the *relativity of values* (such as truth or democracy) which have the logical form of *concepts*: saying that these values (concepts) are *relative* (and not absolute) means that they are too vague and general, and that descriptive (for truth) or prescriptive (for moral values) statements or principles cannot be evaluated according to these values;

– the *relativization of a statement or principle* (such as "my head is above my feet", "John is wicked" or "we should obey God's will"), which have the logical form of *judgments*: saying that these statements or principles must be *relativized* means that their meaning must be specified or qualified relatively to a "frame of reference".

But for commodity (and following Frank) I will not always make this linguistic distinction.

Frank also talks of the relativization of language (whether scientific or ordinary). This is a metonymical way of speaking. One should rather say that the language in itself remains the same, but the statements (which are formulated in this language) are relativized or qualified, *i.e.* additional information (expressed in the same language) is appended to the statement, to specify its "frame of reference".

(21). On the whole, absolute principles are only "good catch-words" for motivating the troops, but can be used misleadingly (55).[21]

9.2.2.3 Absolute and Relative Values

Frank distinguishes "absolute" from "objective values": "relativism" weakens the belief in absolute values, but reinforces the belief in objective values. "Absolute" is in fact synonymous with "unqualified" (28) and opposed to "relative" (which is synonymous with "objective"),[22] in the sense of "'relative to the existing conditions'" (52). An absolute value is thus an *unqualified* value, which therefore (without this qualification process) remains vague and open to any concrete interpretation.[23] This is why the relativists "refuse to recognize 'freedom' or 'democracy' or similar 'values' as 'absolute values'", but only as "*relative* values, ones which have to be striven for 'relative to the existing conditions'" (52).

Thus when Frank talks of "relative truth", he does not relinquish the value of truth as such, and still wants to formulate true statements (compare to footnote 20). To use his words, these statements "are 'absolutely' – or better, dependably – true" (24): what this paradox means is that they are still "absolutely" true (in the sense that they have become "definite assertions"), once they have been relativized *i.e.* once their conditions of validity have been specified. There would be less ambiguity if Frank talked of the relativity of the truth *of a statement* (and not of truth *tout court*), because *truth as a value* is preserved (this is precisely the point of his entire book).[24] Only the statements are specified – and this is precisely what makes them (definitely) true.

[21] On the next page, there is a strange sentence, somehow contradicting what has just been said: "The inspiring [*i.e.* absolute] slogans can *and must* lead our decisions, but they cannot and must not dominate our search for truth." (56, emphasis added). It seems that Frank acknowledges that our actions are not always rationally motivated (this is the "can"), but the "must" introduces a normative component which is at odds with the rest of his conception.

[22] The use of this latter wording also suggests (in addition to "relative") a kind of inter-subjective agreement on these values, which Frank apparently takes to exist (see e.g. 46, where Frank insists on "the 'objectivity' of human values", whose denial would "imperil the foundations upon which decent human conduct is built").

[23] Thus, "[t]he more we discourage the scientific attitude, which is based on 'relativism', the more we run the risk that the ethical or political system which we claim to be 'absolutely true' can be superseded in one rush by the opposite creed which other people believe to be 'absolutely true'" (21).

[24] Nemeth's (2003, 120) formulation, according to which "[Frank] argued [that] science calls into question all absolute knowledge claims and therefore in principle weakens the willingness to believe in absolute values" inherits the ambiguity of Frank's vocabulary, in spite of her adding that "disbelief in absolute values does not mean disbelief in the objectivity of values". It may give the impression of mixing the descriptive (with respect to knowledge claims) and normative (with respect to values) levels. But one may be open-minded with respect to knowledge (which entails factual or theoretical statements subject to potential revision), without necessarily putting into question, or relinquishing, one's values (whether epistemic or not). For example, open-mindedness

9.2.2.4 Relativism, Agnosticism, and Skepticism

We can now go back to Frank's introductory remarks about the roles of the "agnostic and sceptical attitudes", the "disbelief in absolute values" and the "relativism" associated with science, in the collapse of the West during World War Two. We have just seen that "absolute values" are considered pejoratively by Frank, thus it is an illusion to think that they would have prevented the "breakdown of military and political morale" (11) of the West in the face of Germany, and acted as a rampart against totalitarianism (20).[25] Less clear is Frank's apparent condemnation of agnosticism and scepticism.[26] Indeed, and contrary to Frank, one would rather think that it is the incriminated "agnosticism" and "scepticism" which are the best ramparts against totalitarianism, in so far as they stem from a critical attitude which puts into question all "absolute slogans" so condemned by Frank. Of course, the latter critical attitude is justly appreciated by Frank (see e.g. 93 where he praises the capacity to distinguish between "what can be and what cannot be proved"), but it does not seem to include "agnostic and sceptical attitudes".[27] This is because Frank has a strict acceptation of agnosticism and skepticism.

Indeed, whereas skepticism in its modern sense might precisely be taken in the (positive) sense of a doctrine of the relativity of truth or knowledge in Frank's sense (or in Merton's (1942/1973) sense as "organized skepticism", i.e. a cautious and critical attitude not taking anything for granted without a valid proof[28] – an attitude Frank would clearly subscribe to), Frank seems to understand skepticism in its ancient, negative sense (according to which it is impossible to know anything with certainty, and which leads to suspension of judgment and permanent doubt[29]).

with respect to the truth of a particular statement does not necessarily entail open-mindedness with respect to the value of truth itself!

[25] Frank clearly condemns this (illusional) belief, as he makes clear e.g. on pp. 19–20. To understand how and why "absolute values" could be considered as ramparts against totalitarianism, it would probably be illuminating to study the other contributions of the Conference, especially those endorsing this opinion (compare with footnote 3), but this exceeds the scope of this article.

[26] See the quote of p. 12 in the introduction.

[27] However, Frank is not consistent in his use of words: see p. 46 where "sceptic", "relativist" and "agnostic" seem to be taken as synonyms. In the same way, on p. 85, Frank criticizes immovable, "firm ethical convictions" and ethical "principles that are of eternal validity, and not dependent on the advancement of science": an attitude which could be qualified as skeptic or agnostic in the positive sense (see below).

[28] Merton (1942/1973, 277) defines organized skepticism as "the temporary [and not permanent] suspension of judgment and the detached scrutiny of beliefs in terms of empirical and logical criteria". Merton (1938, 334) explains that "organized scepticism involves a latent questioning of certain bases of established routine, authority, vested procedures and the realm of the 'sacred' generally. [...] Most institutions demand unqualified faith; but the institution of science makes scepticism a virtue."

[29] That Frank understands skepticism in this ancient, pejorative sense is clear from *e.g.* p 30 (where it is associated to the "bankruptcy of science" and the "despair ever to penetrate into the realm of 'absolute truth'") or p. 106 (where skepticism in science is implicitly defined as a "lawless doubt of the principles of science").

Although he mentions less often agnosticism, Frank, by associating it with skepticism, also seems to take it in a negative sense (stating the impossibility of knowledge). On the whole, and in spite of some inconsistencies in the use of the term "sceptic", it is relativism (and not agnosticism or scepticism, which are considered pejoratively) that endorses the role of rampart against totalitarianism.

9.2.2.5 Two Sorts of Relativization

Finally, our analysis of relativism makes clear why, according to Frank, general principles (whether theoretical or practical) must be relativized or qualified. For this one has (only) two possibilities (53, 90–91):

- One can keep an absolute principle and complement it with operational definitions of the terms in which the principle is formulated, which qualify or relativize the principle. For example, one can keep the absolute principle "Thou shalt not kill", provided there are operational definitions which describe "what sort of action is meant by killing" in each particular case (90).
- Or one can relativize the principle itself, in which case there is no need for operational definitions. In this case, there is no more general, absolute principle such as "Thou shalt not kill": for each particular case, killing is forbidden or allowed "relative to the circumstances" (91).

In the end, the choice is only linguistic: relativization always occurs. One might, of course, object that the choice does not have to be exclusively (as Frank argues) between these two alternatives, and that principles or values can remain absolute and do not have to be relativized (either with operational definitions or themselves). I will return to this objection in Sect. 9.3.4.

9.2.3 Relativism, Pragmatism and Operationalism

9.2.3.1 The Pragmatic Theory of Meaning

In order to understand the place of relativism in our system of knowledge, one must, according to Frank, go back to the various attempts to clarify the "vague and misleading" statements (such as the ones seen previously) generated by "an over-simplified language" (31). Such attempts try to suppress ambiguities in language by basically distinguishing "'meaningful discourse'" from "'empty talk'", and have been undertaken by "pragmatism, operationalism, semantics, logical empiricism and so forth" (31).

The basic idea behind these movements, which according to Frank have their common source in Peirce's pragmatism, is that, to define the meaning of a statement and to distinguish it from another, one must exclusively consider the *concrete* consequences of this statement, i.e. how it influences *action*. This "pragmatical theory of meaning" tries, to use James' words, to identify "the truth's cash value [of an idea or belief] in experiential terms" (33), i.e. to identify how the truth of a statement concretely influences reality. Note that (31):

- the pragmatic conception of meaning is not limited to science but applies to all areas of human life, including ethics and religion;
- in this conception, once a truth has been acknowledged, it reduces the room available for other truths, which must be compatible with it: in particular, theological beliefs cannot contradict beliefs already acknowledged by science.

Another important feature of this conception is that, according to Frank, "the pragmatic spirit of science is a force for democracy and against totalitarianism", because "the meaning of [the] principles [of science, and more generally of any principle] is explained as consisting only in the experiential facts following from them" (106). Indeed, "[t]his indifference to the words in which the highest principles are formulated proves to be a powerful weapon against totalitarian ideologies", which are based on principles which are mere "banners" whose formulation is, on the contrary, very important, and must absolutely be preserved, "regardless of what follows in real life" (102). This is a very strong, and convincing, feature of Frank's philosophy: he rightfully draws attention to the *concrete* consequences of a general principle – whether in science, ethics or politics –, and not to the words in which it is formulated, however beautiful or seducing they may be.

However, Frank's analogy with relativity in physics is not fully relevant. Indeed, Frank criticizes the fact that totalitarian ideologies promote absolute (*i.e.* absolutely formulated) principles disconnected from the facts, in contrast to his relativistic and pragmatist conception. It is true that, in physics, the "words" (*i.e.* the symbols) in which the laws are formulated do not matter. However the *form* of the laws (i.e. the structure of the mathematical equations expressing them) does matter. It is precisely the point of the theory of relativity to state that the form of physical laws is invariant in any reference frame (this is covariance), as Frank is of course perfectly aware of: in other words, that their formulation is, indeed, absolute. Frank's insistence on the variable – and not the absolute – features of the theory of relativity may have to do with his endorsement of Bridgman's semantic and operationalist interpretation of it (see e.g. Frank 1949, 291). I shall return to this point in Sect. 9.3.3.

9.2.3.2 Operationalism

Indeed, Frank mentions with approval Bridgman's operationalism, based on "operational definitions" which equate a concept with the corresponding set of operations (for example, the scientist describes the concept of length by describing the physical operations by which he measures a length) (34). In other words, operational definitions link the abstract words of theory and the observable facts of our everyday experience (37). According to Frank, the operationalist theory of meaning, which is a conception not limited to science but also applies to abstract concepts like freedom, democracy, religion, etc., bears a strong resemblance with the pragmatist one.[30] Frank then goes on to assert that

> It is obvious that the application of the operational or pragmatic theory of meaning is closely connected with the 'relativistic conception of knowledge' [...]. Bridgman says[31]: 'Relativity in the general sense is the merest truism if the operational definition of concept is accepted, for experience is described in terms of concepts, and since our concepts are construed of operations, all our knowledge must unescapably be relative to the operations selected.' (36-37)

However, since Frank does not give further elements for vindicating this claim, this connection between the operational or pragmatic theory of meaning on the one hand, and the relativistic conception of knowledge on the other, is not so "obvious" as it might seem according to Bridgman's short quotation. Indeed, it rests on Bridgman's operational – and highly disputable – premise that "[all] our concepts are construed of operations" (and nothing else). More generally, one could argue that it is not the same to qualify a theory of meaning as:

- pragmatic: in which case it is mostly interested in the *concrete* consequences of the truth of a statement;
- operational: in which case it is mostly interested in how to relate the concepts used to *sets of operations*;
- relativistic (if one is allowed to extend this appellation to a theory of meaning): in which case it is mostly interested in how the truth of our statements is *relative* to other statements or information (not necessarily "concrete" or formulable as sets of operation: it might be a frame of reference or a theoretical principle).

Now that we have described Frank's relativism, let us contextualize and criticize it.

9.3 Discussion

Before discussing Frank's conception, I shall briefly recall its historical context.

9.3.1 A Brief Contextualization

First, the fact that Frank insists, as we have seen, on the relativity of measurements in physics, and more generally of statements in everyday language (rather than on the invariance of physical laws, or the absolute formulation of ethical principles),

---

[30] "The meaning of a concept is defined not by an 'idea' of which it is to be the image [like in the 'idealistic theory of meaning'], but by the observable consequences that we can draw from sentences in which this concept occurs. [...] a concept is identical with the set of operations from which it has been derived" (36).

[31] This is a quote from Bridgman (1958, 25).

should be read keeping in mind his critique of "misinterpretations" (whether physical or philosophical) of the theory of relativity. Indeed, Frank has written numerous articles on the misinterpretation of physical theories (for the theory of relativity see especially Frank 1938/1953, 1949, Chs. 14 and 15; 1957b, Ch. 7). For example, Frank (1938/1953) criticizes misinterpretations which attribute a greater role to the "observer" in the theory of relativity than in classical physics, or which introduce new psychological conceptions of space and time on its basis, or which in general try to support any philosophical, or rather "metaphysical", conception (such as "fatalism", *i.e.* the view that everything is predetermined and that time does not flow).

All these misinterpretations have in common the fact that they are based on unverifiable statements (which are isolated from the verifiable, observational sentences of the theory, and are therefore "unscientific"). Such "metaphysical" or "meaningless" statements lead to "the belief that they are systems of statements about a realm of things", and this belief "constitutes the foundations of the philosophical misinterpretations of physical theories" (*ibid.*, 216), which can serve non-scientific, "intellectual" or "social" purposes. Here Frank's main concern is to warn against any ideological misuse of scientific theories in other, non-scientific realms. To do so, Frank essentially relies on a *verificationist* conception: statements or concepts have to be formulated in appropriate scientific language, and connected to empirically observable facts. The focus here is not on theoretical considerations (such as invariance), but on practical matters (avoid misinterpretations by empirically probing the meaning of each statement or concept).

More importantly perhaps, Frank's insistence on the relativity of statements should also be replaced in the context of his fight against totalitarianism and political absolutism. An example of this fight is given by Frank's (1957a) article in the *Bulletin of the Atomic Scientists*. There Frank studies philosophical uses and misuses of science, and distinguishes a "technological use", which is not concerned with philosophical interpretations and has historically been devoted to the scientists by the political, religious or educational authorities; from a "moral" or "philosophical use" (which fails to provide operational definitions of the physical symbols used), which purports to use science to serve ideological, political or religious goals, and has been traditionally made by non-scientific authorities. In the latter category, one finds historically the (Catholic as well as Protestant) Church, as well as, in the twentieth century, the totalitarian regimes of Fascist Italy, Nazi Germany and the Soviet Union. All these regimes either rejected the scientific theories which they deemed were not compatible with their philosophical "creed", or kept their "technological" use because of the practical profits which they could gain (*e.g.* in the domain of nuclear energy), while forcing their own philosophical interpretations onto them for their "moral" use. For example, Einstein's theory of relativity was first completely rejected by Soviet ideologues such as A. A. Maximov, then (after the death of Stalin) accepted with respect to its "technological use", but still rejected with respect to its "moral use", and deemed "bourgeois science" (*ibid.*, 129). Similarly, the National Socialist Party in Germany rejected, from the "moral" point of view, Einstein's Jewish origin, the abstract mathematical character of the theory and its counter-intuitive consequences which contradicted "the 'naturalness' and 'directness' of the German mind" and its "living intuition of nature", the "touch of relativism, subjectivism and skepticism" associated with the theory, while at the same time welcoming its technological use and practical results, notably for the development of new explosives (*ibid.*, 130).

In this context of his fights against abusive interpretations, or totalitarian appropriations, of relativity, one can easily understand Frank's insistence on "relativism", based on his operationalism and verificationism (which prevents for example the emotional appeal of political absolutism). Indeed, for Frank relativism "has been for centuries the only effective weapon in the struggle against any brand of totalitarianism" (20).[32] And the CSPR itself was precisely organized in order to help fight totalitarianism: "The precise aim of this Conference was to establish a common understanding of democratic principles that would help to overcome the high pressure propaganda of totalitarian values" (11). But whereas "the members [of the Conference] were anxious to prove that the danger of 'relativism' [...] could be avoided by democratic methods", Frank purported to show that "relativism" represented no danger, but on the contrary represented a means to reach the very goal of the CSPR. As Reisch (2017, 235 sq.) explains (quoting J. B. Gilbert), the CSPR was intended, in the face of Hitler's military successes and the ensuing doubts about the solidity of western civilization, to provide "basic ideas to which people of intelligence in the different spheres of life would pledge" in order to fight totalitarianism and help defend democracy. Frank's conceptualization of relativity was intended to provide such a "basic idea" to serve as a weapon in the "battle of ideas and ideology" which, according to L. Finkelstein (one of the organizers of the CSPR), the battle against Nazism most importantly was (*ibid.*). According to Reisch, "Frank presented the relativity idea as an example of the *unifying truth* of methodology of the sort that Finkelstein and his conferees aimed to locate", as a "global", "unifying principle" which could "connect and unify different kinds of knowledge", and which would come from (physical) science, not from religion or metaphysics (*ibid.*).

Now that these elements of context are in place, we should not, however, be deterred from criticizing Frank's philosophical conceptions as such. But let us first replace Frank's position in the contemporary debate on relativism.

9.3.2 Characterization of Frank's Relativism in the Contemporary Philosophical Debate

As is well known, relativism is a controversial doctrine (for an overview of the concept of relativism, see Baghramian and Carter 2019, and of moral relativism Gowans 2019). *Prima facie*, Frank's relativism, or more precisely the relativization of statements which it advocates, appears like a rather banal, common-sense requirement: in a nutshell, it requires that the meaning of the terms one uses (such as "above" or

[32] See also footnote 23.

"wicked") be specified or qualified, by providing their "frame of reference". One can thus say that Frank has a semantic conception of relativism (which is confirmed in other writings, *e.g.* Frank 1957b, 143), which is about meaning, not ontology.[33] On first sight, Frank's position looks like a mild, respectable version of relativism, in contrast to, for example, the "cheerful anti-realist relativism" of Barnes and Bloor (1982), as Uebel (2000, 148) qualifies it. Let us try to characterize Frank's relativism more precisely with respect to typical relativistic stances.

9.3.2.1 Frank's Relativism of Truth

Apparently, it would seem exaggerated to label the Frank of *RRT* as a relativist of truth in the contemporary sense of Baghramian and Carter (2019, Sec. 4.3), namely holding "the claim that what is true for one individual or social group may not be true for another, and there is no context-independent vantage point to adjudicate the matter. What is true or false is always relative to a conceptual, cultural, or linguistic framework". In *RRT*, Frank rather gives the impression to keep a notion of "definite" truth (recall the "definite assertions" on p. 24), as long as it is "qualified" – "absolute" being only synonymous with "unqualified". However, in other works Frank is well known for having studied, and to a certain extent legitimated, the inclusion of external, non-scientific (in the strict sense of natural science)[34] criteria in theory choice (see *e.g.* Uebel 2000). More precisely, Frank uses the underdetermination thesis of scientific theories by empirical data to allow extra-scientific values to influence theory choice:

> The theories that are built up by scientific methods, in the narrower sense, are pictures of physical reality. Presumably they tell us the truth about the world. If a theory built up exclusively on the ground of its agreement with observable facts tells the truth about the world, it would be nonsense to assume seriously that a scientific theory can be influenced by moral or political reasons. However, we learned that agreement with observed facts does not single out one individual theory. We never have one theory that is in full agreement but several theories that are in partial agreement, and we have to determine the final theory by a compromise. The final theory has to be in fair agreement with observations and also has to be sufficiently simple to be usable. If we consider this point, it is obvious that such a theory cannot be the truth. In modern science, a theory is regarded as an instrument that serves toward some definite purpose. (Frank 1956b, 14)

So far, one may still think that Frank keeps a kind of theoretical notion of *the* (unique and universal) truth, strictly scientifically determined (*i.e.* without any extra-scientific factor), but that in practice this cannot be achieved and therefore the truth arrived at in the selection (by humans) of scientific theories is a "compromise" with their extra-scientific preferences. But Frank goes as far as to equate the notion of

---

[33] It is not about the truth conditions of statements neither (such as in Harman and Thomson 1996).

[34] But in fact, these external criteria can also be considered scientific in the large sense of the social sciences (psychology, sociology) (see *e.g.* Frank 1951; 1956b 1954–55/1956a).

truth with the notion of theory choice, thereby effectively relativizing it in the previously delineated way:

> In what sense does science search for the truth about the universe? This truth certainly does not consist of facts but of general hypotheses or theories. What we call facts in the strictly empirical sense, without any admixture of theory, are, in the last analysis, sense impressions between which no connection is given. Hence, the question of what we have to regard as the truth about the universe has to be formulated rather as follows: What are the criteria under which we accept a hypothesis or a theory? If we put this or a similar question, we shall see soon that these criteria will contain, to a certain extent, the psychological and sociological characteristics of the scientist, because they are relevant for the acceptance of any doctrine. In other words, the validation of theories [or, in other words, the "truth"] cannot be separated neatly from the values which the scientist accepts. This is true in all fields of science, over the whole spectrum ranging from geometry and mechanics to psychoanalysis. (Frank 1956b, vii–viii)

Therefore, it seems quite fair to describe Frank's relativism or "doctrine of relativity of truth" as, indeed, a relativism about truth in the sense of Baghramian and Carter (2019, Sec. 4.3). Now as the latter explain, relativism about truth "is the most central of all relativistic positions since other subdivisions of the philosophical theses of relativism [...] are in principle reducible to it". For instance, "ethical relativism can be seen as the claim that the truth of ethical judgments, if such truths exist, is relative to context or culture" (*ibid.*). Indeed, Frank' relativism illustrates this claim, since it applies to virtually any statements, whether scientific, ordinary, moral or political ones, as we have seen. Therefore, Frank's relativism could be labeled "global relativism" in the sense of Baghramian and Carter (2019, Sec. 1.4.1), namely in the sense that "all beliefs, regardless of their subject matter, are true only relative to a framework or parameter". Importantly, the scope of Frank's global relativism also includes science. For instance, Frank (1957b, 144) seems ready to accept the Ptolemaic system instead of the Copernican one on grounds of social "usefulness" (as providing *e.g.* "a feeling of security"). In the same way he seems ready to accept talk of "absolute motion" (*ibid.*). Here the belief of the world picture is relative to what society deems best useful. Therefore we can talk of epistemic relativism following Baghramian and Carter (2019, Sec. 4.4) (i.e. "the thesis that cognitive norms that determine what counts as knowledge, or whether a belief is rational, justifiable, etc. could vary with and are dependent on local conceptual or cultural frameworks and lack the universality they aspire or pretend to"), and more specifically of relativism about science (Baghramian and Carter 2019, Sec. 4.4.3), where the underdetermination thesis is used in support of relativism,[35] as indeed Frank does.

[35] "Relativists about science have argued that only with the addition of auxiliary hypotheses could the scientist choose between various theories and that such auxiliary hypotheses are colored by socially and historically grounded norms as well as by personal and group interests" (*ibid.*). In Frank's relativism the norms and interests are common (social and political) rather than personal.

9.3.2.2 Frank's Relativism Vs. the Relativism of the Strong Programme

So far Frank can fairly be labeled a relativist in the above mentioned senses. In what sense, however, is his position different from the relativism of the Strong Programme of the sociology of scientific knowledge (see Bloor 1976), as, according to me, Uebel (2000, 149) rightfully claims (*contra* Bloor 2011, 428)?

On first sight, the relativism of Barnes and Bloor (1982) resembles somewhat that of Frank inasmuch as it denies the possibility of formulating "the justifications of [one's] preferences [...] in absolute or context-independent terms" (27). But it also goes much further. It implicitly denies that "certain form of knowledge [have] a privileged status" (22), that there are "distinctions between true and false, rational and irrational belief" (25) and claims that reality, "being a common factor" in "all the vastly different cognitive responses that men produce to it", "is not a promising candidate to field as an explanation of that variation" (34). Reality, then, should not be considered as the main reason why scientists, and people more generally, have such or such belief.[36] Rather, only the local social and cultural context can explain them, since any "principle" or "matter of fact" that we may ultimately refer to to justify them "only has local credibility" (27–28). In the same way, this relativism postulates that "all beliefs are on a par with one another with respect to the causes of their credibility. It is not that all beliefs are equally true or equally false, but that regardless of truth and falsity the fact of their credibility is to be seen as equally problematic" (23). This "'symmetry'" or "'equivalence' postulate" (22) in effect evacuates the truth value of a belief as a reason for having this belief. To conclude, according to this type of relativism neither reality (more precisely, the interactions of the scientists and people with it) nor truth (more precisely, the truth value of our beliefs) can and should explain our beliefs.[37]

Such extreme relativism is clearly not the position of Frank, who does not subscribe to such a "doctrine of equal validity", as Boghossian (2006a, 2) calls it, according to which "there are many radically different, yet 'equally valid' ways of knowing the world, with science being just one of them".[38] Indeed, Frank clearly distinguishes between rational and irrational belief (and between true and false if "true" is taken in its restricted sense of empirically adequate); he considers science as a very special, and the most superior, form of knowledge; and he takes into

[36] Here we understand why Uebel (2000, 148) talks of "cheerful anti-realist relativism".

[37] See Barberousse (2011, 220–223) for a systematic critique of the methodological principles of the "Strong Programme". I will pass over the contradictions inherent in Barnes and Bloor's (1982) conception, for example the fact that their article (*absolutely*) assumes the rules of logic (by using the principle of non contradiction p. 22, or with respect to its very argumentative structure), or uses other concepts or terms absolutely (*e.g.* "primitive cultures" p. 26).

[38] Boghossian presents this conception as a consequence of the "social dependence conception of knowledge" (according to which "whether a belief is knowledge necessarily depends at least in part on the contingent social and material setting in which that belief is produced (or maintained)" (pp. 6–7). However, such a social dependence conception of knowledge does not necessarily lead to a "doctrine of equal validity", as the conception of Frank typically illustrates (which endorses the former, but not the latter claim).

account physical reality in the explanation and justification of scientific theories, and more generally human beliefs. One could say that, in his appraisal of the validation of scientific theories, Frank retains and distinguishes a core truth (*i.e.* facts)-based, scientific conception, and that *in addition* he takes into account extra-scientific (in the strict sense of natural science) factors (i.e. social, psychological and political).

In spite of these differences, the later Bloor (2011) demonstratively endorses Frank's relativism in his study on aerodynamics. There Bloor seems to keep only the core, more respectable idea, of relativism, according to which there are no absolutely valid, context-independent, statements or principles, whether in the theoretical or moral domain. Indeed, Bloor (2011, 439) distinguishes "foolish version[s]" of relativism" (440) from the following "relativism by contrast" (Baghramian and Carter 2019, Sec. 1.2) defined by contrast with absolutism: "if the concept of 'relativism' is to be used with precision, it can only mean one thing: a denial that there are any absolute truths. This is a necessary and sufficient condition for an account of knowledge to be identified as a form of relativism. A relativist can be comfortable with knowledge that is conjectural, inconsistent, expedient, and partial, that is, with everything that science and technology really is" (439). For Bloor the "relativity of knowledge" is a consequence of its "social character" (442). Apart from the strange qualification of "inconsistent", it is indeed difficult not to agree with Bloor on this humble characterization of scientific knowledge.

Nevertheless, even the basic relativistic claim (endorsed by Frank as well as the late Bloor) according to which there are no absolutely valid statements, the "denial that there are any absolute truths" (which by the way is itself contradictorily *absolute*) can be problematic both in the theoretical and practical domains, as I will now show. Although "absolutism" may seem simplistic or inaccurate in the former, and intolerant or dangerous in the latter, I believe there are cases where it is justified. In science, there are for example statements of physical principles (such as that the laws of physics are the same in all inertial reference frames), laws (such as that two bodies with a mass attract each other) or facts (such as that a molecule of water is made up of two atoms of hydrogen and one atom of oxygen) that appear absolute and context independent. Similarly, in ethics, there are, in my opinion, principles (such as the respect of human rights, or the absence of death penalty) which should be held absolutely.

9.3.3 Critique of Frank's Interpretation of the Theory of Relativity

Let us first make clear that Frank is, of course, perfectly aware of the importance of the concept of invariance (in the large sense, including covariance) in Einstein's theory of relativity: he writes for example elsewhere that "[f]or some particular changes of the system of reference the laws of motion remain unaltered. This fact turns out to be the most important property of these laws, and is the very basis of the theory of relativity" (Frank 1949, 239). There is no question that Frank is a master of the theory of relativity.[39] This, however, should not hinder us from criticizing his philosophical *interpretation* of the theory, and especially the way he applies its concepts (and first of all, the concept of relativity itself) outside the realm of physics.

We have seen that Frank uses Einstein's theory of relativity to illustrate his general claim that truth, or better objectivity, is reached, not in an absolute but in a "dependabl[e]" (24), relativized or qualified way. Frank's focus here is on the relativity of *measurements* (of time and space in the special theory of relativity), in conformity with his empirical and operationalist philosophy of science: measures are what is observable. On the contrary, the received view usually insists on the invariance of (the form of) physical laws (i.e. their covariance) with respect to any system of reference: the focus is on the "absoluteness" of physical laws (as well as of the principle of relativity itself), so to speak.[40] As is well known, the reason for this is that the relative features of the theory of relativity (relativity of time and space measurements) *derive* from its absolute features (absoluteness of laws), which are logically prior. The theory of relativity derives the relativity of physical quantities (such a length, time, velocity, mass) with respect to the considered reference frame, from the (observed, and then elevated to the rank of absolute principles) invariance of physical laws and (in the case of the special theory of relativity) of the velocity of light (see e.g. Einstein 1905, 891 sq.).

Frank, for his part, insists on the validity of each particular statement once it has been referred to its relevant reference frame (but not on the overall connection), instead of insisting on the overall objectivity gained by the theory of relativity, whose correspondence rules connect statements in different reference frames. Now although such an insistence is, again, understandable given Frank's previous work and philosophical orientation, it may not be the best way to reach his own goal of debunking philosophical misinterpretations of the theory of relativity. Indeed, such an insistence may favor misinterpretations attributing a greater role to the observer in the theory of relativity than in classical physics, or introducing new psychological conceptions of space and time on its basis, which Frank (1938/1953) himself had to fight, as we noted above. It is worth noting that Einstein himself did not think that his special theory of relativity supported relativism in ethics or epistemology, because although time or distance intervals are relative to reference frames, the physical laws expressing such relativity are absolute and in no sense relative. He thought that the name *Invariantentheorie* was more appropriate for his theory, as Holton (2006, 269) explains, and preferred it to the "so-called relativity theory" (a name proposed by Planck in 1907), which he deemed "unfortunate", because "it has given occasion to philosophical misunderstandings". As Sachs (2005, p. 4) summarizes:

[39] See his many writings on the subject in his bibliography in Frank (1998, 290–296).

[40] Because of its focus on invariance, Holton (2006, 269) writes that "modern relativity is simple, universal, and, one may even say, 'absolute'".

> The principle of covariance asserts that the laws of nature must be totally objective – meaning that their forms must be independent of the space-time reference frame in which they are expressed, from any particular observer's point of view. The space-time language itself is relative to the reference frame in which it is expressed – hence the name of this theory! Still, the primary focus of the principle of covariance is on something that is absolute rather than relative – this is the invariant law of nature itself.
>
> In the early stages of this theory, its name led to the erroneous impression that this approach in science is based on the philosophic view of 'relativism' – the idea that all knowledge is relative only to the 'knower'[41] – i.e., that there is no objective knowledge to talk about. Of course, Einstein never had this view in mind – his approach was just the opposite, where one focuses on the invariant (objective) law of nature. To avoid the confusion, Einstein tried to rename his theory 'invarientheorie' (theory of invariants), implying a focus of this theory on absoluteness rather than relativeness. However, he eventually rejected the name change because of further confusion he thought it might entail.

9.3.4 Critique of Frank's Moral Relativism

Many contemporary relativists distinguish the epistemic from the moral domain because of fundamental differences between natural facts and ethical values (e.g. Boghossian 2006b or Harman and Thomson 1996), but this is not the case of Frank, who extends his relativism to ethical statements. As Bloor (2011, 429) summarizes it, for Frank "[t]he danger to rational thought and moral conduct came [...] not from relativism but from absolutism." In particular, "[r]elativism, he argued, is the only effective weapon against totalitarianism" (*ibid.*). However, I will argue that in the ethical realm, Frank's relativism is also debatable. Although Frank claims to preserve the objectivity of moral values, his conception of relativity does, in fact, open the door to a kind of moral relativism in the contemporary, metaethical sense of Gowans (2019, Introduction), according to which "the truth or justification of moral judgments is not absolute, but relative to the moral standard of some person or group of persons".[42] Frank's conception of the principle "thou shalt not kill" illustrates this.[43] In this case, it is easy, *contra* Frank:

[41] It should be clear by now that this acceptation of relativism is not Frank's.

[42] Frank does not treat the empirical thesis of moral relativism noticing the existence of "deep and widespread moral disagreements" (*ibid.*). Although less visibly, Frank indirectly defends a metaethical thesis of moral relativism (*e.g.* when claiming p. 90 that "every ethical system consists of the principles and the operational definitions of the terms"), in addition to a semantic thesis based on meaning ("only if a principle is accompanied by operational definitions does it specify a definite way of life", *ibid.*).

[43] Principles like Kant's categorical imperative or "obeying God's will" may be better examples in support of Frank's position and less vulnerable to my critique below, insofar as they are more complicated to interpret and respect.

- to define (on the semantic level) the meaning of this principle *absolutely*, without the need for further operational definitions of its terms: for example we can simply define killing as taking someone's life (possibly adding: against her will);
- to argue (on the metaethical level) that this principle must be respected *absolutely,* without the need for further specification, including in Frank's examples: by avoiding at all cost to kill in self-dense, by refusing death penalty, by refusing to go to war, etc.[44]

For Frank, a moral principle (or value) does not have to be always respected, and *exceptions* are always possible, either through operational definitions, or through the formulation of the principle itself (in our example: it is allowed to kill in some special circumstances). But it could easily be argued that there are, on the contrary, some principles (like, precisely, not killing; or, say, not have recourse to torture) or values (such as human rights) which can, and must, be upheld absolutely (i.e. universally, at all places and times), without the need for further specification.[45] What is more, Frank's relativization of moral principles (either through operational definitions or through the formulation itself) depends on the moral "reference frame" that is chosen: but he does not give a clue as to which one to chose.[46] And contrary to relativity in physics,[47] the choice of the reference frame in ethics does have an importance: this is another weakness in Frank's analogy of relativity in physics and in ethics.

[44] If we take the shorter definition of killing as taking someone's life (not necessarily against her will), then there are more obvious cases where the commandment of not killing must be discussed – in particular, if someone else (*e.g.* a terminally ill patient) *asks* to be killed (to end his unbearable pain, etc.).

[45] For example – and this is a conception I fully endorse – human rights can be considered as standing on top of any other legal (whether national or international) norm, and must therefore be respected *absolutely* (*i.e.* whatever the circumstances, and contrary to Frank's conception).

[46] This is what Siegetsleitner (2017, 221) calls "standard relativism": "If we use 'evil' as an evaluative term, morality is indeed relativized to a certain standard. If the standard is held by a certain group, it is also relativized to a certain group—for example, a cultural or religious group. Here, relativization adds the moral standards of said group, the relevant moral system, so to speak. We can call this kind of relativism "standard relativism". Standard relativism does not imperil objectivity relative to shared standards, but objectivity in this case does not encompass criteria for the right standards. Although Frank does not endorse moral skepticism, standard relativism allows for the perspective that fundamental moral principles are unknowable." Thus "relativization in morality may also mean a relativism of standards, and where there is no agreement on shared standards, the door is left open for subjectivism and skepticism" (*ibid*., 225).

[47] Even if, of course, some reference frames are more convenient than others for expressing a physical quantity, depending on the problem at hand. But this is just a matter of mathematical formulation, which can be more or less complicated, but can always be equivalently transformed from one reference frame to another. On the contrary, in morality it is not the same thing to choose one reference frame or another. As Frank rightfully remarks, in the case of "John if wicked", we may chose "God's will", or Kant's categorical imperative, or the "voice of our own conscience" as the "supreme authority" with respect to which wickedness is defined, and the result will vary accordingly.

I am ready to share, to some extent, Frank's view that "the more abstracted and 'absolute' our values become, the less meaning they have and the less practical help they can provide", as Reisch (2005, 222) summarizes it. The need to attend to the concrete meaning and consequences of our values and principles, powerfully underlined by Frank, is indeed fundamental, on pain of falling into void discourse and verbalism. But I am not ready to share Frank's view according to which, as Reisch puts it, "the only good value was a relativized, contextualized, and revisable value" (*ibid.* 362), or that "[a]bsolute moral values are self-defeating [...] because genuinely absolute values and fixed values must be so abstract and general that they are practically meaningless" (*ibid.*, 379). I think that there are some values which do *not* require any relativization (for example human rights, non-recourse to torture, etc.), and that the line of conduct to follow accordingly is perfectly clear and meaningful nonetheless. If one finds these examples too anachronistic for Frank,[48] let us go back again to his example of killing: even in this case, the definition of killing is in fact straight-forward, and does not require any further specification. Even the cases of civil war, or sabotage, or execution of a criminal, still deserve the label of "killing" if the life of someone is taken away against her will.[49] What is more, it is easy to see, *contra* Frank, that relativism can serve the purpose of totalitarianism just as well as absolutism: remember how "killing" someone can, in Frank's view, easily be *re-qualified* according to the circumstances. Think of people in a demonstration killed by police forces: their killing could be re-qualified as "legitimate defense" by the police forces, for example – whereas in fact, they *were killed* by these forces, whatever they did before.

Against Frank, there are two possible lines of argumentation, one truly absolutistic (and probably too idealistic), and one more pragmatic and realistic. The first line would be to uphold the principle "thou shall not kill" absolutely, without the need for further qualification (whatever the aim pursued, including *e.g.* a noble aim of sabotage against Nazi occupation, it shall not be reached by killing someone). The second line would be, following Frank's second option, to "relativize" the principle itself, *i.e.* to respect it or not according to the circumstances (*e.g.* I will allow myself to kill Nazi soldiers because of the importance of my sabotage action). But I wouldn't call this "relativization" properly speaking since, again: (1) on the semantic level, what is meant by killing is very clear; and (2) on the metaethical level, the "moral standard" (as in Gowans 2019) does not need to be specified for the principle to hold – rather, the respect of the principle is relative to the "circumstances", as

[48] But, since the Enlightenment, it could hardly be so: universal values (*e.g.* in the French Revolution) and absolute demands (*e.g.* in Kant) have become common place philosophy.

[49] The formulations of Frank (1946, 23) are particularly striking in this respect, where he endorses statements such as "Bombing an enemy city is not killing", or claims that killing a dictator does not deserve the term "killing", but rather "liquidating". As in *RRT*, he claims that absolute ethical principles "fail to define an exact line of action just in cases in which moral advice is needed the most", and that "the most ardent advocates of 'absolute truth' avail themselves of the doctrine of the 'relativists' whenever they have to face a real human issue" (1946, 24). I certainly do not agree, as I show in this section.

Frank himself makes clear (91). More precisely, I would insert the principle in question in a larger *hierarchy* of ethical principles (thus to be agreed upon), in which each principle is enforced as long as respecting it doesn't breach a higher principle (if I am not mistaken, such a hierarchy of ethical norms bears strong resemblance with the juridical hierarchy of laws). In other words, the *hierarchy as a whole* can, if well designed, be enforced *absolutely*. In my example, if I don't perform the sabotage while potentially killing Nazi soldiers, more innocent lives will be lost, thus I am ready to kill Nazis and breach my principle in this case.

9.4 Conclusion

As we have seen, Frank repeatedly underlines the need not to draw abusive interpretations of physical theories – and in particularly the theory of relativity – outside the realm of physical science, and warns against the uncritical application of physical concepts – such as relativity – to philosophical or political considerations.[50] However, Frank himself is eager to switch the meaning of relativity from one field (physical science) to another (ethics, politics or everyday experience), as we have seen. He advocates a generalized "relativism" and "scientific spirit" in all domains of human thought and activity, in *RRT* and elsewhere.[51] Interestingly, it seems that the same philosophical conception – namely, Frank's operationalism and pragmatic theory of meaning – lies at the basis of both his (legitimate) critique of philosophical misinterpretations of relativity, and his own (illegitimate) application of the concept of relativity outside physics. Frank's moral relativism thus appears as a kind of negative, unintended effect of his overarching operationalism and pragmatic theory of meaning extended outside science.[52]

To conclude, my point in this article was not to bluntly criticize Frank's conception, which is not only perfectly understandable and defensible, but also convincing in many respects (for example his attention to the concrete consequences of general, abstract claims). Frank is of course right to criticize philosophical misinterpretations of physical theories, as well as political absolutism and the dangers of its emotional appeal. My point was rather to show that Frank's overarching

[50] See also his review (Frank 1938a, 79 sqq.) of Ernst Cassirer's (1936) *Determinismus und Indeterminismus in der modernen Physik*, where Frank praises Cassirer's critique of the use of quantum mechanical concepts (such as indeterminacy) to justify moral considerations about freedom of the will.

[51] Illustrated for example by Frank's (1949, 52) summary of *RRT*, presented as an attempt to "prove that the 'relativism of science' has also penetrated every argument about human behavior." (see footnote 4).

[52] It is consistent with Frank's expanding conception of logical empiricism, as he wrongly incorporates Cassirer's conception in it (Frank 1938a).

conceptualization of relativity can itself be seen as a kind of scientistic[53] absolutization of a scientific concept applied in other areas of human life, and that, in certain cases, such a conception may not be desirable. In particular, my claim was that the relativization of values may not always be desirable or necessary, and that absolute ethical principles are not always too abstract to derive "rules for human conduct" from them, as Frank puts it.

Acknowledgements I thank Ádám Tuboly for very helpful comments. This work was supported by the MTA Lendület Values and Science Research Group.

References

Baghramian, M., & Adam Carter, J. 2019. Relativism. In *The Stanford Encyclopedia of Philosophy*, ed. E. N. Zalta. https://plato.stanford.edu/archives/win2019/entries/relativism/

Barberousse, Anouk. 2011. Philosophie Des Sciences et études sur la science. In *Précis de philosophie des sciences*, ed. A. Barberousse, D. Bonnay, and M. Cozic, 206–231. Vuibert.

Barnes, Barry, and David Bloor. 1982. Relativism, Rationalism and the Sociology of Knowledge. In *Rationality and Relativism*, ed. M. Hollis and S. Lukes, 21–47. Cambridge, MA: MIT Press.

Bloor, David. 1976. *Knowledge and Social Imagery*. Routledge & Kegan Paul.

———. 2011. *The Enigma of the Aerofoil: Rival Theories in Aerodynamics, 1909–1930*. Chicago: University of Chicago Press.

Boghossian, Paul. 2006a. *Fear of Knowledge: Against Relativism and Constructivism*. Oxford: Oxford University Press.

———. 2006b. What is Relativism? In *Truth and Realism*, ed. P. Greenough and M. Lynch, 13–37. Oxford: Oxford University Press.

[53] I originally wanted to write a few words about Frank's scientism, but this would take me too far and will be the object of another article. In a nutshell, the idea is that Frank's taking into account of extra-scientific (in the strict sense of the natural sciences), *i.e.* psychological, social and political, factors in scientific theory choice, is itself supposed to be resolved scientifically (in the large sense of a "science of man"). This "naturalistic" understanding of extra-scientific factors, as Howard (2003, 62) puts it, leads to the ambition of a "general science of human behaviour" (Frank 1954–55/1956a, 6). This, together with the claim that the "scientific spirit" has permeated all areas of human life, could be taken as illustrative of a certain (mild) scientism, where for example "social conditions [...] produce the conditioned reflexes of the policymakers" (Frank 1954–55/1956a, 15). However, Frank's position is complex, as he also carefully considers "physics, and science in general, as part of our general pattern of thinking and acting", "as one of the means of expressing man's attitude towards the world, the small world of society and politics and the large world that is our astronomical universe" (Frank 1949, 51–52), and warns that "[s]cience cannot be used as a model for the construction of a 'super-science' that teaches us on a 'higher level' what is 'good' in the same way as ordinary science teaches us what is true" (p. 93). Thus Uebel (2003, 162), mentioning the self-reflection of (unified) science on its own theories' choice which Frank advocates (and which he calls "active positivism") instead of letting philosophy perform the reflective work from the outside, writes: "It is important that 'active positivism' does not defend this autonomy [of unified science from philosophy] for the price of the naturalistic fallacy, by prescribing social goals on purely scientific grounds. Rather, it does so by pointing out the wider contexts into which questions of theory choice are embedded and how the relevant means-ends relations are to be assessed, by providing a scientific analysis of the choice situation."

Bridgman, Percy W. 1927/1958. *The Logic of Modern Physics*. New York: Macmillan.
Cassirer, Ernst. 1921. *Zur Einsteinschen Relativitätstheorie: Erkenntnistheoretische Betrachtungen*. Berlin: Bruno Cassirer Verlag.
———. 1936. *Determinismus und Indeterminismus in Der Modernen Physik: Historische Und Systematische Studien Zum Kausalproblem*. Göteborg: Göteborgs Högskolas Årsskrift.
Einstein, Albert. 1905. Zur Elektrodynamik Bewegter Körper. *Annalen der Physik* 322 (10): 891–921.
Frank, Philipp. 1938a. Bemerkungen Zu E. Cassirer: Determinismus Und Indeterminismus in Der Modernen Physik. *Theoria* 4 (1): 70–80. Translated as "Determinism and Indeterminism in Modern Physics" in Frank (1941, ch. IX) and Frank (1949, ch. 9).
———. 1938b. *Interpretations and Misinterpretations of Modern Physics*. Paris: Hermann.
———. 1938/1953. Philosophical Interpretations and Misinterpretations of the Theory of Relativity. In *Readings in the Philosophy of Science*, ed. H. Feigl and M. Brodbeck, 212–231. Appleton-Century-Crofts. Originally published in Frank (1938b, pp. 34–58).
———. 1941. *Between Physics and Philosophy*. Cambridge, MA: Harvard University Press.
———. 1946. The Relativity of Truth and the Advance of Knowledge. *Analisi: Rassegna Di Critica Della Scienza* 1: 18–24.
———. 1949. *Modern Science and its Philosophy*. Cambridge, MA: Harvard University Press.
———. 1950/1951b. *Relativity – A Richer Truth*. London: Jonathan Cape.
———. 1951. The Logical and Sociological Aspects of Science. *Proceedings of the American Academy of Arts and Sciences* 80 (1): 16–30.
———. 1952. *Wahrheit – Relativ oder absolut?* Zurich: Pan-Verlag.
———. 1954–55/1956a. The Variety of Reasons for the Acceptance of Scientific Theories. In Frank 1956b: 3–18.
———. 1957a. Philosophical Uses of Science. *Bulletin of the Atomic Scientists* 13 (4): 125–130.
———. 1957b. *Philosophy of Science. The Link Between Science and Philosophy*. Englewood Cliffs: Prentice-Hall.
———. 1998. *The Law of Causality and its Limits*. Edited by Robert S. Cohen. Springer.
———, ed. 1956b. *The Validation of Scientific Theories*. Boston: The Beacon Press.
Gowans, C. 2019. Moral Relativism. In *The Stanford Encyclopedia of Philosophy*, ed. E. N. Zalta. https://plato.stanford.edu/archives/sum2019/entries/moral-relativism/
Harman, Gilbert, and Judit J. Thomson. 1996. *Moral Relativism and Moral Objectivity*. Oxford: Blackwell.
Holton, Gerald. 2006. Einstein and the Shaping of our Imagination. In *E=Einstein: His Life, his Thought, and his Influence on our Culture*, ed. D. Goldsmith and M. Bartusiak, 263–286. New York: Sterling.
Howard, Don. 2003. Two Left Turns Make a Right: On the Curious Political Career of North American Philosophy of Science at Midcentury. In *Logical Empiricism in North America*, ed. G.L. Hardcastle and A.W. Richardson, 25–93. Minneapolis: University of Minnesota Press.
Merton, Robert K. 1938. Science and the Social Order. *Philosophy of Science* 5 (3): 321–337.
———. 1942/1973. The Normative Structure of Science. In *The Sociology of Science: Theoretical and Empirical Investigations*, 269–278. Chicago: University of Chicago Press.
Nagel, Ernest. 1943. Malicious Philosophies Of Science. *Partisan Review* 10 (1): 41–57.
Nemeth, Elisabeth. 2003. Philosophy of Science and Democracy. Some Reflections on Philipp Frank's *Relativity – A Richer Truth*. In *Wissenschaftsphilosophie Und Politik. Philosophy of Science and Politics*, ed. F. Stadler and M. Heidelberger, 119–138. Wien/New York: Springer.
———. 2010. Wissenschaftsphilosophie Als Kulturelle Aufgabe. Überlegungen Zu Philipp Frank und Ernst Cassirer. In *Logischer Empirismus, Werte Und Moral*, ed. A. Siegetsleitner, 219–247. Wien: Springer.
Reisch, George. 2005. *How the Cold War Transformed Philosophy of Science: To the Icy Slopes of Logic*. New York: Cambridge University Press.
———. 2017. Pragmatic Engagements: Philipp Frank and James Bryant Conant on Science, Education, and Democracy. *Studies in Eastern European Thought* 69 (3): 227–244.

Sachs, Mendel. 2005. *Relativity in our Time: From Physics to Human Relations*. London/Washington DC: Taylor; Francis.
Siegetsleitner, Anne. 2014. *Ethik Und Moral Im Wiener Kreis. Zur Geschichte Eines Engagierten Humanismus*. Wien: Bölhau.
———. 2017. Philipp Frank on Relativity in Science and Morality. *Studies in East European Thought* 69 (3): 215–225.
Stamenkovic, P. 2015. *The Transcendental Method in Kant and Cassirer. Application to Quantum Theory*. PhD thesis. Paris: Paris Diderot (Paris 7) University. https://hal.archives-ouvertes.fr/tel-02292687.
Stamenkovic, Philippe. 2017. Retour Sur La Philosophie de La Physique Quantique d'Ernst Cassirer. *Revue d'histoire Des Sciences* 70 (2): 397–426.
Toulmin, Stephen. 1951. Review of Philipp Frank. *Relativity – A Richer Truth. The Philosophical Quarterly* 1 (2): 180–181.
Uebel, Thomas. 2000. Logical Empiricism and the Sociology of Knowledge: The Case of Neurath and Frank. *Philosophy of Science* 67: 138–150.
———. 2003. Philipp Frank's History of the Vienna Circle: A Programmatic Retrospective. In *Logical Empiricism in North America*, ed. G.L. Hardcastle and A.W. Richardson, 149–169. Minneapolis/London: University of Minnesota Press.

Chapter 10
Alternative Facts, Fake News, Pseudoscience – New Challenges for a Scientific World-View. An Essay

Hans-Joachim Dahms

Abstract This paper was given in the first year of Donald Trumps administration. It seemed timely to ask, whether this situation required a response from philosophy of science too. So the second section tries a definition and classification of fake news and then gives some historical information about their use up to the present day. But those "news" often don't come in isolation, but in chains and clusters. So the question is, what differentiates them from pseudo-science(s). In the third section I enter the field of the history of philosophy: have there been any philosophers, who contributed to current irrationality and the relativization of science? I concentrate here on Paul Feyerabend and his – up to now – neglected treatment of witchcraft-theory. I then turn to a more positive outlook, namely the scientific world-conception of the Vienna Circle: what is to be learned for todays situation from their fight against metaphysics in the 1920th and 1930th? And in which respects was (and is) their conception perhaps not enough in the struggle with today's irrationalism?

Keywords Fake news · Pseudo-science · Paul Feyerabend on witches · The Relotius case · Scientific worldview today

10.1 Introduction

I would like to start with an introductory remark. When Adam Tuboly invited me in early June of 2017 to the conference about "The Socio-Ethical Dimension of Knowledge: The Mission of Logical Empiricism" for the 12th and 13th December of that year, I proposed for my contribution the title given above which appeared to me timely and actual. Donald Trumps inauguration as president of the United States

H.-J. Dahms (✉)
Institute Wiener Kreis, Universität Wien, Vienna, Austria
e-mail: hans-joachim.dahms@univie.ac.at

C. Damböck, A. T. Tuboly (eds.), *The Socio-Ethical Dimension of Knowledge*,
Vienna Circle Institute Yearbook 26, https://doi.org/10.1007/978-3-030-80363-6_10

had taken place only some months earlier on the 20th of January. Having followed most of the debates between Republican candidates and also those between Hillary Clinton and Trump in 2016, I was convinced that he had a chance to win the presidency and that, if elected, he would not quit after a short time, as many still hoped. With him as president not only the USA, but the whole world would enter a new age, in which fake news would become (again) a "normal" tool of politics.

Already the inauguration ceremony gained special fame immediately, since Trumps speaker Sean Spicer told that the attendance number superseded the ones of his forerunner, president Barack Obama. Trumps advisor Kellyanne Conway coined the term "alternative facts" when Spicers "news" was challenged in an interview with her. Since then one could witness a steady outpouring of fake news from the White House. According to the Washington Post's counting, their number reached 10.000 in late April 2019 and 30,573 at the end of Trumps tenure.

Already at the time of the Budapest conference a number of articles and even booklets on those phenomena had appeared (Neiman 2017; Snyder 2017), in fact on the day of my talk an opinion was voiced in the *New York Times* under the title "Truth Still Matters". Nowadays of course nobody can be surprised by the appearance and revelation of more fake news on an almost daily basis, because the phenomenon is so massive and obvious.

In reaction to this situation I changed my paper. It will not much rely on examples from 2017 of course, but more recent ones. On the whole I will try now to systematize our knowledge of the phenomenon of fake news: a definition, a possible classification, the balance between their quantity and quality, their producers and recipients. That will happen in the first chapter of the contribution.

In Sect. 10.2. I turn to the question, what that all has or might have to do with philosophy. Have perhaps certain philosophical viewpoints contributed to pave the way for the new wave of irrationalism, perhaps with a relativization of science and the propagation of alternative worldviews, which then led to a contempt or denial of its findings? That is discussed in Sect. 10.3. In the last one, Sect. 10.4., I pass to a more positive outlook, namely to the question: What can we learn from science-friendly, even scientistic traditions and the battles their adherents fought against the rise of irrationalism in their time. Special emphasis will be placed on the "scientific world-conception" and the Vienna Circle in their fight against, what they termed – perhaps somewhat vaguely – "metaphysics".

10.2 Alternative Facts and Fake News

I start with a sort of warm-up, namely "alternative facts", which seems to be the easiest part and I will follow in the beginning the Wikipedia entrance. The expression is a relatively recent invention. "Alternative facts" is a phrase used by U.S. Counselor to the President Kellyanne Conway during a *Meet the Press* interview on January 22, 2017, in which she defended White House Press Secretary Sean

Spicer's controversial statement about the attendance numbers of Donald Trump's inauguration as President of the United States. When pressed during the interview to explain why Spicer "utter[ed] a provable falsehood", Conway stated that Spicer was giving "alternative facts". The interviewer responded, "Look, alternative facts are not facts. They're falsehoods."

Conway's use of the phrase "alternative facts" to describe demonstrable falsehoods was widely mocked on social media and sharply criticized by journalists and media organizations. The phrase was extensively described as Orwellian. Within 4 days of the interview, sales of George Orwell's famous book "1984", where a "ministry of truth" itself produced and overlooked falsehoods, had increased by 9.5%, which *The New York Times* and others attributed to Conway's use of the phrase, making it the number-one bestseller on Amazon.com.

What is less well known is, that Conway also made an attempt to explain what she had meant by her new invention: "Two plus two is four. Three plus one is four. Partly cloudy, partly sunny. Glass half full, glass half empty. Those are alternative facts." But later on under the weight of criticism Conway slightly changed her choice of words, when she explained "alternative facts" as "additional facts and alternative information".

"Partly cloudy, partly sunny" is perhaps not a very good example, because on that logic every mixed situation would constitute an alternative fact. But with the other one Conway went right on into the philosophical sphere. The equation reminded me of Gottlob Frege's treatment of identity. He introduced his doctrine of sense and reference in order to solve the problem, why certain equations are completely trivial, whereas other ones may contain valuable information. The equations used by Conway are of the latter kind. They contain different "Gegebenheitsweisen" of one and the same thing, namely the truth-value "The True", according to Frege. That treatment, by the way, is not restricted to arithmetic alone, but also to empirical identity statements, like "the evening star is the Venus" and "the morning star is the Venus", so that one gathers from those two statements the identity "the evening star is the morning star". Whereas "the Venus is the Venus" is trivial, the former "the evening star is the morning star" can formulate an empirical achievement. Nevertheless all those statements denote all the same thing, the truth, whereas Conway's "alternative facts" denoted also the same thing, the falsehood (in Frege's terminology). The other example "glass half full, half empty" is easier, because it seems to indicate only the emotional attitude, with which the optimist and the pessimist perceive the same fact. They would not necessarily disagree about measurement results of the content of the glass.

Conway could have adduced also a more difficult example, namely the different images common in Gestalt-psychology. From the middle of the 1950s onwards they were used also by philosophers and historians of science like Norwood Russell Hanson, Thomas Kuhn and Paul Feyerabend in order to exemplify, how seemingly the same observations would depict different realities after a paradigm (or "patterns of discovery", Hanson's expression) change.

10.2.1 A Definition of Fake News

Now: What are fake news? They belong to the vast class of false statements. In order to qualify as fake *news*, they must fulfill some additional requirements:

1. They must report some recent event, episode etc. or a recent finding about an older event. Otherwise they are not news.
2. The persons reporting the event must know that the report is not true (so they are lying about the "described" event). The report may be about an event which never took place or describe an actual event in a way that is false. (I will show examples later).
3. Those statements need to be made in the public sphere (by governments, political parties or other political agencies and their speakers, in newspapers, in social media etc.).

Now lies are perhaps as old as man uses language. Fake news on the other hand occurred only since there existed a public sphere. It may be disputed whether the Agora in Athens in the times of Socrates, Plato and Aristotle constituted a "public sphere". In any case of course there has been a rise in the possibility of spreading fake news with the invention of print in the fifteenth century. Around 1600 the first newspapers were founded. And from then onwards the possibilities for spreading fake news rose with every mayor invention in the media technology: radio, television, the internet.

10.2.2 Non-Fake-News: Some Problematic Cases

In the press and public discourse we find also many examples of alleged fake news, which according to my definition are something else. I have in mind utterances which refer to the future, be it forecasts or promises (especially, when proclaimed in political parties manifestos before an election). Forecasts are in a special class. One is inclined to think that they cannot be fake: Scientific tests often fail to confirm a theory, but the scientists did not propose the theory or the portion to be tested in bad faith. But when e.g. polls about future voting behavior or scientific results (say: about the future pace of global warming and its effects on sea-level rise) are falsified, they come close to fake news. In any case one cannot judge these forecasts with the same degree of accuracy as present or past events, and therefore I would not count them as fake news. A special case would be, if someone made forecasts which could not come true under any plausible circumstances.

Promises are an especially difficult case, because they somehow combine a forecast (about ones own behavior) with a moral obligation to act to make this forecast come true. But promises are sometimes not kept, even if the persons in question did not give them in bad faith. Only when the author of a promise knows in advance that he does not want to keep it or even knows that it is impossible to keep it, it is morally bad to proclaim these pledges. Furthermore it can well be that the author of false promises learns only after a while (e.g. after a "look into the books" after an election) that it is impossible to keep them. That is why new governments start their work almost uniformly with a cash-check and then declare that money for certain programs promised during election campaigns is sadly not available, because the former government acted financially incompetent and hid the economic catastrophe created by it before the public.

A good example for a failed promise is Boris Johnson's pledge – placated on the side of a big bus - before the Brexit referendum on 23rd of June 2016 to spend 350 million of pounds saved by a future withdrawal from the EU for the NHS every week. Was this a forecast, a promise or what? Did he know that the sum was exaggerated in the first place and did he perhaps never intend to spend any saved money for the NHS? If so, it was certainly a morally despicable attempt to manipulate a referendum. But fake news it was not in my opinion.

10.2.3 A Classification of Fake News

Almost every day one can find examples of fake news in the conventional or the new social media. Is it possible to classify them in order to get a grip on their occurrence and function? In a paper on "Fake News vs. Satire" a vast volume of examples were searched (Golbeck et al. 2018). The authors, Jennifer Golbeck et al., broke these down under the heading "Themes of fake news vs. satire" to seven "codes", a sort of classification of the themes of fake news. To each category they added – real or in some cases invented – examples, as follows:

1. Hyperbolic position against one person or group (e.g. Trump, Clinton, Obama, Islam refugees)

   Example headline: "Obama Signs Executive Order Banning The Pledge of Allegiance in Schools Nationwide"
2. Hyperbolic position in favor of one person or group (e.g. Trump, Clinton, Obama, Islam, refugees)

   Example: "Because Trump won the presidency, Ford shifts truck production from Mexico to Ohio"
3. Discredit a normally credible source

   Example: "MIT researchers: Global Warming Data is Complete Bunk"
4. Sensationalist crimes and violence

   Example: "George Zimmerman Found DEAD Just Hours After Bragging About Killing Trayvon Martin"
5. Racist messaging

   Example: "Trump has Fired Muslim Sharia Judge Arrested and Charged"
6. Paranormal theories (e.g. Aliens, Flat Earth)

   Example: "Donald Trump Says the Earth is Flat"
7. Conspiracy theories

Example: "Hillary Clinton Busted in the Middle of Huge Pedophilia Ring Cover Up at State Department".

I doubt that all "paranormal" and "conspiracy theories" can be labeled as fake news, because that would require that the proponent of those theories don't believe in them themselves. And that is sometimes not the case. I will have to say a little bit about these sorts of irrational ideas in the paragraph about "pseudo-science" below.

I also need to point out that the list is not exhaustive: one "code" is left out, although it is one of the most important ones. So I take the liberty of adding

8. Reasons for going to war and news about the development of a war;

   Example: "Saddam Hussein had weapons of mass destruction".

Fake news were present in many big events in world history. When we only look at the twentieth century, it springs to mind that many wars originated with an propaganda effort using fake news: almost every war is preceded or accompanied by fake news about the actions and intentions of the enemy. During wars of course many lies are told by government authorities about the course of the war. I will come back to examples below.

10.2.4 Some Historical Remarks

The word "fake news" itself seems to have been used for the first time in 1890. Hundred years earlier happened a major event in part triggered by fake news, namely the French Revolution. It was successful, because the agrarian population sided with the revolutionaries. They were deceived by fake news, namely the rumors that the nobility had hired soldiers (in part from foreign countries) in order to oppress the farmers and also to kill them. Georges Lefebvre, the famous French historian and co-founder of the Annales school of history, described the whole process in his book *La Grande Peur de 1789* (Lefebvre 1932). It is of special interest, because the spreading of those false rumors was not affected by printed media at the time. So Lefebvre followed the spread of those fake news, using as source old church-books. That way he was able to show, from where the news originated and where and when those news arrived elsewhere.

Sometimes the truth or falsehood of news can only be established much later, when new sources are discovered or the access to closed sources is made possible. Think of the Katyn-massacre of polish officers and intellectuals in 1941, which was falsely blamed on the Nazi troops by soviet Russia or of the Hitler-Stalin treaty with the suppressed information about the division line of spheres of interest in the European East between Germany and Russia. Most of these fake news were spread by state agencies, and some of them were only revealed much later (the Soviet examples only by Boris Jelzin in 1993). Military defeats or revolutions are the decisive events which lead to revelations of historical fake news covered up by previous regimes.

What is now the difference between fake news then and now? One has to take the technical means to communicate them into account: the French revolutions "grande peur" was spread only by mail or oral communication. The fake news surrounding preceding and during the Second World War used the new medium of radio on a massive scale (think of Hitler's peoples radio, the "Volksempfänger"). The alleged Polish attack on the German radio station Gleiwitz (itself a fake-event, since it was staged by SS personal put into Polish army uniforms, the German victim being a KZ inmate) was transmitted immediately by German radio, including Hitler's declaration "from now on we will shoot back". The fake news preceding the Vietnam War (the Tonkin event) came in the time of television as well as those preceding the Second Iraq War of 2003 (Saddam Hussein's alleged possession of weapons of mass-destruction). And nowadays it is the internet and social media.

These new media have an impact on the quantity of fake news and their repetition for hammering them into the heads of the recipients. Of course also their speed plays a large role. Since the recipients are either not able to detect them or in part believe whatever they are told by those media, the degree of believe in fake news has risen dramatically. Many don't care what other media publish and stay in their social bubble or echo-chamber. Social media strengthen that tendency, when they offer links to information of similar kinds. Social bots or whole armies of (in part paid) trolls were also of course not present in earlier times.

10.2.5 Quantity and Quality of Fake News

The Washington post kept track of the fake news published by American president Donald Trump since his inauguration. Now I think: just counting is misleading. This atomistic approach to the problem overlooks two things:

- some fake news are (often: much) more important than other ones,
- fake news come sometimes in chains, which connect with former "news" of that type or build whole clusters.

The former fact is quite obvious: false number of attendants at Trumps inauguration ceremony are a bad thing. But telling, that Saddam Hussein had weapons of mass-destruction is far worse, since this "news" was the basis of the Second Iraq War in 2003 and the cause for several hundred thousand deaths among the people of Iraq and the later beginnings of ISIS.

That fake news come in chains or whole clusters is more tricky. This actually happens quite often: news of the same type and theme – with some pause in between – come again and again. In order to have some minimal credibility they need to mix up with real news. Just one example: On New Years Eve 2016 there were hundreds of cases of molestations, thefts and rapes of women in front of Cologne's (Germany) cathedral. In order to build on that image of public disorder caused by a "Muslim invasion of Germany" Breitbart news published a piece about the alleged burning of "German oldest church" on New Years eve exactly a year later. But: the church, where the "event" happened (St. Rheinhold in Dortmund), was not Germany's oldest church (this is the cathedral in Trier, begun in the fourth century and so at least 700 years older). And there had not been a big fire ignited by Muslims destroying the church, but only a minor fire on a scaffolding, caused by firework and extinguished in 10 min by fire fighters. In due course the narrative about the destruction of churches by Muslims took a different direction. Now it was France, where thousands cases of vandalism and destruction were reported by the same press-organ. To be sure: there have been cases, were damage in French churches have been caused by Muslims, for instance: a man from a north African immigration background broke into the old church of St. Deny near Paris, destroyed a window of the church, stayed there for a night and damaged the organ, where he slept. That there should be thousands more cases of that type is not established at all. Furthermore in France there is a strong anti-clerical tradition in place. So people from the most different backgrounds could be authors of similar damages. Nevertheless in far-right discussion forums one finds the firm conviction that also the fire at Notre Dame in Paris was caused by Muslims.

But the destruction and vandalism of churches is often only part of a cluster of fake news. Take as an example the "Great Replacement-theory". According to its inventor Renaud Camus and its adherents there is a big population replacement going on which in time will wipe out Western Europe's former population and replace it with Muslims from the Near East. This narrative tells,

- that there is a migration crisis going on, which brings millions of Muslim "invaders" to Europe as alleged asylum-seekers,[1]
- when landed in European countries like Germany, Sweden and Austria (which have the largest numbers per capita of asylum-seekers) they kill many of the "indigenous" people and rape the women,
- they destroy what is left of the former European culture like Christian religion, churches etc.,
- they start a "demographic jihad", i.e. they will "outbreed" the former people and so will in due course arrive at the majority.

In the end they will establish a new caliphate, where sharia law reigns.

That whole theory also comes with a conspiracy part: behind all these activities are "the elite" and their organizations, like the EU and the UN. Since the migrations crises of 2015 and the following admission of more than a million asylum-seekers from Syria, Iraq and Afghanistan especially the German chancellor Angela Merkel is named as one of its representatives and she is alleged to be "in the pockets" of the Hungarian-US-American George Soros.

[1] Breitbart showed a very nice example of illegal entrance from Africa to Europe with a special device, namely someone crossing the Gibraltar straight with a sort of hover-boat. In Breitbarts reader forum people lamented, how the depicted passenger could afford that device and even was able to buy expensive sunglasses. Now the thing is, that in fact that man was Germanys soccer international player Lukas Podolski, and the photo was shot during a free day during the training of Germanys team in Brazil's world cup tournament in summer 2014.

That whole theory relies on vast exaggerations (e.g. concerning the numbers of immigrants), a wrong construction of causes for the immigration, and really fantastic projections of developments to the future. The conspiracy part is pure nonsense: why should Angela Merkel, the daughter of a protestant pastor, would like to live under sharia law and why should Soros with his Jewish background have an interest to deliver Europe to Muslims?

In sum: it is much more important to point to more complex chains, clusters and these broader ideological constructions of fake news than simply to count them in isolation, irrespective of their importance and size.

10.2.6 Producers and Recipients of Fake News

There is a whole range of producers of fake news: people who want to elevate their social standing by giving wrong information about their ancestry, education, former occupations, achievements, wealth etc., or people who want to demolish the reputation of actual or perceived enemies by denying any of those points. Of course they need the media for spreading those fake news. But those media also sometimes produce fake news of their own, for example in order to promote an economic or political agenda. An outstanding source of fake news was revealed in 2018 in German journalism, when the famous journalist Claas Relotius was unmasked as author of dozens of fake reports he had produced over the time-span of 5 years for the prestigious German magazine Der Spiegel and other journals. These covered everything from the alleged start of the civil war in Syria to the alleged clandestine actions of militias at the Southern border of the USA in their hunt of illegal immigrants.[2]

But of course the most important producers of fake news are politicians in their struggle to gain or retain power and influence. This phenomenon has been described and analyzed exemplary by Hannah Arendt in her famous article about the function of lies in politics. She wrote the article "Die Lüge in der Politik" (Arendt 1971/2013) in reaction to the revelation of the Pentagon Papers[3] by Daniel Ellsberg, published jointly by the *New York Times* and the *Washington Post*. The Papers showed the complete disaster of the Vietnam War and contributed to its end. Surprisingly a similar spectacular publication about the War on Terror in Afghanistan made much less furor (Beaumont 2019; Whitlock 2019). Now in Arendt's analysis the thesis plays a important role that in time politicians tend to believe in their own lies and as a consequence are no more able to tell truth from falsehood.

---

[2] Relotius' fake-stories were in fact exposed by Juan Moreno who had coauthored the report. See his book: Moreno (2019); *Der Spiegel* itself reported the Relotius case (*Der Spiegel*, 52, 2018), investigated intensely and published an extensive report on the whole scandal and changed its editorial policies (see the concluding report by Fehrle et al. 2019).

[3] Their original title was "United States-Vietnam Relations, 1945–1967: A Study Prepared by the Department of Defense".

The most important recent example for this phenomenon is president Donald Trump's treatment of the Corona pandemic. In the beginning he constantly played down the danger of the virus, telling the public that it would go soon away "like a miracle". Later on he recommended really "alternative" treatments like bleach.[4] One would assume that he was too incompetent or unable to listen to his scientific advisers to comprehend what was happening. But with the publication of Bob Woodward's book *Rage* this impression received a thorough falsification: Trump knew from the start about the danger of Covid-19, but kept this knowledge to himself. These are his words:

> It goes through air. [...] That's always tougher than the touch. You don't have to touch things. Right? But the air, you just breathe the air and that's how it's passed. And so what's a very tricky one. That's a very delicate one. It's also more deadly than even your strenuous flus. (Woodward 2020, XIX)

He told Woodward as an attempt at an explanation for his divergent public utterances: "I wanted to always play it down. [...] I still like playing it down, because I don't want to create a panic" (Woodward 2020, XVIII).

It seems that a representative of a party which often complains about a "nanny state", when it comes to public healthcare, free education, better social security, now plays the role of a anti-panic nanny (which costs tens of thousand of additional deaths, by the way). Now one would think that the downplaying tactic went obsolete with Woodward's revelations. But no: Trump during his rallies before the general election in November returned to this downplaying when he said that the US is "turning the corner" in its fight against Covid, while in fact in almost every federal state infection numbers were rising. It seems that he heavily succumbed to the phenomenon Arendt had described. That of course applies also to his complaints of a "stolen election" from November 2020 onwards to the present day.

But of course such a tactic can only work, if big parts of the public either don't care, what the facts are, or in time are used to believe in whatever fake news are dished out to them. And that seems to be the case. Furthermore: whenever in media like Breitbart news an article on Barack Obama or his family appear, you find in the readers columns not only the usual repetitive falsities of the birther campaign, according to which "the Kenyan" Obama is not an American citizen and so was not eligible as president, but in addition that he is homosexual, his wife is a transsexual, their children either adopted or simply stolen somewhere etc. Those people most probably know what a sort of offensive nonsense they are producing. But they are so in tune with fake news about their enemies, that they freely add to it and even new ones, when the occasion arises.

Sometimes the consumers of fake news are not only consuming and propagating (seemingly harmless) nonsense, but start counter-action against the "deep state". The case of Edgar Welch, who entered the Washington-based Comet Ping Pong

[4] His subsequent effort to depict the recommendation as satire are not convincing, since he had immediately before contact with a bleach producer.

pizzeria, from where a ring of child-traffickers allegedly operated under the order of Hillary Clinton, with an assault rifle to free those children, is a telling example.[5]

10.2.7 A Quick Look on Pseudo-science and the Current Rise of an Anti-scientific Attitude

Here we no longer speak about isolated events (or chains of events) and their truthful or wrong descriptions, but about whole interrelated groups of facts, like theories, scientific laws, principles and their use in explanations and forecasts. But proponents of pseudo-science often don't know about the falsehood or implausibility of their theories. So here in some cases one finds examples where the proponents do not lie intentionally. And that marks the difference to fake news.

Pseudo-sciences are those theories which don't stand up to the test of observation, but nevertheless make claims for a scientific status. Often they represent simply a former stage of knowledge, later on superseded by scientific advance.

The use of the concept of pseudo-science is sometimes also used by enemies of established science, when their findings threaten to damage powerful interests (be it "big tobacco" or the fossil fuel industry). The history of science is also full of such fake-sciences. A collection of case-studies edited by Dirk Rupnow et al. (2008) showed examples from alchemy in the early modern times all the way through spiritism in the nineteenth century till the race-theories which preceded in part (and flourished in) the Third Reich.

But there are also important examples of pseudo-science which are still relevant today. I have in mind "Intelligent Design"-theory about the evolution of species and man on the one hand and the denial of man-made climate-change on the other. Christian Circles invented Creation Science and more recently Intelligent Design theory in order to present an alternative theory of the universe, the earth, the emergence of life on earth, of mankind etc. based on the first chapters of the Bible. Some of them take the biblical description literally, that everything was created by God in 6 days (allowing only that a day during creation is translated into a thousand years in history).

In Kentucky you find a museum, where statues of men walk along dinosaurs and small children are riding minor dinosaurs with saddles attached to them. Even against that anti-scientific nonsense a group called "Christians against dinosaurs" rose in protest saying that "those beasts (dinosaurs) lack any Christian values and even lack existence" and so they should not be included in the school curricula nor shown in museums.

Then we have the theory of man-made climate-change. Its critics deny its scientific status, some even deny global warming at all and not only the man-made

[5] See Horton (2020) with a report on the documentary "After Truth: Disinformation and the Cost of Fake News", where "pizzagate" plays a big role.

influence on it. As is well known, among climate scientists there is a large consensus about the correctness of the theory. But in the USA and here and there elsewhere it is still denied. Naomi Oreskes and Erik Conway (2010) pointed in their book *The Merchants of Doubt* to the fact, that a very small group of physical scientists like Bill Nierenberg, Fred Seitz and Fred Singer were able with the backing of well-funded institutions to misled the public about the dangers of tobacco, especially passive smoking. In chapter 6 the actions of those "merchants of doubt" is described with respect to man-made climate change.

Now I certainly think that this conflict is not so easy to solve as the other examples. One should take into account that quite a number of scientific breakthroughs were achieved against the opposition of established science and so consensus among scientists is not always a criterion of truth. The examples of Gregor Mendel, Johann Carl Fuhlrott and Alfred Wegener come to mind, who all had to struggle as outsiders with their groundbreaking innovations against the scientific establishment and its consensus. Bernard Barber's (1961) wonderful article "Resistance of Scientists to Scientific Discovery", which inspired Thomas Kuhn's *Structure of Scientific Revolutions*, offers many more examples.

To come back to the example of climate change: here we find problems like

- the question of different methods of temperature measurement (by terrestrial and sea stations or via satellites) and their sometimes seemingly divergent results,
- the problems of sorting out relevant factors for simulation models,
- the question how big the man-made factor should be estimated in comparison with other contributing causes. The whole theory of global warming is a quite complex thing, which still presents challenges for climate researchers. At the same time it offers fascinating material for philosophical reflection too. Philip Kitcher has been a pioneer in this discussion. I think that more needs to be done here by our profession.

10.3 Did Philosophers Contribute to the New Surge of Irrationalism?

Now what have "alternative facts, fake news and pseudo-science" to do with philosophy? I hinted at some interrelations in passing already. But now I will have some more general things to say. My lead question in this section is: did perhaps some philosophers contribute to this rise of irrationalism?

My thesis is, that some philosophers certainly contributed to irrationalism. Susan Neiman (2017, 48 ff.) in her booklet *Widerstand der Vernunft. Ein Manifest in postfaktischen Zeiten* named Michel Foucault and postmodern philosophers with their denial of a universal truth, their talk of "narratives" and their thesis, that questions of truth and morals are hidden question of power. There are authors who claimed that philosophical postmodernism with authors like Lyotard, Baudrillard, Derrida etc. perhaps even anticipated the post-truth attitude of Donald Trump. To me it seems difficult to verify such claims. It is well known that Trump seems to have a slight aversion against books and theories and said once that he had no time to read them (possibly with the exception of books co-"authored" by himself). So there needs to be a bridge which brings us from philosophy to perhaps some people in Trump's environment who have an outspoken ideology and influenced him. Or one has to rely on something difficult to grip like the "Zeitgeist" or the "spiritual atmosphere" which were influenced by philosophical postmodernism, relativism and all that and in turn impinged the political sphere in Trump's government.

Of the philosophers who somehow personified the relativistic Zeitgeist since the 1970s I have someone else in mind than the above-mentioned French postmodernists, someone who is in addition closer to philosophy of science. I will discuss in the following some utterances by Paul Feyerabend showing an anti-science stance (in his anarchistic phase from the late 60s onwards). It is of course absurd to construct a direct impact of his writings on Donald Trump or his environment, so much so that a Berlin-based caricaturist could depict him, together with Vladimir Putin and (then) Italian prime minister Silvio Berlusconi, all dressed in typical intellectual outfits of the French Existentialists of the post World War II period and heavily smoking, as a group reading and discussing Feyerabend's *Against Method*.[6]

Now what do we find in Feyerabend that could have informed nevertheless the anti-science, relativistic worldview of today's populists around the world? I concentrate on the most extreme example of anti-science revolt in Feyerabend, namely the theme of witchcraft, because that seems up to now not have received sufficient attention.[7] The example of witchcraft played different roles during Feyerabend's career as a philosopher of science. It seems that he used it up to the middle of the 1960s in order to undermine the positivistic differentiation between observational and theoretical language. He maintained the theory-ladenness of observation: that there is no neutral observation, that on the contrary the theories structure and affect in advance what can be later on observed. Witchcraft theory is a good example. It has indeed different layers, one more observational one, where all the deeds of the witches and their interactions with devil and what individual witches had done and were still doing after their intercourse with him belonged to, and a more theoretical one, namely how the devil comes into contact with the witches, how he made them dependent on him, whether offspring was originated by their intercourse and eventually what type of offspring etc. Feyerabend was very much interested at that time also in the question, how witch mania was overcome in the eighteenth century. His thesis was, that this was not due to rational reasons during the Enlightenment, but to an ongoing cultural revolution.

---

[6] See the reprint of Studio Pong's wonderful caricature "Die postfaktische Universität", first published in the German journal *Die Zeit* (15 December, 2016 edition), in Hagner (2017)

[7] I was born in the small town of Lemgo near Hannover. That town had been an important member of the Hanseatic trade-organization over many centuries, but reduced its populations in half during the thirty Years war and three subsequent terrible prosecution waves of alleged witches and their witch-masters and never recovered from this backlash. See Meier-Lemgo (1962, 147–203) for a description of the "century of decline" and the four waves of witch-hunt with hundreds of victims.

After this early period, where the example of witch-"theory" was used in arguments against a special philosophy of science he played with the idea that witchcraft was no less rational than the sciences and so implicitly turned against science itself. The correspondence with Imre Lakatos is telling in this respect. So he writes him in autumn 1972:

> I just arrived in time at Hamburg airport… I gave a rousing talk before all the German philosophers, they sat there in their suits, with their ties, I walked up in front of them in my pullover and my bellbottom trousers talking to them about incubi, succubi, the problem of the cold member of the devil (sensed as such by all witches), the question whether the devil could create seeds or whether he had to steal them from some man to inseminate women. (Lakatos and Feyerabend 1999, 300)

All that is *not* contained in the published version of Feyerabend's talk to the congress of the society of German philosophy in Kiel (near Hamburg) under the nice title: "Wissenschaftstheorie: eine bisher unerforschte Form des Irrsinns?" ("Philosophy of Science, a up to now Unknown Kind of Insanity?", Feyerabend 1978a), where he identified mental illnesses with certain schools in the philosophy of science: logical empiricism with schizophrenia, Popper's critical rationalism with hysteria and Paul Lorenzen's Erlangen program of "proto-physics" with catatonic rigidity (Feyerabend 1978a, 310). But as a participant of that congress confirmed to me, Feyerabend indeed used witch-theory extensively in the discussion afterwards.

In Feyerabend's *Against Method* (1975a; and *Wider den Methodenzwang*, 1975b, its German counterpart), we find many places, where this idea is developed. Even Pope Innocence VII and Sprenger's and Institoris' "Hexenhammer", the most outspoken proponents of witch-theory, are mentioned, discussed, and positively evaluated. So from then on witchcraft went on a par with science. Later on a whole zoo of non-scientific "theories" were assembled: voodoo, astrology, traditional Chinese medicine and so on.

It seems that Feyerabend had planned to write a bigger article or even a book on witchcraft, because he asked correspondence-partners like Wolfgang Stegmüller to tell him more about the phenomenon. It comes as a surprise then, that the subject is only mentioned shortly in his autobiography. Perhaps Feyerabend had only used it in the 70s as a provocation. But no: he gave also lectures at (some of?) the various universities, where he held chairs in the early 70s (Feyerabend 1995, 173).

Now some critical remarks on that example, which are meant to show that Feyerabend's use of it were from the start completely irresponsible:

1. The "observational" part of witch theory stands on very shaky basis: the tests for being a witch, especially the water and the needle-test are only a pretext for imprisoning the victims, torturing them and committing them to the flames (see Soldan-Heppe 1843/1911, 380–386). In the case of the water test the victim was put (in a cage) into water. If she survived, she belonged to the devil, if not, she had not been the devils companion. So she would perish in any case. In the case of the needle test the victim was pinched with needles all over her body in search of an insensitive spot the devil might have left as a mark for her belonging to

him. If such a mark was found, that counted as proof that the inmate was indeed a witch and had intercourse with the devil. But if no mark was found, that showed, that the subject was all the more a witch, because the devil has the habit not to mark his truest followers. The inmate was "convicted" in any case. So the different test-procedures for a persons being a witch can be refuted on logical grounds alone and have nothing to do with any philosophical outlook which may have informed observations.
2. There were some critics and even critical publications about witchcraft theory at the time. But it was very dangerous to speak out, because then it could happen, that the critic found himself accused of being a sympathizer of the devil and his witches (similar to what happened under national-socialism to critics of the anti-Semitic race-theory). So for instance one of the best critical books against witch mania, namely Friedrich von Spee's book "Cautio Criminalis oder Rechtliches Bedenken wegen der Hexenprozesse" had to be published anonymously in 1632 (see von Spee 2000).
3. According to todays estimates around 3 million people became victims of witch-hunt from the fourteenth to the eighteenth century in all parts of Europe (irrelevant, whether catholic or protestant, by the way); there are 70.000 registered cases of death through burning alone. This is perhaps a bit different from the other cases Feyerabend had in mind: Chinese medicine and astrology were not intended to destroy people. So witch theory is comparable more to national-socialist race-theory and its practical "application" in forced sterilization and the Holocaust. And everything Feyerabend said about witch theory could be exemplified by race-theory as well. So one might ask: is race-theory perhaps also an alternative view of the world?

But Feyerabend did not stop there. In his fight against first the philosophy of science and then against the authority of science itself and the authoritarian traits of the scientific mandarins he then went on to extend his relativistic propaganda to the education system: why should schools and universities teach only science, why not alternative world-views as well? He proposed to subject the school curricula to public democratic voting and wrote:

> For example democratically committees must investigate, whether the theory of decent is really so well justified, as the biologists persuade us, and whether Genesis is really finished, so that only the first, but not the second should be taught at our schools. (Feyerabend 1978b, 364, my translation)

Alternatively he could have demanded that private schools, each with their favored worldview, be it Darwinian theory of selection (including human descent) or Creation Science and Intelligent Design could be taught not only as alternatives, but even exclusively, each according to the preference of the parents and school-boards.

To sum up: Feyerabend demands became in part reality: there is a proliferation of "alternative worldviews" and a growing relativization of scientific findings (opening the door ultimately to a neglect and even a contempt of science). The layman now feels competent to condemn climate science and one of its findings, man-made global warming, and thinks that is only a "hoax from China". So yes: there have been philosophers who contributed to the rise of irrationalism, even in the ranks of philosophy of science. Of course it has to be seen, to which degree: whether they only furthered a cultural atmosphere, the Zeitgeist, in which the propaganda of "anything goes" could thrive or whether their views informed also the views of todays major right-wing ideologues like Steve Bannon (or Mark Spencer).

Steve Bannon remains even after his ouster from Trumps White House adviser function a faithful follower of his master and propagates everything the president has in mind, like building the wall, abandoning all the international trade-agreement, the climate-accord etc., including a daily fight, or, as he calls it, "the war" within the Republican party against more liberal minded people, the "rinos". But what is *his* ideological background, which he perhaps transmitted in part to president Trump? He published in mid-December 2017 a list of his dearest 6 books (in addition of the Bible, of course). There were many classical scriptures like Thucydides account of the Peloponnese war, Plutarch about the Roman wars against Hannibal, Gibbon's work about the decline of the Roman Empire. To this list one must add books like the Bhagavad-Gita or the book of the Chinese theorist of war Tzu. These choices show that Bannon sees world-history as a coming and going of empires and an uninterrupted war between them.

My understanding of this list is that he is now concerned to protect the USA from suffering a fate similar to the decline of the Roman empire:[8] his idea seems to be, that after the build-up of a (world-wide) US-empire one could strengthen it through a huge military in a way that no dependence on any allies is needed, hinder the incoming enemies from outside (building the wall and restrict access especially from Muslim countries) and driving out all the enemies inside, in this case the illegal immigrants and silence the opposition.

But the list of his "classics" includes no contemporary history, let alone philosophy. But there is a book which informed Bannon's (and seemingly his masters) view on the epoch in which the world's leading power finds itself now. I have in mind Neil Howe and William Strauss (1997) book *Generations*: *The History of America's Future, 1584 to 2069*. This is a special theory of societal development, according to which a society and culture goes every 80 years (approximately the timeframe of a human life) through four different phases: blossoming, becoming adult, crises and finally decay (a picture similar in many respects to Oswald Spengler's *Decay of the West*, although of course with a completely different time scale than Spengler's Culture Circles with a life-span of a thousand years). As with Spengler's diagnosis of the *Decay of the West* with its tacit aim of revealing the fate of Germany after the First World War it is important to know in which of those phases the USA is currently standing according to the generation theory. The answer is that it entered with the big economic crash of 1908 its ultimate phase. Unless someone took decisive leadership, strengthen its status on the international scene with drastic investments in military power (albeit not necessarily much of military interventions), defend itself against more immigration, and drove out its internal enemies, the USA would

[8] Compare the relevant pages of his biography: Koffler (2017, 111 ff.).

vanish as a world power. Bannon read Howe and Strauss's book and adopted their worldview (Koffler 2017, 58 ff.). He even produced a film in 2010, "Generation Zero" (Koffler 2017, 60 ff.), along these lines depicting the rise of the present situation, the "fourth turning".

Those authors and their follower Bannon had no ear for philosophies of science (see Howe 2017), be they relativistic or otherwise. So the question, how those views of science might have contributed to paving the way for today's irrationalism, especially in the personal of the current US-administration, remains largely open.

10.4 What Can We Learn From the History of Scientific Philosophy (Especially: Logical Empiricism) for Today's Political Problems?

I think that Logical Empiricism stands relatively strong in its fight against irrationalism, even though its insistence on verification or at least conformation of every declarative sentence on the one hand and on its verification principle as a meaning theory cannot be upheld literally and was abandoned a long time ago.[9] But its principal questions: "What do you mean? How do you know?" were not only of cultural and political relevance in its time, but remain relevant today.

But: there seem to be relatively few examples for the application of said questions on politically relevant issues of the 1920s and 1930s. Of course Otto Neurath and Edgar Zilsel contributed a number of articles to the theoretical journal of the Austro-Marxist party SDAP "Der Kampf". Zilsel attacked Nazi ideology fiercely in his "SA philosophiert" and "Das Dritte Reich und die Wissenschaft".[10] Maybe one can find more contribution of this sort in the party's "Arbeiterzeitung".

But what about other members of the Vienna Circle? In Philipp Frank's book *The Law of Causality and its Limits* from 1932 we find a constant criticism and polemic against worldviews who positioned themselves against the formulation of laws, be it in natural sciences or the humanities (Frank 1932/1998). Especially Othmar Spann, who was called in 1922 to Vienna, the same year as Moritz Schlick, is singled out for criticism, when he recommended going back to romantic natural philosophy in order to detect the essence of nature. More specific is Frank's attack on the anti-causal attitude, when it demands super-naturalistic wholes, be it in Biology or in Economics and philosophy of the state. I mean especially the concept of entelechy, favored by the neo-vitalistic schools and revived by people like Hans Driesch.[11] And one should mention the concept of the whole (*Ganzheit*) with its applications

---

[9] See the classical article describing this development: Hempel (1964).

[10] Both articles – published under the pseudonym Rudolf Richter – are reprinted in Zilsel (1992, 153–166 and 167–178).

[11] Frank (1932/1998, 133–144) discusses Driesch's alleged proofs for his neo-vitalistic theory in astonishing detail in his book.

in biology, psychology and especially sociology and political theory. Frank now is very anxious to differentiate between the application of the Ganzheit-terminology in different sciences. So he acknowledges that Driesch restricts the formula "the whole is more than the sum of its parts" to Biology and Psychology, but is a vehement critic of is use in the social sciences, whereas adherents of the fascist Ständestaat (and later on of National Socialism) like Othmar Spann lay special emphasis on subordination of the individual under the relevant Ganzheit, the fascist state.

It is remarkable that Moritz Schlick treated the very same problem in his contribution to the VIII. International Philosophical Congress of Philosophy in Prague 1934 in his lecture "Der Begriff der Ganzheit". He did not care to differentiate between leftist and rightist theoreticians of Ganzheit with their divergent application of the concept. Hans Driesch in the discussion made just the same distinctions as Philipp Frank in his book, which Schlick seemed to have overlooked.

In Carnap's case the fight against metaphysics is especially interesting. One only needs to follow his ideas about the misuse of thought and language from his talk about "The Misuse of Language" at the Bauhaus 1929 through his famous paper "Overcoming of Metaphysics by the Logical Analysis of Language" of 1932 up to his talks to students and progressive audiences in 1934 in Czechoslovakia. The Bauhaus-lecture is politically still more or less harmless. The examples for the misuse of language are drawn from every-day life and the common-sense sphere. Metaphysics is still not the target. That changed a lot when Martin Heidegger had published his famous booklet "Was ist Metaphysik?" in 1931. Because then Carnap seems to have presented Heidegger's nice sentence "das Nichts selbst nichtet" (the nothing itself nothings) to a private party of his close friend Franz Roh. There he asked the participants: who could have said this? And a hearer, Edith Tschichold answered: Kurt Schwitters (the Dadaist artist, who indeed had invented a sense-free Ursonate a little bit earlier). So from then on Carnap used the Heideggerian example in his fight against metaphysics, and with a fine instinct: Heidegger presented himself in May 1933 as prophet of National Socialism. In a number of talks to leftist student organization in Prague, Brno etc. in 1934 Carnap then sharpened his fight against metaphysics, saying for instance that it was "the opium of intellectuals" (a variance of Marx's talk about religion as the opium of the people). These talks have only been transcribed very recently and I can present no details at the moment.

Now some remarks on the problem of pseudo-science and how it was tackled by the Vienna intellectuals!

When we take the Vienna progressive scene of the 20s and early 30s (in a wider sense, not only comprising the Vienna Circle), we find that one of them put special emphasis on the fight against pseudo-science: Karl Popper named in his early work "Die beiden Grundprobleme der Erkenntnistheorie" (which preceded his famous *Logik der Forschung* of 1935) the demarcation of science and pseudo-science as the most important problem of the theory of knowledge (Popper 1979).

He prided himself to have found the solution to this problem in the falsifiability criterion.[12] His criterion had not the drawback of the positivistic verifiability criterion of meaning, which excluded statements of wider generality (like natural laws) from the status of scientific respectability. Various solutions for this self-inflicted problem given by Vienna Circle members are presented in Popper's book and shown not to be satisfying. Now Popper extended the reach of the criterion to specific examples, like the Marxian theory of history or Freudian psychoanalysis. I found this criticism (contained in his lead article: "Science: Conjectures and refutations", Popper 1963) not convincing. Of course, Marx's theory of history and also Freud's psychoanalysis contain parts which *are falsifiable. They contain law-like general statements. A prominent example is Marx's central thesis of historical materialism: "all history (up to now) has been a history of class struggle". The falsification would consist in finding very early or completely different societies from ours. Likewise* Freud's theory of dreams contains the law-like statement "every dream is a whish fulfillment" (not necessarily of a sexual kind, by the way). So one can try to falsify this general statement too by adducing non-whish-fulfilling dreams (for instance nightmare dreams). There is a vast literature about these topics and I have here no space to delve into it. Popper – by the way – never applied his criterion to a more obvious contemporary candidate: Nazi race-theory.

Now: how far does one get with the verifiability theory of meaning or the falsifiability criterion of science, when one discusses Intelligent Design or climate change denial?

Intelligent Design points to gaps in cosmological and Darwinian theories which up to now could not be filled with astronomical, petrographical etc. findings and uses the authority of the bible as the basis of its "science". The first argument demands a fixed, secure and never changing foundation for any respectable science. This is something that no science at all can deliver. The second brings in an agent, whose existence and functioning is even much more improbable than any amount of gaps in cosmology and evolution theory.

A false idea about what science is and can deliver also informs much of the criticism of current climate science and global warming. There are not many who completely deny the phenomenon of global warming. What they mostly deny are the man-made causes of it over the last 200 centuries since the start of industrialization. The "merchants of doubt" have detected that gaps, imprecisions etc. are inherent in the climate scientists' theories. Also they complain on the quality of the data and especially the models which are used in building climate-models and forecast based on simulations. These problems pose a challenge not only for climate scientists, but for philosophers of science as well. But the alternative: climate change denial (and doing nothing against it) are not part of any science, but of its opposite: pseudo-science.

---

[12] I leave here open, whether he borrowed that criterion not from others (especially Leonard Nelson) and only applied it to different examples. See Dahms (2019) for details.

To sum up: the demarcation-problem brought up by Popper in the 1930 is a real problem. Popper later on developed his criterion into a comparative and even a numerical one, creating thereby the illusion, as if one could "simply" know before any confrontation of theories in any field of scientific research with reality, whether they were serious candidates for scientific theories, and even give a number for that. Other philosophers presented different criteria. But it seems that no satisfying candidate has been found to fill this role for every field and subject of research (see Hagner 2008). Nevertheless Popper's problem remains as urgent today as it has been during his time in Vienna.

10.5 Concluding Remarks

So: what can we learn from all these examples for the task of philosophy in the necessary fight against current irrationalism?

1. There needs to be a discussion in the philosophers' own ranks. Discussion about truth-theories have to take into account that truth is something that is used in journalism and law as well as in philosophy and need to adjust their discussion to this fact.
2. Past sins of philosophers in undermining rationality have to be exposed (like the piece of Feyerabend about witch theory and other over-relativistic tendencies).
3. Discussions about science and pseudo-science need to be intensified with a view to current examples like intelligent design and climate change.

There are Philosophers like Philip Kitcher (2008), who do the latter since a long time. But more people should enter the field. They will find that arguments which some Vienna Circle People left out as metaphysics, namely those of values and applied ethics, will play a large role in some of these discussions (like international and intergenerational justice in the case of global warming). Another topic for applied ethics would be in the actual Covid pandemic the discussion about the burden different age-strata, social classes and ethnicity could or should shoulder. That question cannot be answered by science alone too. It has to be decided by politics, but also needs to be discussed by philosophers as moral and political problems. But it happens so that the neglect of former philosophical fields like ethics, political philosophy (as well as aesthetics, by the way) is one of the drawbacks of the historical "scientific worldview" of the Vienna Circle and similar groups and movements.

Furthermore there remains the task of bringing all these results to the general public. Historians Timothy Snyder's 20 lessons are very instructive when one wants to know what can be done for what he terms "the resistance" (Synder 2017). Logical empiricists of the Vienna Circle and other Circle underscored on many occasions that theirs was a this-worldly philosophy. Personalities like Rudolf Carnap, Philipp Frank and also Hans Hahn were prepared to give public speeches in – then still republican – Czechoslovakia in order to fight metaphysical irrationalism in the 1930s.

One would be a "brave empiricist" (Feyerabend's term) in name only when one would not do something similar against the forces of irrationalism and evil today.

Acknowledgements I thank Adam Tamas Tuboly (Budapest/Pécs) and Daniel Kuby (Konstanz) for corrections of this paper and useful suggestions.

References

Arendt, Hannah. 1971/2013. Die Lüge in der Politik. In *Wahrheit und Lüge in der Politik*. München: Piper Verlag.

Barber, Bernard. 1961. Resistance by Scientist to Scientific Discovery. *Science* 134: 596–602.

Beaumont, Peter. 2019. US lies and deception spelled out in Afghanistan papers' shocking detail. *The Guardian*, December 9.

Dahms, Hans-Joachim. 2019. Karl Poppers philosophische Anfänge. In *Karl Popper Handbuch*, ed. G. Franco, 189–204. Wiesbaden: Springer.

Fehrle, Brigitte, Clemens Höges, and Stefan Weigel. 2019. Der Fall Relotius. Abschlussbericht der Aufklärungskommission. *Der Spiegel* 22 (25 May, 2019): 130–146.

Feyerabend, Paul. 1975a. Against Method. *Outline of an Anarchistic Theory of Knowledge*. London: Verso.

———. 1975b. Wider den Methodenzwang. *Skizze einer anarchistischen Erkenntnistheorie*. Frankfurt am Main: Suhrkamp Verlag.

———. 1978a. Die Wissenschaftstheorie – eine bisher unerforschte Form des Irrsinns? In *Der wissenschaftstheoretische Realismus und die Autorität der Wissenschaften*, 293–338. Braunschweig/Wiesbaden: Springer.

———. 1978b. Die Wissenschaften in einer freien Gesellschaft. In *Der wissenschaftstheoretische Realismus und die Autorität der Wissenschaften*, 351–367. Braunschweig/Wiesbaden: Springer.

———. 1995. *Killing Time*. Chicago: University of Chicago Press.

Frank, Philipp. 1932/1998. *The Law of Causality and Its Limits*. Dordrecht: D. Reidel.

Golbeck, Jennifer, et al. 2018. Fake News vs. Satire: A Dataset and Analysis. *Websci* 18: 27–30. Amsterdam (Netherlands).

Hagner, Michael. 2008. Bye-Bye Science, Welcome Pseudoscience? Reflexionen über einen beschädigten Status. In Rupnow et al. (2008): 21–50.

———. 2017. Wider den Populismus. Paul Feyerabends dadaistische Erkenntnistheorie. *Zeithistorische Forschungen* 14 (2): 369–375.

Hempel, Carl G. 1964. Empiricist Criteria of Cognitive Significance: Problems and Changes. In *Aspects of Scientific Explanation and other Essays in the Philosophy of Science*, 101–121. New York: The Free Press.

Horton, Adrian. 2020. After Truth: How Ordinary People Are 'Radicalized' by Fake News. *The Guardian*, March 19, 10–31.

Howe, Neil. 2017. Where did Steve Bannon get his worldview? From my book. *The Washington Post*, 24 February, 2017.

Kitcher, Philip. 2008. Darwins Herausforderer. *Intelligent Design oder: Woran man Pseudowissenschaftler erkennt. In Rupnow et al. (2008): 417–34.*

Koffler, Keith. 2017. *Bannon: Always the Rebel*. Washington, DC: Regnery Publishing.

Lakatos, Imre, and Paul Feyerabend. 1999. *For and Against Method: Including Lakatos's Lectures on Scientific Method and the Lakatos-Feyerabend Correspondence*, ed. Matteo Motterlini. Chicago/London: University of Chicago Press.

Lefebvre, Georges. 1932. *La grande peur de 1789*. Paris: Armand Colin.

Moreno, Juan. 2019. Tausend Zeilen Lüge. *Das System Relotius und der deutsche Journalismus*. Berlin: Rowohlt.

Neiman, Susan. 2017. Widerstand der Vernunft. *Ein Manifest in postfaktischen Zeiten*. Salzburg/München: Ecowin.

Oreskes, Naomi, and Erik M. Conway. 2010. *Merchants of Doubt: How a Handful of Scientists Obscured the Truth on Issues from Tobacco Smoke to Global Warming*. London/New York: Bloomsbury.

Popper, Karl. 1963. Science: Conjectures and Refutations. In *Conjectures and Refutations. The Growth of Scientific Knowledge*, 33–65. London: Routledge.

———. 1979. *Die beiden Grundprobleme der Erkenntnistheorie*. Tübingen: Mohr Siebeck.

Rupnow, Dirk, et al., eds. 2008. *Pseudowissenschaft*. Frankfurt am Main: Suhrkamp.

Snyder, Timothy. 2017. *Über Tyrannei. Zwanzig Lektionen für den Widerstand*. München: C-H Beck.

Soldan, Wilhelm G. 1843/1911. *Geschichte der Hexenprozesse*. 2 vols. Newly edited by Max Bauer. Hanau: Müller and Kiepenheuer. (Cited as: Soldan-Heppe).

Strauss, William, and Neil Howe. 1997. The Fourth Turning. *What the Cycles of History Tell Us About America's Next Rendezvous with Destiny*. New York: Three Rivers Press.

von Spee, Friedrich. 2000. *Cautio Criminalis oder Rechtliches Bedenken wegen der Hexenprozesse*. Rinteln.

Whitlock, Craig. 2019. At War with the Truth. U.S. officials constantly said they were making progress. They were not, and they knew it, an exclusive Post investigation found. *The Washington Post*, December 9.

Zilsel, Edgar. 1992. *Wissenschaft und Weltanschuung. Aufsätze 1929–1933*, ed. Mit einem Vorwort von Karl Acham and introduction by Gerald Mozetic. Wien/Köln/Weimar: Böhlau.

Part II
General Part

Chapter 11
Viennese Lessons: Wittgenstein, Carnap, and Schlick

2nd Arthur Pap-Lecture 2019

Scott Soames

Abstract The logical empiricism of the Vienna Circle was the second great step in the development of analytic philosophy, following the turn to logic and mathematics imitated by Frege and Russell, and the practice of analysis illustrated by Moore. Although Wittgenstein was not a member of the Circle, he greatly influenced those who were. The paper will articulate three important contemporary lessons to be gained by identifying, reconstructing, and developing under-appreciated insights of Wittgenstein, Carnap, and Schlick concerning the nature of propositions, modality, and ethics.

Keywords Meaning · Truth · Propositions · State descriptions · World-states · Necessity · Apriority · Value · Moral reality

The logical empiricism of the Vienna Circle was the second great step in the development of analytic philosophy, following the turn to logic and mathematics imitated by Frege and Russell, and the practice of analysis illustrated by Moore. Although Wittgenstein was not a member of the Circle, he influenced those who were. Having moved beyond the conception of philosophy shared by his predecessors, Wittgenstein took philosophy's task to be that of finding the essence of representational thought and language, which, he believed, required discovering the scope and limits of intelligibility. This is what he meant when he said, in the notebooks he kept while writing the *Tractatus,* that his whole task was to solve the one great problem of philosophy, to wit, explaining the nature of the proposition (Wittgenstein 1914–1916, p. 39). He realized, of course, that this was paradoxical. Believing that philosophy's only real problem was to answer the question *"What are the limits of*

S. Soames (✉)
University of Southern California, Los Angeles, CA, USA
e-mail: soames@usc.edu

C. Damböck, A. T. Tuboly (eds.), *The Socio-Ethical Dimension of Knowledge*,
Vienna Circle Institute Yearbook 26, https://doi.org/10.1007/978-3-030-80363-6_11

intelligibility?", he also believed that the answer couldn't be stated by any proposition. Worse, he thought that propositions traditionally regarded to be philosophical simply don't exist. Hence, if philosophy was to continue, it had to be radically reconceived.

In this, Wittgenstein was (for a time) at one with Schlick and Carnap, despite their other differences. Today, I, along with many others, think their shared big idea about the limits of intelligibility and the nature of philosophy was fundamentally mistaken. But they weren't wrong root and branch. On the contrary, there are aspects of their work of nearly a century ago from we still have much to learn. I will illustrate with Wittgenstein on propositions, Carnap on modality, and Schlick on scientific ethics.

11.1 Wittgenstein

I begin with the *Tractatus*. Like Frege and early Russell, Wittgenstein took sentences to be the primary meaning-bearing units, but unlike them he didn't take meanings to be propositions. Rather, he denied that any entities were sentence-meanings. Although *propositions* were bearers of truth, Wittgenstein took them to be something like uses of meaningful sentences, rather than imaginary sentence meanings. For him, sentences were *linguistic facts* consisting of expressions standing in syntactic relations. For them to be meaningful is for them to be governed by conventions. For example, the sentence 'John loves Mary' consists in the two names standing in a certain syntactic relation R – which involves, among other things, *the first name being followed by the phrase 'loves' which is followed by the second name.* The sentence is the fact that the names stand in that relation. The conventions governing it stipulate (i) that the names are used to designate John and Mary, respectively, and (ii) that structures in which two names stand in R are used *to represent* the referent of the first name as loving the referent of the second. One who uses the sentence in this way represents John as loving Mary. This tells us what the world must be like if one's use of the sentence is to be true.

It is tempting to think that the bearer of truth is the sentence – i.e. the syntactic structure – in which the two names stand in R. After all, the sentence *is* governed by the conventions I gave. But it didn't have to be. Had it been governed by other conventions, it would have meant something different, and had different truth conditions. So there is something, the sentence, that *is* used to represent John as loving Mary, but *could have been used* differently, and had different truth conditions. However, there is also something that has those truth conditions essentially. Necessarily the *proposition John loves Mary* is true iff John loves Mary. This wouldn't be so, if propositions were sentences. Wittgenstein agrees; they aren't.

Perhaps propositions somehow incorporate both sentences and conventions governing them. Perhaps the proposition that John loves Mary is *a use* of the sentence in accord with the conventions I stated. *What, then, is this thing–a use of a sentence S in accord with conventions?* Since to use an atomic sentence S is to do something, *a use of S* is a cognitive doing, an act or operation of some sort. It is the act of using the two names to designate the two people, while using the relation R to represent the referent of the first name as loving the referent of the second. This *repeatable act type* represents John as loving Mary, because *for any agent to perform it is for the agent to represent them that way.* So, elementary propositions are representational act or operation types. For them to be true is for them to represent things as they really are. Since they are *repeatable*, they are a kind of abstract object, distinguished from concrete events in which they are performed.

This reconstruction fits several tractarian themes. (i) It explains the meaningfulness of a sentence without positing an independent entity as its meaning. (ii) It identifies the truth-bearer, *a meaningful use*, as something the truth of which is defined in terms of its representational accuracy. (iii) It preserves the idea that the constituents of the sentence are isomorphic to the constituents of the atomic fact that makes a use of it true. (iv) Since conventions governing use are those governing the sentence's constituents, no extra convention governing the sentence as a whole is needed. (v) The proposition has its truth conditions essentially because any *possible* agent using the sentence in this way represents the same thing as being the same way.

This is encouraging, but it isn't exactly what Wittgenstein had in mind. Uses of sentences do represent, or, *picture*, reality. But Wittgenstein's pictures are supposed to be *facts,* not *acts*. Still, this doesn't mean that sentences, which *are* tractarian facts, are the primary bearers of truth after all. Wittgenstein distinguishes propositions from sentences, which he calls *propositional signs*.[1]

3.11 *We use the perceptible sign of a proposition (spoken or written, etc.) as a projection of a possible situation.* (The method of projection is the thinking of the sense of the proposition.)

3.12 The sign through which we express the thought I call the propositional sign. And the proposition is the propositional sign in its projective relation to the world.

3.13 *A proposition includes all that the projection includes, but not what is projected.*

Therefore, though what is projected is not itself included, its possibility is.

A proposition, therefore, does not actually contain its sense, but does contain the possibility of expressing it.

3.14 The propositional sign consists in the fact that its elements, the words, are combined in it in a definite way. The propositional sign is a fact.

Propositional signs are syntactic structures which, though meaningful, aren't individuated by what they mean. A proposition combines the sign with a projection. Recall, "The method of projection is the thinking of the sense of the proposition." A proposition *represents* objects, which are projections of the names in the propositional sign. The proposition represents these objects as standing in the relation projected from the linguistic relation that unites the names in the propositional sign. Wittgenstein says, the proposition *"includes all that the projection includes, but not what is projected." What is projected,* is the sense of the proposition – the possible fact, whose actuality would make the proposition true. It isn't "included"

[1] Italicized quotations are from the Pears and McGuinness translation; others are from the Ogden translation.

in the proposition because *there are no merely possible facts* in the tractarian ontology for real, existing false propositions to include. Nor are the objects and relations projected from constituents of the propositional sign *included* in the proposition. But, Wittgenstein tells us, the rest of the projection, namely, the conventions governing uses of the constituents of the sentence, are somehow included in the proposition. They determine the fact that would have to exist if the proposition were true.

How are conventions included in the proposition? Wittgenstein isn't clear. Having identified the propositional sign as a purely syntactic structure, he says the proposition is *the propositional sign in its projective relation to the world.* Unfortunately, this language, *the sentence S in its relation to the world,* doesn't pick out an entity other than S – any more than the phrases *Soames-in-his-relation-to--USC, Soames-in-his-relation-to-his-wife,* or *Soames-in-his-relation-to-Vienna* pick out entities other than me of which I am an essential part. There aren't several me's, or me-complexes, here, just misleading ways of talking about the fact that I teach at USC, live with my wife, and lecture in Vienna. The same is true of Wittgenstein's talk of propositional signs *in their projective relations to the world.* Fortunately, we can remedy this. Propositions are *uses of sentences in accord with their conventions.* Although this view departs from the tractarian doctrine that propositions are facts, not acts, the modification preserves important tractarian themes while avoiding well-known tractarian and contemporary puzzles about propositions (See Soames 2015, 2016, 2018, chapter 2).

In taking this step, we remain true to the tractarian idea that although propositions aren't sentences, talk of propositions is talk about sentences. But this too must be modified. It is essential to thought that agents represent things as being certain ways. It's not essential what, if any, artifacts they use in doing so. Any organism whose cognitions can be true or false represents things as being various ways. Sometimes it may do so by using symbols. But we needn't think that agents *always* use symbols when *thinking of* something as being some way. When they perform representational cognitive acts linguistically, the propositions they affirm may be uses of symbols. When they non-linguistically represent things as being various ways, the propositions they affirm don't involve symbols. Thus we needn't accept 4.0312 of the *Tractatus, "The possibility of propositions is based upon the principle of the representation of objects by signs."*

The tractarian account of *elementary* propositions is an incomplete realization of three valuable insights. (i) Declarative sentences are representational, not because they express primitively representational Fregean or Russellian propositions, or because they name possible facts, but because of how they are used. (ii) They are bearers of truth because they are used to represent things as bearing properties. (iii) A use of an atomic sentence is true at a world-state w iff were w actual things would be as anyone who *used the sentence in that way* would represent them.

This approach allows different propositions to be true at all the same world-states. Thus we may ask, *What use of which sentence is identical with the proposition that John loves Mary?* There is no more reason to take it to be a use of a sentence of one language than there is to take it to be a use of a sentence of another. The proposition we are after is something all *representationally identical* uses of candidate sentences have in common. Now, consider the act type of using *some sentence or other* to represent John as loving Mary. Anyone who uses a *particular* sentence *S* in this way thereby performs a representational act type one can perform without using S. If acts of using particular sentences are propositions, then, it seems, this general act type should also be.

What about the act of predicating *loving* of John and Mary, cognizing the two *by any means at all*. It's the best candidate for being the proposition that John loves Mary. If it's *not* possible to perform this representational act *without using a sentence*, then the proposition is the act of using some *sentence or other* to represent John as bearing that relation to Mary. If, as I believe, it *is* possible to perform the general act without using a sentence, then it alone is the proposition we seek. However, even that is too simple. There really isn't just one proposition with the content *that John loves Mary*; there are many. Thus, we should reject the assumption that sentential clauses, *that S*, are fine-grained enough to pick out unique propositions. This, I contend, greatly expands the solution space for Frege's puzzle, Kripke's puzzle about belief, and many others.[2]

If we identify propositions as representational cognitive acts, we must individuate propositions in the way we individuate other act types. Consider *driving to work* and *traveling to work*. Anyone who performs the first act performs the second as well. But since there are many ways of getting to work, one can perform the second without performing the first. So, the acts are different. *When we apply this lesson to propositions, we generate representationally identical propositions that are cognitively distinct, because they place different demands on agents who entertain them.* These propositions are not merely true at the same world-states. They are representationally identical; they represent the same things as being the same way. Still, they are cognitively distinct, agents can often bear a propositional attitude to one without bearing it to the other. All of this follows if we take propositions to be cognitive acts including, but not restricted to, *uses of sentences* in accord with conventions.

This modification and reconstruction of the tractarian account of elementary propositions can be extended to non-atomic propositions, while avoiding the problems caused by the very different tractarian analysis of non-atomic propositions. We also avoid the tractarian conclusion that necessarily equivalent propositions are identical, which obscured the breakthrough that its conception of atomic propositions might otherwise have been. True, Wittgenstein himself wouldn't have seen things this way. Without his identification of propositions with their modal truth conditions, the *Tractatus* couldn't have been the tool for transforming philosophy's self-conception in the way he wanted. But had he properly generalized his account of atomic propositions to all propositions, our philosophy of mind, language, and psychology would have been better off.

[2] See chapters 3, 4, and 5 of Soames (2015).

11.2 Carnap

Having extracted philosophical gold by reconstructing tractarian propositions, I will now combine them with Carnap's account of modality in *Meaning and Necessity* (1947). The best current accounts of possible-world states identify them, not with concrete universes disconnected from ours, but with *properties the world can coherently be conceived to have.* These include both metaphysically possible world-states and those that are only epistemically possible. The actual world-state is a property the universe does instantiate. Metaphysically possible states are those that could have been instantiated. *Epistemically possible states are properties we can coherently conceive to be instantiated, which we can't know apriori not to be instantiated.*[3]

For insight into these properties, I start with Carnap's notion of *a state description*, which is a complete, consistent, set of atomic sentences or their negations. Truth values of complex sentences at state descriptions are determined using familiar accounts of quantifiers, truth functions, and modal operators. To update this, I replace Carnap's atomic sentences and their negations with *propositions* that are *uses* of atomic formulas, and their negations, *in accord with conventions governing them* -- including allowing variables to be used to name any object. Complete, consistent sets of such propositions are used to define world-states. Let D be a domain of objects under consideration. Let B be the set of properties expressed by simple predicates of a Carnapian language, including an existence predicate. A world-description S_W is a set each member of which is either an *atomic proposition* that predicates a property in B of objects in D or the negation such. S_W is complete iff for each atomic proposition, either it or its negation is a member of S_W. It is consistent iff its members can't be known apriori not to be jointly true. A world-state w is *the property of making the propositions in the state-description S_W true*. To conceive of w as being instantiated is to conceive of every member of S_W as true, *while taking the objects in the universe to include only those in D.* Other propositions are true at w iff they are apriori consequences of the claim that w, so conceived, is instantiated.

A structure of world-states is a set of epistemically possible states; the one that is instantiated is *actual*. A state is epistemically possible iff it can't be known apriori not to be actual. Those that could have been instantiated are *metaphysically possible*. [Possibly S] is true at w iff S is true at some state *metaphysically possible* from w. World-states metaphysically possible from one state may differ from those possible from another. The truth values, at w, of modal propositions are determined by w's position in the space of states. The actuality operator picks out the world-state @ of any context at which it is used. If S expresses proposition p in context C, a use of [Actually S] in C expresses the proposition *that p is true at @*. Its truth value at any world-state is its truth value of p at @.

To sum up, a world-state is a property that determines a complete story of what the universe would be like if w were instantiated. Since it's not part of that story to specify what the universe would be like if other world-states were instantiated, the

[3] Soames (2005, chapter 8), (2010, chapter 6), (2011).

propositions in terms of which w is defined don't contain explicit information about other states. Nevertheless, for any state w* and proposition p, we can evaluate the truth value of the proposition *that p is true at w** at any world-state. We need only remember that a proposition can be true at a world-state without being one of the propositions that define it.

The space of states is relativized to *contextually varying inquiries.* Those relative to an inquiry are all possibilities relevant to it. For each inquiry, there may be another inquiry requiring more fine-grained states, in which case no state may be a maximally informative story about the universe that answers every conceivable question and evaluates every conceivable proposition. Rather, world-states are *treated as maximal* for particular purposes. This doesn't mean they are unreal. The properties are real. It is the use to which we put them that is relative to us.

Next consider instances of the Kripkean necessary aposteriori that predicate essential properties of objects they can be known to possess only empirically. The function of the needed evidence is to rule out epistemically possible, but metaphysically, impossible states at which the propositions are false. With this in mind, let S express a contingent truth p and [Actually S] expresses the proposition *that p is true at @*. Since it is true at every world-state, including those that are only epistemically possible, empirical evidence is never needed to establish it. Thus, it's not necessary aposteriori, despite the fact that [Actually S] is trivially inferable from S, and may not be known by us in any other way. How can this be?

The mystery can be dispelled by imagining a tiny universe consisting of two blocks side by side, with a third on top. This world-state, *Tiny*, is *the property of containing blocks 1 and 2 side by side, with block 3 on top*. Since we can easily comprehend this property, we can know, just by thinking about it, that if it were instantiated, block 3 would be on blocks 1 and 2. So, if p is the proposition that one of them is sitting on the other two, it is knowable *apriori that p is true at Tiny.* The point generalizes. If the world-states relevant to an inquiry are finitely specifiable, then, for every such state w, and every proposition p the truth of which is calculable from the propositions in the state description S_w defining w, the proposition *that p is true at w* is knowable apriori. This applies to the actual world-state (relative to an inquiry), as much as any other. Thus, the proposition expressed by a use of [Actually S] is typically *knowable* apriori, even if it's not *known* apriori. Since the actual state @, relative to an inquiry may be complex, we may not know the world-story in terms of which @ is defined, and even if we did, we might not have calculated all its relevant consequences. In such cases, our only practical way of learning *that p is true at* @ is by inferring it from p. So, when p is aposteriori, the knowledge we express by saying [Actually S] is typically aposteriori, even though *what we know* can, in principle, also be known apriori.

So far, so good. Now a puzzle. The empirical evidence required to know Kripkean necessary aposteriori truths is needed rule out metaphysically impossible, but epistemically possible, world-states at which they are false. This may seem to suggest that *if p is false at an epistemically possible state then p isn't apriori*. Thus, it may seem, that apriori truths are true at all such states. But that conclusion threatens the *contingent apriori*. How can p be false at a metaphysically possible state (which contingency requires), while being true at all epistemically possible states? Surely no *metaphysically possible* state is *epistemically impossible*. Since a metaphysically possible state is a property the universe could have had, it shouldn't be incoherent to suppose the universe does have it, in which case metaphysically possible states *are* epistemically possible. Thus we must either give up the contingent apriori, or explain why apriority doesn't require truth at all epistemologically possible states.

We can't give up the contingent apriori. Suppose agent X at @, knows apriori *that S iff S*. From this X can reason, without evidence, (i) that *it is true at this very world-state that S iff S* (said demonstrating @), (ii) that (*it is true at this very world-state that S*) *iff* (*it is true at this very world-state that S*), and (iii) that *S iff it is true at this very world-state that S*, which X knows apriori. Since this *is* the *contingent* proposition *that S iff actually S*, it is knowable apriori.

To solve our puzzle we must explain how it *can be known apriori to be true*, despite being *false at some world-states we can't know apriori not to be actual.* Surely, it seems, if it's knowable apriori (i) *that p is true* and (ii) *that if w were actual, p would be false*, then it must be knowable apriori *that w isn't actual* (in which case w isn't epistemically possible). After all, apriority is closed under apriori consequence, isn't it? No, in fact, it isn't. Why not?

Let world-state w be the property of making propositions $q_{1...}q_{100}$ true (where these are expressed by $S_1 \ldots S_{100}$). Let 'w' name w and let 'PW' be the property name ⌈the property *making it true that* S_1*,..., that* S_{100}⌉, which also names w. Finally, let p be some further proposition. For it to be true at w (i.e. at PW) is for p to be an apriori consequence of the claim *that PW is instantiated.*

Now consider (1a,b).

1a. Proposition p is true at PW – i.e. at the property *making it true that* S_1*,..., that* S_{100}.
1b. Proposition p is true at w.

The relationship between these propositions is like that between propositions (2a,b), where 'logicism' is a proper name of the proposition that arithmetic is reducible to logic.

2a. Russell attempted to establish *that arithmetic is reducible to logic.*
2b. Russell attempted to establish *logicism.*

In both cases, the (a) and (b) propositions predicate the same property of the same things; the (a) proposition differs from the (b) proposition only in what it takes to entertain it.

To entertain proposition (2a) is to predicate *attempting to establish* of the pair Russell and the proposition *that arithmetic is reducible to logic,* identifying that proposition by entertaining it. By contrast, to entertain proposition (2b) merely requires predicating *attempting to establish* of Russell and the proposition that arithmetic is reducible to logic, *without placing any special constraints on how one identifies it.* It follows that anyone who entertains or believes proposition (2a) thereby entertains or believes proposition (2b), but not conversely.

Similar comments apply to (1a) and (1b). To entertain proposition (1a) is to predicate *being true at* of the pair p and world-state w, cognizing w, i.e. *being such that* S_1*, that S2,..., that* S_{100}, by entertaining its propositional constituents. By contrast, to entertain proposition (1b) requires predicating *being true at* of p and world-state w, without constraining how one cognizes w. If follows that anyone who entertains, believes, or knows proposition (1a) entertains, believes or knows proposition (1b), but not conversely. So, when p is a contingent aposteriori truth which is an apriori consequence of the claim *that PW is instantiated*, proposition (1a) is knowable apriori. Since knowing it without empirical evidence *guarantees* knowing proposition (1b) without evidence, (1b) is also *knowable apriori*, even if the only knowledge we actually have of it is aposteriori.

What makes this possible is that sentences (1a) and (1b) express cognitively different but representationally identical propositions; knowledge of the first guarantees knowledge of the second. Although there are ways of entertaining proposition (1b) that don't allow you to recognize its truth without appealing to empirical evidence, there are other ways of entertaining it – namely by entertaining proposition (1a) – that do allow you to establish it apriori. For a proposition to be apriori, there must be some way of entertaining it that allows one to recognize its truth apriori; this is so even if there are other ways of entertaining it that don't allow this.

Next, consider a use of sentence (3), which predicates *being true at* of p and the actual world-state @.

3. *Actually* P/The proposition *that P* is true *at this very world-state.*

This use (3) involves cognizing the actual world-state @ as *this very world-state* (the one we experience). When @ is identified in this way, we can know proposition (4) apriori, even if the proposition p, expressed by sentence P, is knowable only aposteriori.[4]

4. Actually P iff P/the proposition *that P* is true at this very world-state iff P.

Thus, propositions (3) and (4) are both knowable apriori. Nevertheless, proposition p – i.e. *that P* – isn't, apriori. Since it *is* an apriori consequence of propositions (3) and (4), *apriority isn't closed under apriori consequence.*

To see why, let *that P* be the contingent truth *that SS is a philosopher.* The proposition that *actually SS is a philosopher* is then true at all metaphysically possible world-states. It is knowable apriori, because it is an apriori consequence of the claim *that PW is instantiated,* and hence of the claim *that @ is instantiated.* The biconditional *that SS is a philosopher is true at @ iff SS is a philosopher* is false at some metaphysically possible states. Still, we can know it apriori, *because there is a way of cognizing it,* using the sentence "*Actually SS is a philosopher iff SS is a philosopher*" such that when we cognize it that way, we don't need evidence for its truth.

The key point is that the cognitions of @ required to come to know (5a) and (5b) apriori are radically different and *can't be recognized apriori* as cognitions of *the same world-state.*

5a. The proposition that SS is a philosopher is true at @.
5b. The proposition that SS is a philosopher is true at @ iff SS is a philosopher

[4] In this sentence, and in examples (3) and (4). 'P' is used as a schematic letter for a sentence that expresses proposition p.

When we are thinking about @ in a way that allows us to come to know (5a) apriori, we can't thereby see that (5b) is apriori, and when we are thinking @ in way that allows us to know (5b) apriori, we can't thereby see that (5a) is apriori.

This doesn't prevent us from knowing the conjunction of (5a,b) apriori. We can, after all, safely accept the conjunction of *sentences* (6a) and (6b), in which 'PW' is the articulated name of the property that is the actual world-state @, which is indexically demonstrated in (6b).

6a. The proposition that SS is a philosopher is true at PW.
6b. The proposition that SS is a philosopher is true at this very world-state iff SS is a philosopher.

Knowing each of these propositions apriori, we can know the proposition expressed by the conjunction (6c) apriori.

6c. (The proposition that SS is a philosopher is true at PW) & (the proposition that SS is a philosopher is true at this very world-state iff SS is a philosopher)

But since knowing the proposition expressed by (6c) without evidence *guarantees* knowing the representationally identical proposition (6d) without evidence, it too is knowable apriori.

6d. (The proposition that SS is a philosopher at @) & (the proposition that SS is a philosopher is true at @ iff SS is a philosopher).

What one *cannot* know apriori is the lambda abstracted version (6e) of (6d).

6e. λw [(the proposition that SS is a philosopher is true at w) & (the proposition that SS is a philosopher is true at w iff SS is a philosopher)] @.

Proposition (6e) is an apriori consequence of (6d), because there is a way of entertaining (6d) that allows one to make the apriori inference. But there is *no* way of *knowing (6d) apriori* that allows one to make the inference. Hence, both (6e) and (7) are apriori consequences of premises that can be known apriori, even though they can't themselves be known apriori.

7. SS is a philosopher

So, apriority is not closed under apriori consequence.

The key Carnapian insight leading to this result is that the propositions defining possible world-states don't include all propositions true at those states. The insight about propositions needed for the result, gotten by modifying Wittgenstein, is that representationally identical propositions may be cognitively distinct.

11.3 Schlick

I will close by looking at Schlick's book *Problems of Ethics*, whose time has, after decades of dismissal and neglect, finally come. Unlike other logical empiricists, who held that moral language has only emotive content, and so makes no contribution to cognitively meaningful statements, Schlick recognized moral truths capable of motivating action. Unlike the British anti-emotivists, Ross and Pritchard, who took knowledge of basic moral truths to be synthetic apriori, Schlick took it to be aposteriori. Ethics itself, like other empirical inquiries should, he thought, aspire to be a science- a science of human nature including psycho-biology and sociology.

For Schlick, ethical behavior is conduct we demand from others and ourselves, conduct we fundamentally desire that relates us to others. This, for him, is a rock-bottom. If our most basic desires of this type can be identified, there is no further question of justification to be raised. It is nonsense to ask, *Is what we most fundamentally value really valuable?* We simply value what we do. He wanted to know what *conduct we most value and why.* Realizing that different individuals might have different moral standards, over and above differences stemming from varying factual circumstances, differences in available actions, and gaps in their non-normative knowledge, he was confident that we would find many common values.

He took the morality to be a hierarchical system of norms specifying good conduct in various circumstances. The claim that something is a group norm is a factual claim about what conduct group members desire and expect. Sometimes, there is highest moral norm, sometimes top norms are independent. Either way, one *justifies* the claim that an act is morally good by citing the norm under which it falls, while justifying lower-level norms in terms of higher level norms. The process ends with the highest norm, or norms, for which no further moral justification makes sense.

Schlick's next step was to subsume moral norms under higher, non-moral norms.

> It might be that *moral* good could be shown to be a special case of a more general kind of good…If [so]…the question, "Why is moral behavior good?" can be answered by "Because it is good in a more general sense…The highest moral norm would be justified by means of…a higher principle of life. (p. 24)

How is such subsumption supposed to help? Thinking of morality as a system of demands – which are often inconvenient or worse – that we place on ourselves and others, Schlick asks why we make these demands. In part because we wish to constrain others. But why do we constrain ourselves? In part because we need their cooperation and can only get it by being seen as conforming to rules we impose on them. But if that were the whole story, we would feel fine about cheating when not detected. Since we don't, we must find value in living up to norms connecting us to others. Schlick wanted to know how and why.

He found the answer in our psycho-biologically based social attachments to others, which he took to be as basic to us as the desire to eat when hungry, while being more important for achieving happiness than many other natural desires. His point is reinforced by our recognition of our own mortality, leading, as we age, to increasing satisfaction derived from contributing to others, whom we have come to love, and who we hope will long outlast us. For Schlick, our social impulses, which lead us to value others, are central to our well-being. Today, social scientific versions of this view maintain that our genetic endowment, our early family experience, and the facts of the human condition provide us with a motivational base that ties us by affection, social affiliation, and mutual interest to our fellows. This is the raw material that generates reasons for other-regarding action, the authority of which is recognized by most human beings.

Because cooperation promotes survival, we have been bred by natural selection to be social animals. It is not just that we need what others can provide, and so are impelled by self-interest to depend on them. We are also disposed to form cognitive and emotional attachments to them. Parents are innately disposed to nurture and love their young. Children naturally bond with parents, while emulating all with whom they are intimate. In our early years we form reciprocal bonds of affection and trust in which our well-being and self-conception is intertwined with others. Entering into games and collective activities, we learn the rudiments of fairness, adhering to common rules and earning rewards proportional to the value of our efforts.

This fusion of natural sentiment with early experience and rational principle gives birth to morality. Sentiment infuses our participation in games and activities with those we like and admire, and whom we hope will return the favor. Often our companions are models of the people we wish to become. The rules governing our shared activities are often impersonal principles that apply to anyone who occupies a given role in the effort. Because they define the accepted terms of participation in a mutually beneficial activity, it is in the self-interest of each to obey them. But the rules are more than prudential. Rule violations by friends bound by ties of social affiliation, carry psychic risks beyond the loss of the self-regarding benefits secured by participation. Volitions are affronts to one's socially affiliated fellows, to one's friendship with them, to one's image in their eyes, and to the person one wants, with their help, to become. With this, instrumentally useful rules obeyed to secure benefits of group action become principles to be honored even when no one is looking. In this way, sentiment, social affiliation, and recognition of mutual interest are incorporated into moral commitments.

To complete this basic picture, we must assign truth conditions to *ought* statements in ways that explain why the truth of such claims normally provides us with reasons for action. Consider, to begin with, cases in which we use a sentence *A ought to do X* to express the *prudential* judgment that A's doing X would advance A's welfare more than doing anything else, from a range of relevant acts. Since agents typically want to advance their welfare, recognizing that performing X would increase it nearly always provides them with a potentially motivating reason to do it. So, if it is true that A prudentially ought to do x, A will have a reason to act as we advise. To derive this prudential *ought* from a factual *is,* we need facts about what A's welfare consists in – which we may take to include, health, safety, companionship, freedom of action, development of A's capacities, enjoyment of sensual pleasures, opportunities for excitement and the pursuit of difficult goals in concert with others, and the ability to contribute to the welfare of others A cares about and to benefit from those who care about A.

Suppose we can derive prudential *ought-statements* from factual truths about these recognized values and the actions that would promote them. Still, the result is limited. If A prefers an action X that would benefit B whom A cares about, even though performing X would diminish A's own welfare, then A *ought not*, prudentially, do X. But that needn't mean that A *ought not* do X, all things considered. If A cares more about B's welfare than A's own, A may still think "*I ought to do X*," because A wants to benefit B more than anything else. This suggests that *the prudential statement* is true iff doing X would most advance A's welfare, while the statement reflecting what A most values is true iff doing X will satisfy A's most important desires. Both may be derivable from factual premises.

What about moral statements? The challenge is to find facts about normal human agents supporting claims that they *morally ought* to do X that provide them with potentially motivating reasons. These must include facts about *other people* agents care about plus *relationships and activities* from which participants derive value that couldn't be had if they couldn't rely on others to play their parts. Moral reasons include facts about the impact of one's actions on the welfare and *activity or relationship-based expectations* of others. If an action would advance the welfare of those one cares about, one may have a moral reason to perform it. Similarly, if an action conforms to expectations of those with whom one voluntarily interacts in an activity providing benefits for all, then one normally has a prima facie moral reason to perform it.

Imagine a voluntary group activity that benefits all if each plays his or her part, but which fails to be beneficial if one or more opt out. Realizing this, and wishing to avoid hostility that would result from discovery that one is shirking, one has a self-interested reason not to cheat. One will also have a moral reason when one cares for the other participants, or when one doesn't want to be the kind of person one would condemn, if one were in the position of another participant. The strength of this moral reason is proportionate to the importance of one's role in the activity, the benefits produced for participants, and the centrality of the activity in the social life of which one is a part. The acts one morally ought to perform are those one has the strongest moral reasons to perform, provided that they don't require sacrifices out of proportion with the benefits they secure.

Weighing these reasons raises deep conceptual challenges. Still, the determinants of the needed calculations – the effects of an action on one's welfare and that of others, and the relation of the action to the activity-based expectations of others – are factual matters. If their relative strengths and their manner of combination are also factual, it may, in principle, be possible to derive *moral oughts* from factual premises about what *is*. At any rate, it's not obvious that that this isn't possible. Thus, we should be open to the idea that moral facts, like other facts, are capable of being investigated and known, even if that knowledge is sometimes hard to achieve. So, I think, Schlick would have thought.

Still, there is a worry. Although moral objectivity is welcome, it may seem that an objectionable moral relativity comes with it. In grounding moral reasons in interests and values of an agent, one must give up the idea that moral obligations are binding on all rational agents, who could, in principle, entirely lack fellow feeling with others. The point is illustrated by facts that would, ordinarily, be regarded as relevant to establishing moral obligations: for example, (i) the fact that lying or breaking a promise subverts the trust that makes one's lie or promise possible, (ii) the fact that free riders in a collective effort ask others to do what one refuses to do oneself, (iii) the fact that benefiting oneself sometimes seriously harms innocent others. Imagine a rational agent lacking concern for others who calculates benefits for himself alone and always acts accordingly. Because facts (i)–(iii) don't touch his interests, he won't recognize them as reasons for acting. Of course, a race of relentless interest-maximizers might sometimes coordinate their actions to achieve mutually beneficial ends. But although they may seem to behave cooperatively, they don't behave morally, because they will cheat whenever they can get benefits without incurring the costs of participation, and because genuine reciprocity will be absent.

This suggests that some facts *we* take to support moral conclusions don't provide reasons for *all conceivable rational agents* to act. How then do they provide *us* with reasons? Couldn't we know both those facts and our own interests, without taking them to provide *us* with reasons to act? If so, then the idea that we have duties that *can't* be shirked is false. This is a challenge to moral objectivity. Surely, reasons for action do depend on potentially motivating values. If these can, in principle, vary without limit, no mere *facts* provide *all rational agents* with reasons to perform other-regarding acts. Thus, no objective morality can bind them all. This has, for decades, been taken to be a conceptual truth by many philosophers and social scientists. If it is such a truth, nothing can override it.

The Schlick-inspired way out is not seek the impossible. An objective morality for *all possible rational agents* is not what he sought, or what we need. Instead, we need a morality grounded in human nature, governing normal *human beings.* As he emphasized, this requires a rich species-wide psychology, or psycho-biology. But how rich is the species-wide psychology, or psycho-biology, that he takes to ground human morality? Although it may be rich enough to get morality started, more may needed to reach the expansive morality he seeks. As the social scientist James Q Wilson has emphasized in *The Moral Sense*, there is evidence that human beings are predisposed to other-regarding moral behavior. Human communities at every time and place have shared our native sociability, our prolonged dependence as infants, the bond between parent and child, social attachments with friends, family, and neighbors, and the intertwining of our self-interest and self-conception with a desire, not only for the good opinion of others, but also for their well-being. Yet, despite this commonality, only a few communities have recognized rich moral obligations beyond family, friends, and tribe. Why, if morality is grounded in our innate endowment, should expansive morality be unusual? Why do we think it is *morally better* to recognize some obligations to those outside our tribe? *What does this even mean?*

We know that patterns of social organization that require agents to extend moral standing beyond family, friends and tribe tend to be better in advancing human welfare than those that do not. Although this is important, it isn't enough. It *is* best for humanity that agents internalize this lesson, But recognizing this doesn't solve the problem, because a concern for humanity is not everyone's cup of tea. So, we ask, *Is it true that those whose conceptions of morality aren't universalistic ought to treat outsiders with sympathy, fairness and reciprocity?,* Schlick can't say they morally *ought* to treat outsiders this way unless doing so would, in the main, advance interests they already have. Would it? Perhaps such a change would advance their culture to the benefit of their descendants, whom they may care about. But such a historical transformation is speculative; there seems to be no guarantee it will ever come. Must we resign ourselves to the thought that these agents don't form a moral community with us?

I think not. Universalistic moral conceptions typically tie fairness to reciprocity. If less universalistic cultures wish to reap the benefits of cooperative interaction with more universalistic cultures, they must realize that reciprocity in dealing with outsiders is expected, and so come to find it to be in their interest to reciprocate. If they do, then, increased participation in mutually beneficial activities may change the attitudes of all parties, expanding the moral universe of each. It was just such reciprocity that led our descendants, and most of us at different stages of our lives, to extend empathetic moral concerns we initially restricted to the near and dear to larger groups. This suggests that the proper response to the those with limited moral horizons is to work to expand them by engaging them in reciprocal social interaction, on the basis of the commonalities they already share with us. It is hard to imagine that Schlick wouldn't agree.

What is true of the relationship between cultures with mismatched moral universes may also be true of mismatched individuals, or groups, within a culture. All societies contain some who fail to accord significant moral status to some of their neighbors. The centrality of reciprocity to our moral lives dictates that morally expansive neighbors offer their more morally isolated fellows opportunities for cooperative interaction, *conditional on the willingness of all parties to engage in genuinely reciprocal behavior.* If the required reciprocity isn't forthcoming, such offers should be withdrawn. There is, after all, a price to be paid for opting out of practices of mutually beneficial concern. If there wasn't, moral virtues would never be acquired.

This is the conception of morality to which we are led by pursuing Schlick's view of psychology as the basis of ethics. It is one in which moral systems are social institutions that evolve into richer systems capable of increasing human welfare by increasing the scope of human cooperation. As social cooperation increases, psychologically and biologically-based values are extended, creating new moral relationships that allow previously underivable oughts to be derived from expanded factual premises about what agents value and are able to contemplate. It is this conception – not of changing moral *opinions* but of expanding moral *reality* – that those who value Schlick's contribution are called upon to elaborate and refine.

References

Carnap, R. 1947/1956. *Meaning and Necessity*. Chicago: University of Chicago Press.

Schlick, M. 1930/1939. *Problems of Ethics*. Trans. David Rynin. New York: Prentice Hall, 1939. Originally published as *Fragen der Ethik*. Vienna: Springer.

Soames, S. 2005. *Reference and Description*. Princeton/Oxford: Princeton University Press.

———. 2007. Actually. *Proceedings of the Aristotelian Society*, Supplementary Vol. LXXXI, 251–277.

———. 2010. *Philosophy of Language*. Princeton/Oxford: Princeton University Press.

———. 2011 "Kripke on Epistemic and Metaphysical Possibility", in *Saul Kripke*. ed., Alan Berger. Cambridge: Cambridge University Press.

———. 2015. *Rethinking Language, Mind, and Meaning*. Princeton/Oxford: Princeton University Press.

———. 2016. Propositions, the *Tractatus*, and "The Single Great Problem of Philosophy". *CRITICA* 48 (143): 3–19.

———. 2018. *The Analytic Tradition in Philosophy*. Vol. 2. Princeton/Oxford: Princeton University Press.

Wilson, J.Q. 1993. *The Moral Sense*. New York: Free Press.

Wittgenstein, L. 1914–1916. *Notebooks. 1914–1916*, 2nd ed., ed. G.H. von Wright and G.E.M. Anscombe. Trans. G.E.M. Anscombe. Oxford: Blackwell. 1979.

———. 1922. *Tractatus Logico-Philosophicus*. Trans. C.K. Ogden. London: Routledge and Kegan Paul. Reprinted in 1999 by Dover; Retrans. D. Pears and B. McGuinness. London: Routledge, 1961.

Part III
Review Essays

Chapter 12
George A. Reisch, *The Politics of Paradigms. Thomas S. Kuhn, James B. Conant, and the Cold War "Struggle for Men's Minds"*. Albany: State University of New York Press 2019, 456 pp., ISBN: 978-1-4384-7367-3

Hans-Joachim Dahms

Abstract This review-article is about George Reisch's second big and impressive volume about the fate of philosophy of science after World War II. This one treats a single philosopher and his most successfull book: Thomas Kuhn's *The Structure of Scientific Revolutions (Structure for short)*. It provides the reader with much new and interesting material about Kuhn's life and career from childhood till around 1970, including the production and reception of Structure. It also tries to show that Structure is affected by the political zeitgeist of the Cold War 1950ties. Without denying the usefulness of an "external" approach to the history of philosophical of science, I criticize Reisch's thesis, that paradigm-changes resemble "conversions" in spheres outside science (like in religion and politics) and that the adoption of a new paradigm can be called a sort of "brain-washing". The book is recommended to all readers interested in the interrelation of philosophy (of science) and contemporary history.

Keywords George Reisch · Thomas Kuhn · Cold War · James Conant · Structure of Scientific Revolutions

George Reisch, the author of the book under consideration, is known as managing editor of *The Monist* as well as editor of the Open Courts Publishing company's series "Popular Culture and Philosophy". Reisch made his first big appearance on the stage of the History of Philosophy of Science in 2005, when he published the

H.-J. Dahms (✉)
Institut Wiener Kreis, Universität Wien, Vienna, Austria
e-mail: hans-joachim.dahms@univie.ac.at

C. Damböck, A. T. Tuboly (eds.), *The Socio-Ethical Dimension of Knowledge*,
Vienna Circle Institute Yearbook 26, https://doi.org/10.1007/978-3-030-80363-6_12

book *How the Cold War Transformed Philosophy of Science* (Reisch 2005). His object was there to explain which forces caused the change of philosophy of science from a socially, politically and culturally interested project, as it had been before in Middle Europe and especially in Vienna, to a formalistic and politically neutral discipline during the post-World War II years in the USA (Reisch 2005, 6). The answer was, roughly, that a group of philosophers, who championed the first part of the alternative (like Otto Neurath, Philipp Frank and also Charles Morris) were superseded by others (like Rudolf Carnap, Herbert Feigl and Hans Reichenbach), who favored the second one (Reisch 2005, XII). The explanation is mostly given in political terms, because it was the political pressure on the first group that led to this result. Reisch did not maintain that his explanation was complete. But he wanted to inject some historical and political context into otherwise mostly "internal" debates. That his book sparked discussion and controversy, had also to do with the fact, that he was the first one to try such an approach.

The Politic of Paradigms is now the second major monograph by Reisch. It picks up where the former left off: It is, as the title suggests, about the seminal work of Thomas Kuhn, *The Structure of Scientific Revolutions*. Both its origins and impact after the publication as penultimate volume in the "International Encyclopedia of Unified Science" (IEUS) in 1962 are described and discussed. Kuhn's *Structure* caused in the opinion of many a complete overthrow of Logical Empiricism's view on philosophy of science. But: although the focus and time-frame is different, the second book starts earlier than one might expect: with Kuhn's upbringing and schoolyears as biographical background and the decisive influence of Harvard's president James B. Conant after Kuhn's change from a career in physics to one in history and philosophy of science. In 1953 Kuhn was offered the task to write the volume for the IEUS by its main editor Charles Morris. Since he was for a while distracted by other projects to write it down, the book appeared nearly a decade later. So a larger time-span is covered already with the beginning of the whole story. A similar thing happens at the end, when Reisch follows the impact of Kuhn's book till around 1970.[1] In between there is some overlap with the first book in the decisive 1950s and early 1960s when Kuhn's *Structure* was conceived and written.

Nevertheless the method employed is similar: also in the new work Reisch follows a distinctively "external" approach. He wants to find out, whether and how the political environment of Cold War had an impact, now not on a whole movement (like Logical Empirism), but on a single outstanding work, namely on *Structure*.

Let me now start the discussion of Reisch's second book with the title page, especially with the subtitle *Thomas S. Kuhn, James B. Conant, and the Cold War "Struggle for Men's Minds"*, which is emphasized on the book-cover by the portraits of the two men with the American flag somehow running as small shadows

[1] Reischs decision to stop there seems understandable, because it would require investigations not only into the histories of philosophy, history of science, but also into "sociology, anthropology, literary criticism, business, law, the self-help industry, leftist politics and other endeavors from 1962 onward" (Reisch 2019, 360). Nevertheless it would be interesting to see someone – Reisch himself or another author – take up that task: Kuhn's (1962) success is completely extraordinary.

over their faces. Reisch emphasizes from the start that he has no biographies of his two main characters in mind. Nevertheless his chapters on Kuhn bring a wealth of new material about Kuhn's life and work up to ca. 1970. His information about Conant can round out Hershberg's monumental biography, where remarks on Kuhn are only made in passing (Hershberg 1993, 409 f.).

Now Conant played without doubt a very important role in Kuhn's life: he hired him to teach the course "Science IV" in the Harvard curriculum and discussed with him intensively and for a long time about what to read in the preparation and what to teach in those lectures. Furthermore he let him know his ideas about the importance of science for modern society in general and during the growing confrontation with the emerging other super-power, the Soviet Union, in particular. Conant reminded him that one needed to show the impact of science on society not only in philosophical terms but relying on original historical material (and not on textbooks!) and in detailed case studies. When Kuhn asked, Conant also introduced him into the Society of Fellows, where he could exchange ideas with senior researchers at a very young age. So it comes as no surprise that Kuhn asked Conant to write a foreword to his "Copernican Revolution" and later on dedicated *Structure* to him.

In these early years around 1947 in post-war Harvard Kuhn had two decisive aha-experiences, which had not much or nothing to do with Conant. Reisch describes the first one, his "Aristotle experience" on a hot summer day in various places.[2] It consisted in a sudden understanding of Aristotle's *Physics*, which before seemed to be mostly false and grossly misleading to him. Now Kuhn learned all of a sudden, that terms like *kinesis* (movement) or *dynamis* (force) had a completely different meaning in classical Greek and carried in Aristotle's text a whole lot of theoretical connotations with them which were completely different from the uses of the terms in modern physics. I would not overrate this experience the way that Kuhn (and also Reisch, following him uncritically) did. Had Kuhn only learned to read the original Greek text and perhaps some suitable commentary, he could have come to a similar conclusion.

On the other hand, Reisch plays down[3] – also following Kuhn's memories – a book that Kuhn read at almost the same time, namely Ludwik Fleck's book *Entstehung und Entwicklung einer wissenschaftlichen Tatsache*. Later on Kuhn recalled that he agreed with Fleck to take a sociological approach in describing science as an organized community instead of concentrating on the singular scientist as unit of historical description. But he also "remembered" to have kept Fleck somehow "at arm's length" (Kuhn 2000, 283): Fleck's German-speaking texts were difficult to read and more importantly theoretical misleading, because Fleck modeled the scientific community after the image of the singular scientist. Now I think that Fleck's German is perfectly in order and it was not Fleck, who made that (theoretical) mistake, but Kuhn himself, when he brought in the analogy of Gestalt-switch in order to describe paradigm-change: Gestalt-switches happen instantaneously to a

[2] see Reisch 2019, 65 f. and passim.

[3] at least when compared to the pages devoted to the Aristotle experience: Reisch 2019, 220 f.

person and can be reversed, whereas paradigm-changes in a scientific community can not. Fleck's book contained many more insights Kuhn took over to *Structure*, i.a. the paradigm-monopoly-thesis.

So a more distanced view at Kuhn's recollections about his time in Harvard would be in order. And it remains to be seen, whether Conant's influence on Kuhn survived into later years: in 1953, Conant retired from Harvard to work as High Commissioner and later on as America's first ambassador to West Germany.[4] When he returned to the US in 1957, Kuhn himself had already left Harvard for an appointment at the University of Berkeley.

Happily we learn in the "Introduction" of Reisch's book, that not only the relation between Kuhn and his mentor Conant are covered, but both the origins and the reception of *Structure* as well (Reisch 2019, XXXVII ff). The account starts in part I ("War and Crisis") with three chapters about Kuhn's upbringing and the political climate that surrounded Kuhn and Conant after World War II. Part II ("The Struggle for Men's Minds") covers biographical developments which were similar to paradigm-changes, like "The Cold War Conversions of Thomas S. Kuhn and James Bryant Conant" or "Brainwashing, or the Structure of Philosophical Revolutions". In part III, Reisch describes the planning and writing process of Kuhn's book under the headline "The Cold War Origins of the Structure of Scientific Revolutions". In the fourth and last part ("The new world of paradigms") the reception of Kuhns *Structure* is covered, starting with reactions of American scientists and philosophers like Paul Feyerabend and Bentley Glass. In this context, Conant's reaction is remarkable, because it can serve perhaps as a test case for the ongoing influence of his view about science on Kuhn. But more important are in this section the descriptions and discussions of the first two public debates of Kuhn's work, both in GB, the first one at the conference on "Scientific Change", organized by A. C. Crombie in 1961,[5] the second one at the LSE in 1965 with the famous encounter Kuhn had with Karl Popper and his followers, later published as *Criticism and the Growth of Knowledge* by Imre Lakatos and Alan Musgrave (Lakatos and Musgrave 1972).

Towards the end of the book and before its epilogue we find an episode (Reisch 2019, 343 ff.), where Kuhn actually "crossed the line" into politics (prompted by Nixon's invasion of Cambodia and the shooting of students protesting against this war crime at Kent State University), namely as chair of a commission charged with the task to look into military-related research at the university of Princeton (and also elsewhere). So Kuhn's early and late (in Reisch's account) years somehow round out the whole narrative of a political engaged and well-informed person, but who was never an activist. And that's what distinguishes Reisch's book from Stephen

[4] By the way: the first West-German chancellor after World War II was Konrad (not Conrad) Adenauer (Reisch 2019, XVII).

[5] published as Crombie 1963: see especially the portion on the Sociology of Science (including Kuhn's paper on "Dogmatism …" on page 347–369, as well as comments and discussion).

Fuller's treatment of Kuhn, where Kuhn is suspected to have been perhaps someone as "the American Heidegger".[6]

The whole book is very well written and offers rich new material for everyone interested both in Kuhn as a person and as a scientist/philosopher. The information about his schoolyears were completely new to me. Imagine 13-year-old Kuhn giving at the progressive Hessian Hills school he attended a passionate speech in favor of pacifism and against spending on armament (Reisch 2019, 3 f.)! Another interesting and innovative part of the book describes Kuhn's early dealings with the sociology of science (Reisch 2019, 202 ff.), especially his connections with former Vienna circle member Philip Frank at Harvard, and then with Robert Merton and Bernard Barber (both from Columbia). One may add to these very valuable informations the episode where Kuhn tried to convince the Social Science Resarch Council at the end of the 1950s to fund efforts in the new discipline of Sociology of Science (Dahms 2016, 118 f.). In view of Kuhn's strong ties to that new research field one might like to know why this project seems to have been given up and by whom.

But of course Reisch doesn't want to tell only a loosely connected story of some noteworthy episodes. His aim is to show the imprint of the Cold War on Kuhn's philosophy of science, which is guided or supported by three main claims, namely:

1. that the term "paradigm" in Kuhn's theory is a place-holder for the former concept of "ideology",
2. that during the reign of a paradigm (in "normal science") its students and adherents are somehow "brainwashed" to accept it, and
3. that the process of transition from a paradigm to the next (in times of "revolutionary science") is a sort of conversion.

Later on I will also add some critical remarks on the "reception" part, which are to a lesser degree "external".

Ad 1): The term "ideology" does not play a special role in *Structure*. But Reisch shows convincingly that it was Kuhn's central concept when he started to work on his IEUS contribution,[7] till it was superseded by "paradigm" at the end of the 1950s. What was the cause of this change in terminology? Already in the correspondence between Kuhn and Morris (the acting editor of the IEUS) the term "ideology" is seen by both as problematic (Reisch, 2019, 298 f.), possibly because it may invoke leftist associations, as already the title of the IEUS had done (with "Unity", "Internationalism" etc.). Reisch suggests that it was the fear to be brought in line

---

[6] Reisch mentions him only shortly in the Introduction, XXXV.

For Fuller's detraction of Kuhn see especially Fuller 2003, chapter 17: "Is Thomas Kuhn the American Heidegger?". The question alone shows that Fuller is not well informed about both: Kuhn *and* Heidegger. See Dahms 2016, 106 f.

[7] Reisch 2019, 197; I doubt, by the way, that Kuhn could have "shared with Shils *and* Wirth" a keen interest in ideology: Wirth had been invited by Otto Neurath in 1937 to write a contribution to the IEUS about the sociology of science, but abandoned the field completely after the end of World War II in order to work as president of the world organization of sociologists (ISA) and afterwards on an unfinished mammoth project on "The City". See Dahms 1999, 204–211. Wirth died already in May 1952.

with these associations which led Kuhn to give up "ideology" and to adopt the new word "paradigm". But it has to be noted, that "ideology" in Kuhn's use was something restricted to scientists already from the start. It was meant to embody all the traditions, theories, skills and achievements that bind a scientific community together. This use did for him not carry specifically political overtones with it – comparable to the use nowadays fashionable in soccer, where a pressing vs. a possession "philosophy" or "ideology" are discussed at length.

Ad 2) Chapter 6 of Reisch's book is entitled "Brainwashing, or the Structure of Philosophical Revolutions".[8] The claim here is that according to Kuhn scientists are somehow brainwashed on entering a (new) paradigm, with all the associations that the practice of this method of social conditioning meant in the 1950s and early 1960s in the USA. I don't find the whole idea convincing. As Reisch notes himself, Kuhn mentions the word nowhere in his writings. Therefore one is led to assume that it slipped into his *Structure* subconsciously, because the Zeitgeist around 1960 was loaded with it. But: when a scientific community adopts a new paradigm, this doesn't happen normally under pressure, torture or other means of coercion (like in brainwashing). Reisch himself remarks (Reisch 2019, 127):

> Of course, any comparison of Kuhn's scientists to [...] captured GIs goes only too far. The scientists Kuhn described are not forcibly captured, manipulated, tortured, or forced to accept a new paradigm during times of revolution.

In order to stick nevertheless to his brainwashing theory, he then adds:

> They are, however, educated rigorously and their professional careers are dependent upon their accepting the reining paradigm that shapes their thought and the ways they perceive the world.

This is true, of course. Being "educated rigorously" is the usual practice in many professions which require an academic degree (like medicine, law etc.). But nobody is forced to enlist in these studies and can quit at almost any time (at least in countries where education is for free and where the former investment in some study semesters has not to pay off immediately). And when finished with their degrees and entering a profession: why should scientists work along a superseded paradigm in order to avoid further "brainwashing" into the new one? If one thinks of the way Einstein's general theory of relativity was accepted in the German physics community, one sees that there was indeed much ideological or political struggle (i.a. by people who later on advocated the Nazi's "Deutsche Physik"), but no one was under pressure to succumb to the new theory. And then stunning proofs from forecasts on the basis of the theory arrived: for instance the observation of light deflection by the gravity of the sun during the solar eclipse of 29th of May 1919 (Fölsing 1993, 490). There are no parallels to these spectacular successes on the basis of precise and also counterintuitive forecasts in political revolutions. But that's what guaranteed the breakthrough of Einstein's theories, especially amongst the younger generation in

[8] Reisch 2019, 111–131; The brainwashing theory is mentioned in passing in many other places of the book too.

the physics community, who in turn spoke out against the traditionalists at spectacular public gatherings.[9]

Ad 3) In contrast to the term "brainwashing" Kuhn often used the concept of "conversion", when he described the transition from one paradigm to another. Some critics seized on it and its irrational overtones, concentrating on *religious* conversions.[10] Reisch now uses a somewhat different interpretation of the term "conversion" in order to defend his political reading of *Structure*: he follows the *political* conversions that Kuhn, Conant (and also Sidney Hook, with whom Conant was for a while in close contact) underwent. So Reisch describes in chapter 5 and 6 the Cold War conversions of Kuhn and Conant and in chapter 6 also the political conversion of Sidney Hook which had started already in the late 1930s and led to his rampant anticommunism in the Cold War period. The basic questions are here in each case: When and why did the conversion start? What were the political standpoints at the beginning and at the end of the conversion-process? As Reisch shows, Kuhn had been in his younger years in the 1930s a convinced pacifist. In left-wing circles in the USA (like in the *Partisan Review*) the debate of participation in a war against Nazi Germany was still going on, when World War II had long started. But: a move away from pacifism during the Second World War should surprise no one. Who could possibly remain a pacifist when your nation is under attack, as it happened to the US with Pearl Harbor and Hitler's declaration of war in December 1941? Conversions of that kind I would simply call "learning from experience". And it is no wonder that even convinced pacifists like Einstein and Russell went out of their way in these special circumstances. With Conant and Hook it is of course more complicated than with Kuhn himself. Hook had come a long way from being the most famous Marxist expert and theoretician in the US to the notorious cold warrior after the war. His original motives were events like Stalin's purges and trials of 1936, the Hitler-Stalin-accord of 1939 etc. Conant went from a liberal scientist and university administrator (during his tenure as president of Harvard university) to a supporter of the US war effort and organizer of the Manhattan Project and later on became an anticommunist after the war mainly because of the spectacular revelations about communist spies like Alger Hiss or Klaus Fuchs (in GB) 'infiltrating' the nuclear weapons programs.[11] To sum up: when Kuhn speaks of "conversions", he means the deep impact of scientific changes and sometimes the short timespan in which they occur, but does not want to say that they are completely non-rational (in the religious or (sometimes) the political sense).

Now some remarks about the reception of Kuhn's *Structure*! Conant's first reactions to the book showed that his picture of science was quite different from Kuhn's. It assumed a evolutionist cumulative process, where pieces of new knowledge were

[9] See the precise and readily understandable account of Fölsing 1993, 488 ff.: I wonder why the debate about *Structure* did not center on the rich material about Einsteins's revolution and the political and cultural fights around it only some decades earlier and instead on scientific revolutions in centuries long gone.

[10] Take as an early examples Watkins 1972, 33; and Lakatos 1972, 93.

[11] Nancy Greenspan published recently a book on Fuchs (based i.a. on the family archive).

added to the existing body of theories. He also depicted the scientists as an open-minded critical person, very similar to Popper's proclaimed view (while not always in line with his behavior).

In another passage, Reisch describes the already mentioned 1961 and 1965 conferences in Great Britain, where Kuhn's *Structure* was heavily debated. In his presentation of these events, Reisch suspends his external approach for a more conventional, "internal" reading: He piles up all the different reactions to Kuhn, one after the other, in order to show that Kuhn was mostly on the defensive. In this context, Reisch has omitted, however, two important contributions: The first is the contribution of Michael Polanyi to the conference in 1961, who had welcomed Kuhn's own presentation about "The Function of Dogma in Scientific Research" with the following words (Polanyi 1963, 375):

> The paper by Mr. Thomas Kuhn may arouse opposition from various quarters, but not from me. At the end of it he says that the dependence of research upon a deep commitment to established beliefs receives the very minimum of attention. I could not agree more, I have tried in vain to call attention to this commitment for many years. I hope that if I join forces with Mr. Kuhn we may both do better.

Kuhn in his closing remarks emphasized the similarities of his views with those of Polanyi, too. But he also named some differences, the most important one being his stress on the scientific community versus Polanyi's individualistic outlook (Kuhn 1963, 394 f.):

> It is not, after all, the individual who decides whether his discoveries or theoretical inventions will become part of the body of established science. Rather it is his professional community, a community which has and sometimes exercises the privilege of declaring him a deviant….I take the "social implication" of my views more seriously than his (= i.e.: Polanyi's) commentary suggests.

Nothing came out of that proposed cooperation. It seems that Kuhn had noticed soon that it was a fault to align himself with Polanyi who had a fame for defending an elitist, authoritarian picture of science dominated by scientific geniuses.[12] When Kuhn realized that he was here on the wrong track, he saw to it, that his paper on "dogmatism", included in the proceedings of the 1961 conference, was not reprinted in his own collections of writings (like Kuhn 1977) and not anywhere else when people approached him for permission to do it.

When we look now at the conference of 1965 at the LSE, it seems that the whole event was set up by the Popperians to destroy Kuhn's allegedly dogmatic picture of science in its "normal" phases. Contributions like Poppers "Normal Sciences and its Dangers " (Popper 1972) and Watkins' "Against 'Normal Science'" show that the participants were talking in part on cross-purposes: Kuhn wanted to construct a more realistic view of the actual functioning and practice of science, whereas the Popperians aimed at a heroic ideal, where scientists invited criticism everywhere, showed the weaknesses of their own theories in advance of tests and experiments

[12] Nye 2011, 259 ff. (especially p. 263).

and were prepared to give them up immediately when the outcomes went the wrong way.

In his analyzes of this later conference, Reisch has left out the important paper by Margaret Masterman, the linguist. Analyzing Kuhn's *Structure* she found the (perhaps a little bit exaggerated) number of 21 different meanings of the word "paradigm" (Masterman 1972, 61–66). As Kuhn later acknowledged (Kuhn 2000, 299 f.), Masterman's paper was the reason why he went from the concept of paradigm to "disciplinary matrix", where mathematical generalizations, models (and analogies), values (including, by the way, also non-scientific values) and exemplars are distinguished (Kuhn 1970, 182 ff.). Reisch seems to think that this shift marks a complete abandonment of his former views. I think that the move is less dramatic: it aims mainly at removing ambiguities. But one thing is right: it signifies a further step away (after having given up on the sociology of science around 1960) from taking its social and political aspects seriously into consideration.

To sum up: Reisch made a remarkable and thought-provoking effort to convince us of the impact of the Cold War zeitgeist on Kuhn's *Structure*. Such an external approach is interesting and important. Criticism of some of his findings shouldn't deter neither him nor other scholars to pursue that road.

References

Crombie, A.S., ed. 1963. *Scientific change. Historical studies in the intellectual, social and technical conditions for scientific discovery and technical invention, from antiquity to the present.* London: Heinemann Educational Books.

Dahms, Hans-Joachim. 1999. Otto Neurath's International Encyclopedia of Unified Science. In *Otto Neurath: Rationalität, Planung, Vielfalt*, ed. Elisabeth Nemeth and Richard Heinrich, 184–227. Wien: Oldenbourg Verlag/ Berlin: Akademie-Verlag.

———. 2016. Thomas Kuhn's *Structure*: an 'Exemplary Document of the Cold War Era'? In *Science Studies During the Cold War and Beyond. Paradigms Defected*, ed. Elena Aronova and Simone Turchetti, 103–125. New York, NY: Palgrave.

Fölsing, Albrecht. 1993. *Albert Einstein. Eine Biographie.* Frankfurt am Main: Suhrkamp Verlag.

Fuller, Steve. 2003. *The struggle for the soul of science. Kuhn vs. Popper.* Duxford, Cambridge: Icon Books Ltd.

Hershberg, James G. 1993. *James B. Conant. Harvard to Hiroshima and the Making of the Nuclear Age.* Stanford, CA: Stanford University Press.

Kuhn, Thomas S. 1962. *The structure of scientific revolutions.* Chicago/London: University of Chicago Press.

———. 1963. The function of dogma in scientific research. In Crombie 1963: 347–369.

———. 1970. Postscript – 1969. In *Thomas Kuhn. The Structure of Scientific Revolutions. Enlarged*, 2nd ed., 174–210. Chicago: University of Chicago Press.

———. 1977. *The essential tension. Selected studies in scientific tradition and change.* Chicago/London: University of Chicago Press.

———. 2000. A Discussion with Thomas S. Kuhn (Aristides Baltas, Kostas Gavroglou, Vassiliki Kindi). In Thomas Kuhn. *The road since structure. Philosophical essays, 1970–1993*, with an Autobiographical Interview, ed. James Conant; John Haugeland, 255–323. Chicago/London: The University of Chicago Press.

Lakatos, Imre. 1972. Falsification and the methodology of scientific research programs. In Lakatos and Musgrave 1972: 91–196.
Lakatos, Imre, and Alan Musgrave, eds. 1972. *Criticism and the growth of knowledge, Proceedings of the International Colloquium in the Philosophy of Science, London 1965*. Vol. 4. Cambridge: Cambridge University Press.
Masterman, Margaret. 1972. The nature of a paradigm. In Lakatos and Musgrave 1972: 59–89.
Nye, Mary Jo. 2011. *Michael Polanyi and his generation. Origins of the social construction of science*. Chicago/London: The University of Chicago Press.
Polanyi, Michael. 1963. Commentaries on part four (= Kuhn 1963). In Crombie 1963: 375–380.
Popper, Karl. 1972. Normal science and its dangers. In Lakatos and Musgrave 1972: 51–58.
Reisch, George. 2005. *How the cold war transformed philosophy of science*. New York: Cambridge University Press.
———. 2019. *The politics of paradigms. Thomas S. Kuhn, James B. Conant, and the cold war "struggle for men's minds"*. Albany, NY: State University of New York Press.
Watkins, John. 1972. Against 'Normal science'. In Lakatos and Musgrave 1972: 25–38.

Chapter 13
Amrei C. Joerchel, Gerhard Benetka (Eds.), *Memories of Gustav Ichheiser. Life and Work of a Social Scientist* (Theory and History in the Human and Social Sciences Bd.1). Cham: Springer 2018, 243 pp., ISBN 978-3-319-72507-9

Maria Czwik and Bastian Stoppelkamp

Memories of Gustav Ichheiser, edited by Amrei Joerchel and Gerhard Benetka, marks a double premiere: It is the first volume of the newly founded series *Theory and History in the Human and Social Sciences* and probably the first anthology about the life and work of Gustav Ichheiser (* 1897 Krakow, Poland; † 1969 Chicago, USA).[1] The book provides a personal and professional biography of Ichheiser as well as an extensive overview of his work. It includes eleven research papers (or chapters) plus an unpublished autobiographical manuscript in which Ichheiser deals with his experience of being in a psychiatric hospital all through the 1950s. By contextualizing his scientific non-career as an outsider of academic life, discussing his prime interests, his most important terminologies, his German, English and Polish works and by showing the relevance of his social-psychological theories for today's sciences, the book supplies various starting points for digging deeper into the work of Gustav Ichheiser.

Although largely forgotten today, Ichheiser was one of the most promising social psychologists and sociologist of the interwar period in Austria. Between 1925 and

[1] Some publications on Ichheiser appeared shortly after his death and some of his texts were published in the 1970s (e.g. Freund 1974; Ichheiser 1972; Ichheiser 1977). Then it became rather quiet around him. In recent years he received some attention, partly in the context of a broader research on scientists in exile (Woller 2013; Müller 1993; Fleck 2015). Earlier some publications pointed out his relevance in different contexts (e.g. Rudmin et al. 2011; Parales-Quenza 2010; Velho 2008; Rudmin 2007; Brissett/ Edgley 1990; Bayer/Strickland 1990) – but overall it seems that there was little recognition of Ichheiser's theories.

M. Czwik (✉) · B. Stoppelkamp
Institut Wiener Kreis, University of Vienna, Vienna, Austria
e-mail: maria.czwik@univie.ac.at; bastian.stoppelkamp@univie.ac.at

C. Damböck, A. T. Tuboly (eds.), *The Socio-Ethical Dimension of Knowledge*, Vienna Circle Institute Yearbook 26, https://doi.org/10.1007/978-3-030-80363-6_13

1938 he published a couple of books and three dozen articles in leading psychological, sociological and philosophical journals (Fleck, chapter 2). His investigations into the phenomena of social success and social mis-perceptions were major contributions to the "scientific counterculture" (Benetka, chapter 1) of Red Vienna and Austrian "Late Enlightenment" (*Österreichische Spätaufklärung*).[2]

As a contemporary of the Vienna Circle, Ichheiser's personal, social and intellectual biography is also of interest for the broader historiography of Logical Empiricism. In an enthusiastic review, written in 1930, Otto Neurath welcomed Ichheiser's *Kritik des Erfolges* as an important step towards a "scientific world-conception" (Neurath 1930). Six years later, Ichheiser published an article for the main outlet of Logical Empiricism, Carnap's and Reichenbach's journal *Erkenntnis*, in which he analyzed the social and individual preconditions of psychological inquiry (Ichheiser 1936).

Although these episodes are not directly addressed in *Memories of Gustav Ichheiser* – we will come back to that later –, the book is still useful for students and scholars of Logical Empiricism, because it takes a closer and sometimes shattering look at the personal and professional struggles of a Jewish socially engaged scientist whose biography bears a striking resemblance to the experiences of many members of the Vienna Circle.

One of the most radical, methodological takes of this new anthology is to show the general importance of biographical studies for the historical understanding of scientific contents. As pointed out by the editors in the preface of the book, Ichheiser's biography delivers a test case for how close scientific work and private experience can interact, because "these two life trajectories are never separate as the traditional academic research is used to portraying" (Joerchel, Benetka, preface). In this sense, Ichheiser is framed as an "outsider" or "underdog" who transformed his own professional and social marginalization into a self-reflexive research program. This methodological program is pretty close to Ichheiser's own point of view expressed in one of his articles: "[...] the born psychologist, sociologist, anthropologist is the marginal man who, split within himself, is fully aware that the world is not as we see it, but that we see it as we do because we are as we are" (Ichheiser 1970, 34).

As the book shows, this idea of the social scientist as "marginal man" can be interpreted in at least two different ways. According to a more optimistic interpretation, Ichheiser was an 'outsider as insider' using his own sensibilities for social mechanisms of exclusion and deception as a constructive force for social self-enlightenment. From another more pessimistic perspective, Ichheiser's work seems like an expression of personal "alienation" (Schor-Tschudnowskaja, chapter 4) and "uprooting" (Dietrich, chapter 11).

The more optimistic narrative is first laid out by Gerhard Benetka (chapter 1) in the opening, "historical" section of the book. From Benetka's point of view,

[2]For the idea of "Austrian Late Enlightenment" and the connections between Austrian Social Democracy, the social reform movement and the Vienna Circle see Stadler 1981.

Ichheiser was a paradigmatic, yet unknown exponent of the social-democratic, "scientific counterculture" of Red Vienna. He was born on 25th December 1897 in Krakow as the second son of a Jewish family. After the death of his father the family moved to Vienna where Ichheiser, around 1919, attended the university to study law and later philosophy and psychology. In 1924, he completed his studies with a philosophical dissertation on the *Object of Aesthetics* (*Gegenstand der Ästhetik*) under the supervision of Karl Bühler (Ichheiser 1924).

As a Jewish student, Ichheiser experienced the strong political wave of antisemitism and German-Nationalism that since the end of the First World War had turned the Austrian universities more and more into 'reactionary strongholds' (*Bollwerke der Reaktion*) and made a continuation of his academic career impossible. "During the interwar period, young Jewish and Socialist intellectuals [...] had no chance for an academic career": It would never had crossed their minds "to strive for habilitation at the university", writes Benetka in his essay comparing the biography of Ichheiser with the tragic experiences of Edgar Zilsel.

Like Zilsel and other progressively-minded Jewish intellectuals, Ichheiser found a new platform for his own scientific ambitions in the public institutions of Red Vienna: He got involved in the adult education movement, worked as a supervisor and researcher at the career services (*Berufsberatung*) of the Viennese Chamber of Labour (*Arbeiterkammer*). In his article, Benetka describes the "innovative power" of the Viennese social-democratic machinery that turned former academic outsiders into scientific insiders by engaging them into the socialistic project of cultural and political reformation. Through his involvement into this project, Ichheiser relocated his scientific identity: The former academic philosopher and aesthetician grew into a "non-university" social scientist who "anchored practical relevant problems in his theoretical psychology or the other way around".

This reflexive understanding of science as a force for social emancipation and transformation, that – as Benetka mentions – can also be found in the *Bildstatistik* of Otto Neurath – is further fleshed out in other contributions to the book. One important topic is Ichheiser's opus magnum *Kritik des Erfolges* (Critique of Success) from 1930 that is tackled among others in the articles of Markus Wrbouschek (chapter 8) and Luca Tateo (chapter 9). As shown by Wrbouschek, Ichheiser's monograph was part of a larger "comprehensive research program", which started with an analysis of Machiavelli's political philosophy in 1927 and resulted – until the mid-1930s – in a psychological and sociological investigation into the modern ideology of success and "meritocracy" (Tateo).

According to Ichheiser, modern meritocracy is based on a categorical confusion between the concepts of "productivity" and "success". While "productivity" refers to the ability to fulfill a certain task (*Leistungstüchtigkeit*), "success" involves the ability to get socially rewarded or recognized for the fulfillment of a task (*Erfolgstüchtigkeit*). Being a successful "shoemaker" – to use one of Ichheiser's examples – is thus something very different from being a productive or skilled shoemaker, because in order to achieve success some additional social qualities and standards must be in place, like for instance a good self-marketing or the proper social connections and opportunities. As Wrbouschek makes clear, the modern ideology of success leads according to Ichheiser into two common and coordinated forms of deception: We (wrongfully) justify social success via personal productivity and measure personal productivity via social success. This ideological identification of success and productivity conceals, and at the end: accelerates their factual disconnection, because – as Luca Tateo puts it in a more timely context: "the lack of meritocracy in real life becomes a general demand for 'more' meritocracy, rather than questioning meritocracy as an ideal".

Another interesting example for this reflexive and praxeological understanding of social science is given by Lucas B. Mazur who has analyzed "Ichheiser's Pre-War Polish Texts" (chapter 7). With the rise of the Austro-fascist *Ständestaat* and the destruction of the tight-knit cultural network of Red Vienna in 1934, Ichheiser found a new scientific safe haven in Warsaw (Poland) where he worked – till 1938 – as an associate of the newly founded Institute of Public Affairs (*Instytut Spraw Spolecznych*, ISS). The main goal of the ISS was to improve "the quality of life of the average Polish citizen" by "carrying out scientific research, propaganda and educational work [...] in the areas of workplace protection, social insurance, the employment market, unemployment, emigration and social services". As Mazur shows, this agenda furthered Ichheiser's own scientific identity: For him proper social science was an act of "intellectual bravery", connecting "knowledge and life" by "analyzing the consequences" of existing or potential social goals and demands. In this vein, Ichheiser wrote a book about the psychology of workplace accidents and a couple of articles on the relationship between "Theory and Practice in Social Life": He also published the first volume of a scientific encyclopedia (letters A to K) entitled *The World of Ideas* (*Świat pojęć*), which, as Mazur writes, was designed to be "easily accessible" for the common reader and should "develop in response to reader comments" (Ichheiser 1939). Inevitably, one is reminded of Otto Neurath's project of an *International Encyclopedia of Unified Science* that was carried out at exactly the same time. This is not the only parallel: Who wouldn't think of Neurath or Zilsel when Mazur sums up the tone of Ichheiser's polish publications: "Although he was writing as one whose academic convictions where currently on 'the outside', he believed that those convictions would soon find themselves on 'the inside'".

This optimistic interpretation of Ichheiser's marginality is now contrasted in the anthology by a second, more pessimistic narrative: The most striking example is delivered by Christian Fleck (chapter 2), whose article – with the telling subtitle "A Person Who Failed Himself And The World" – meticulously reconstructs Ichheiser's later life as a scientific refugee: After the occupation of Austria and Poland through Nazi-Germany, Ichheiser turned first to Switzerland and England (1939) before he – at the end of 1940 – found a permanent refuge in the United States, where he lived in a variety of places till his suicide in 1969: "His eventful, three-decade-long life in the US was marked by no highs but numerous lows", writes Fleck, pointing to a certain tragic or even ironic quality of this episode: In difference to other European refugees at the time, Ichheiser had not only the professional capabilities, but also an unusual amount of opportunities to make a working career as a scientist in Amerika: During his time in Vienna and Poland he had established an impressive track record of publications that continued to grow steadily even under the precarious and difficult circumstances in the US. In the course of the 1940s, Ichheiser wrote two books – one about antisemitism, the other about *Misunderstandings in Human Relations* – and published more than a dozen articles in prestigious journals like *Philosophy of Science* or *American Social Review*. With the help of American supporters, among them the sociologist Louis Wirth and the political scientist Hans Morgenthau, Ichheiser obtained a variety of employments: first a non-academic job at a publishing house, then a professorship at Talladega College in Georgia and a research-position at the University of Chicago. All these hopeful job-opportunities were given up "frivolously" after a short period of time, which – according to Fleck – was overall typical for Ichheiser's personal character: He was "often smug and arrogant", expressing even in his articles and reviews a certain sense of superiority and condescension towards his peers in the scientific community, which proved at the end to be pure poison for his academic career. "All in all", writes Fleck in his final, damning assessment, "one must come to the conclusion that someone like Ichheiser would not have been any more successful in any other scientific world".

Apart from these character issues there is probably a deeper explanation for Ichheiser's lifelong professional struggle hinted at by a quote from Hans Morgenthau included in Fleck's article. According to Morgenthau, the "trouble with I[chheiser] was, that he lived his social theories". In other words: What made Ichheiser great as a social scientist, made him fail on a personal level. As pointed out in the papers of Anna Schor-Tschudnowskaya (chapter 4), Lisa-Teresa Woller (chapter 5), Ian Burkitt (chapter 6) and Ivana Marková (chapter 10), Ichheiser's overall worldview was fueled by a deep skepticism towards society. This skepticism set him apart not only from his contemporaries in the Vienna Circle, but also from most of the other leading social philosophies of the day: In difference to the views of Emile Durkheim, Georg Simmel, Edmund Husserl or Alfred Schütz, who otherwise had a tremendous influence on him, Ichheiser rejected the (transcendental) idea of society as a "shared realm" of mutual experiences, attitudes and goals (Schor-Tschudnowskaya). For him, the social world was a source of constant misunderstandings, deceptions and disappointments that "dominated the interpersonal space". All these problems were anchored in human consciousness fractured and torn apart by the individual and social factors of its existence. With the help of dyadic concepts (Marková) like "the Self versus Other(s), appearance versus reality, success versus failure, visible versus invisible", Ichheiser tried to analyze the alienation and isolation of the modern mind that was hindered from personal self-expression by the diverse social roles and requirements it had to fulfill. "At last", writes Ichheiser in one of his earlier essays cited by Woller, "all ideology of 'inner freedom' fails due to the inexorable fact that we as an intrinsic constitution of our mental life remain dependent on the You and a community and thus – as paradox as it might sound – one side of the I is lost to the environment".

In his later years of American exile, Ichheiser's lifelong professional obsession with the modern alienated mind reached a pathological level: He was diagnosed with "schizophrenia, paranoid type" and brought to a psychiatric hospital where he stayed as an inmate for twelve consecutive years from 1951 to 1963. In a manuscript, entitled *Was I Insane – Or was I 'railroaded' to a State Hospital?* written in

1966, Ichheiser later tried to come to grips with this unfortunate episode in his life: The main aim of the manuscript, now published for the first time as a supplement to the anthology and analyzed in a separate article by Stephan Dietrich (chapter 11), was, as Ichheiser puts it himself: "to achieve a public vindication as well as a personal catharsis" by countering the diagnosis which had initiated his psychiatric commitment: What, from the perspectives of the involved public institutions, seemed at the time as a clear case of a "mentally ill person incapable of managing his own estate", was according to Ichheiser in reality an act of social marginalization and oppression directed against him, as an outsider whose political and professional beliefs were not in alignment with the "conventional-traditional scheme". In this sense, as Dietrich describes it, Ichheiser's personal, psychotic experiences turned into a self-fulfilling prophecy "proving his own theoretical works about misinterpretation in the judgement of other people". Or to paraphrase Morgenthau: Ichheiser not only "lived his social theories", he suffered them.

One of the most extraordinary features of *Memories of Gustav Ichheiser* is without a doubt the way in which nearly all of its articles follow through on the idea of marginality in its different aspects. Another strong element of the book is its comprehensive character. Both aspects are highly unusual for an anthology. That being said, there is still a lot of work to do. One of the episodes, which is not mentioned in the book and would have been a further illustration of its central narrative, is Ichheiser's short interaction with the Vienna Circle.

In 1930, after Ichheiser's *Kritik des Erfolges* (Critique of Success) was published, Otto Neurath appreciated his work in *Der Kampf*, a leading theoretical journal of Austrian social democracy. Neurath took up Ichheisers specification of social constellations and mechanisms determining success or failure and their impact on the individual ability to be productive (*Leistungstüchtigkeit*) and achieve success (*Erfolgstüchtigkeit*). Neurath discovered with Ichheiser, that success, social recognition and opportunities cannot be explained through performance and quality[3]: We have the impression that we judge people by their performance, but we already judge them by their success – a deception which would be dissolved by a consistent scientific worldview. For Ichheiser as well as for Neurath success is socially conditioned, not a part of the personality itself. The fact that the social mechanisms for success and failure are hidden and disguised leads to pseudo empiricism and keeps science from getting its way. Neurath presented Ichheiser's theory of success and the resulting need for uncovering deception and restricting value judgments as a desirable ingredient of a new socialist social order based on a scientific worldview: "Will the deception persist in a large socialist organization? There seems to be much to suggest that the revolutionary process of transformation of our days will prove effective here too, …".[4]

[3] See also Wilsmann 2004, 67.

[4] *"Wird in einer sozialistischen Großorganisation die Täuschung" beharren? Uns scheint vieles dafür zu sprechen, daß der revolutionäre Umgestaltungsprozeß unserer Tage auch hier sich wirksam erweisen wird, […]*", Neurath 1930, 397.

Neurath also highlighted another aspect of Ichheiser's analysis of the antinomy between the ability to achieve success and the capability of being productive (Neurath 1930, 396): Personal characteristics, that enabled a person to be outstanding in a certain field, could at the same time prevent her from being successful and socially recognized in the existing society (like an increased sensitivity). Ichheiser himself referred to this point in his later article "*Über einige Voraussetzungen der psychologischen Begabung: Ein Beitrag zur Psychologie des psychologischen Wissens*" (About some requirements of psychological talent. A contribution to the psychology of psychological knowledge), published in Reichenbach's and Carnap's *Erkenntnis* in 1936. In the text, he analyzes the personal preconditions for psychology as a science and as a specific form of knowledge. Ichheiser's key term is here the concept of "psychological talent", which refers to the mental *possibility* and *probability*, anchored in a personality, for having psychological "intuitions" (*Erlebnisse*) and "insights" (*Erkentnisse).* Besides certain conceptual skills (*Formulierungsfähigkeit*), psychological research requires the mental sensitivity and "instability" (*Psychologische Sensitivität*) to perceive even small disturbances, structures and processes – a non-rational "talent", that often goes along with a neurotic personality (Ichheiser 1936, 118–124). A second important element of psychological talent is an impartiality from psychological conventions (konventionelle Unbefangenheit): To effectively replace the conventional, mythological-metaphysical image of the human personality with a scientific one, psychologists need to free themselves from the biases and preconceptions of folk psychology (Ichheiser 1936, 124–139). In this sense, Ichheiser's paper further elaborates his provocative thesis, that the ideal social scientist is the marginal man: the social or psychological deviant.

Regarding Ichheiser's time in Vienna some other questions remain open in this book. As Gerhard Benetka states, Ichheiser was an "autodidact" and "nobody's scholar for good reason" (chapter 1). But are there really no content-related or theoretical connecting points? What influence did his work have on other Viennese researchers? Did he give some input to them? Benetka and Fleck treat this question only marginally – the references in Ichheiser's German texts indeed seem to be rather vague. Further questions concern his time at the University: What courses did he take and whose lectures did he attend? Did he maintain contacts to the various academic and extra-academic circles at the time? The book does not give much information on this period.

Concerning the supervisor of his dissertation Karl Bühler, Benetka refers to Reinhard Müller and his study on Ichheiser from the 1990s (Müller 1993). Müller pointed out, that Ichheiser took only one lecture from Bühler during his study-period, furthermore that his dissertation on the *Gegenstand der Ästhetik* (Object of Aesthetics) would not fit into the general program of the Bühler's – neither methodically nor theoretically or in terms of content. This is not entirely convincing: In 1924 Ichheiser graduated during Bühler's first years in Vienna. Further works by other students on related themes followed. It would be interesting to compare Ichheiser's approach with that of other students as well as with Bühler's own

Viennese works. In his dissertation Ichheiser at least touched on Bühler's *Gestaltwahrnehmung.*[5]

In general, it could pay off to evaluate the dissertation more precisely especially as it seems that several elements of Ichheiser's later work were already established in there.One example is his holistic interpretation of consciousness and experience, mentioned among other thought patterns by Anna Schor-Tschudnowskyja (chapter 4). To Ichheiser the only aesthetical "real" is the concrete and indivisible situation of consciousness (*Bewußtseinssituation*). In his dissertation, Ichheiser proposes a four-sided analytic model to capture the whole aesthetic situation by reconstructing its different layers of meaning in a describing process as a quasi-interim solution or as "points of passages in thinking". At the end of this process only the sum of steps in specifying the whole subject of interest is able to come close to the described aesthetic situation.

Another element already established as sticking point in his dissertation seems to be his criticism of a confusing use of the term "impression" (*Eindruck*) in the field of aesthetics, which later can be found in his works on "*Ausdruck und Eindruck*" (Expression and Impression), in a sociological context (Ichheiser 1932, 1938).

Between 1929 and 1931 Ichheiser worked as a lecturer at the pedagogical institute of the city (*Pädagogisches Institut der Stadt Wien*) where he gave lectures on *Psychological Problems of Careers Counseling*; *Social Psychology*; *Facts and Problems of Social Psychology*; *Sources of deception in the psychological assessment of fellow human beings (students).* This has also not been evaluated yet.

Overall the impression remains that what makes Memories of *Gustav Ichheiser* such a fascinating read is the way it highlights how close biography and theory, science and life can interact. This interaction, as the example of Ichheiser clearly shows, can take a productive or a pathological turn: it can be a source of self-reflection or turn into a force of self-destruction.

References

Bayer, Betty M., and Lloyd H. Strickland. 1990. Gustav Ichheiser on sociopsychological misperception in international relations. *Political Psychology* 11 (4): 699–719.

Brissett, Dennis, and Charles Edgley. 1990. *Life as theater: A dramaturgical sourcebook*. New York, NY: de Gruyter.

Bühler, Karl. 1913. *Die Gestaltwahrnehmungen. Experimentelle Untersuchungen zur psychologischen und ästhetischen Analyse der Raum- und Zeitanschauung*. Stuttgart: Spemann.

Fleck, Christian. 2015. Who is Ichheiser? Ein an sich und der Welt Gescheiterter. In *Etablierung in der Fremde: vertriebene Wissenschaftler in der USA nach 1933*, 295–332. Frankfurt, M: Campus Verlag.

Freund, Peter. 1974. The visible and invisible: A look at the social psychology of Gustav Ichheiser. *Journal of Phenomenological Psychology* 5 (1): 95–111.

Ichheiser, Gustav. 1924. *Gegenstand der Ästhetik. Eine kritische Untersuchung*. Wien: Uni-Diss.

[5] On this topic see: Bühler 1913.

———. 1930. *Kritik des Erfolges. Eine soziologische Untersuchung*. Leipzig: Hirschfeld.

———. 1932. Ausdruck und Eindruck. *Kölner Vierteljahresschrift für Soziologie* 11: 59–64.

———. 1936. Über einige Voraussetzungen der psychologischen Begabung. Ein Beitrag zur Psychologie des psychologischen Wissens. *Erkenntnis* 6: 114–132.

———. 1938. Ausdruck und Eindruck. In *Onzième Congrès international de psychologie. Paris, 25. juillet 1937. Rapports et comptes rendus*, ed. Henri Piéron and Ignace Meyerson, 376–379. *Agen*: Impr. Moderne.

——— (1939). *Swiat Pojec. Slownik encyklopedyczny pojec z dziedziny filozofii, psychologii, filozofii przyrody, socjologii, etc. Czesc 1 [The world of ideas. Encyclopedic dictionary of ideas from the fields of philosophy, psychology, natural philosophy, sociology, etc. Part 1]*. Warszawa: Mathesis Polska.

———. 1970. *Appearance and reality*. San Francisco, CA: Jossey-Bass.

———. 1972. Blind to the Obvious. In *Down to Earth Sociology. Introductory Readings*, ed. James M. Henslin, 4–7. New York: Free Press.

———. 1977. Image of the other. In *Bridges, not Walls. A Book about interpersonal Communication*, ed. John Steward, 158. London: McGraw-Hill.

Müller, Reinhard. 1993. Teilprojekt, Gustav Ichheiser. Ein Zwischenbericht. *Archiv für die Geschichte der Soziologie in Österreich* 9: 8–11.

Neurath, Otto. 1930. Der Erfolg. *Der Kampf* 20: 394–397.

Parales-Quenza, Carlos José. 2010. El error fundamental en psicología: reflexiones en torno a las contribuciones de Gustav Ichheiser. *Revista Colombiana de Psicologia* 19 (2): 161–175.

Rudmin, F., and P. Boski. 2007. Ichheiser's theories of personality and person perception: A classic that still inspires. *Journal for the Theory of Social Behaviour* 19 (3): 263–296.

Rudmin, F., R. Trimpop, I. Kryl, and P. Boski. 2011. Gustav Ichheiser in the history of social psychology: An early phenomenology of social attribution. *British Journal of Social Psychology* 26 (2): 165–180.

Stadler, Friedrich. 1981. Spätaufklärung und Sozialdemokratie in Wien 1918–1938. Soziologisches und Ideologisches zur Spätaufklärung in Österreich. In *Aufbruch und Untergang. Österreichische Kultur zwischen 1918 und 1938*, ed. Franz Kadrnoska, 441–473. Wien, München, Zürich: Europa-Verlag.

Velho, Gilbert. 2008. Goffman, misunderstandings, and international risks. *Revista Brasileira de Ciências Sociais* 23: 145–147.

Wilsmann, Stefan. 2004. Lebenschancenkonzepte. Von Theodor Geiger über Otto Neurath bis zu Amartya Sen. In *Kritische Empirie. Lebenschancen in den Sozialwissenschaften. Festschrift für Rainer Geißler*, ed. Horst Pöttker and Thomas Meyer, 61–78. Wiesbaden: Verlag für Sozialwissenschaften.

Woller, Lisa-Teresa. 2013. *Sein und Erscheinen. Leben und Werk von Gustav Ichheiser*. Wien: Unpublished Bachelor-Thesis.

Index

C. Damböck, A. T. Tuboly (eds.), *The Socio-Ethical Dimension of Knowledge*, Vienna Circle Institute Yearbook 26, https://doi.org/10.1007/978-3-030-80363-6

The manufacturer's authorised representative in the EU is Springer Nature Customer Service Centre GmbH, Europaplatz 3, 69115 Heidelberg, Germany. If you have any concerns regarding our products, please contact ProductSafety@springernature.com

Printed and bound by CPI Group (UK) Ltd, Croydon, CR0 4YY
30/09/2026
02225650-0007